High-Performance Construction Materials
Science and Applications

Engineering Materials for Technological Needs

Series Editor: Deborah D. L. Chung
(University at Buffalo, State University of New York, USA)

Vol. 1: High-Performance Construction Materials: Science and Applications
edited by Caijun Shi (Hunan University, China) &
Yilung Mo (University of Houston, USA)

Preface

This is the inaugural volume of the book series "Engineering Materials for Technological Needs". The technological needs relevant to this series include those that are related to the electronic, optical, magnetic, thermal, electrochemical, energy, environmental, structural, construction, aerospace, automobile, security, manufacturing and medical industries. Each book in the series addresses materials that serve various functions in a particular area of technological need by describing the relevant scientific concepts, the status of the technological need, the limitations of the technology, the pros and cons of the materials available, how the materials are applied, evaluated and fabricated, how various materials are integrated to form an assembly that can be used in practice, the performance and durability of the competing materials for each specific application, and the future directions.

The books in this series differ from other books in their emphasis on materials for technological needs. Existing books in the materials area largely focus on the fundamental science of particular classes of materials. The material classes in these existing books tend to be distinguished from the viewpoint of science rather than the viewpoint of applications. In contrast, from the viewpoint of a particular application, the relevant materials may include metals, polymers, ceramics, composites and semiconductors, which constitute the traditional classes of materials. The impetus of scientific research hinges on the success of applications. It is important to develop materials for applications rather than just studying materials for the sake of advancing the science of materials. In spite of the obvious importance of the linkage between science and applications, this linkage has long been weak, due to insufficient communication between the science community and the application community. This book series is intended to alleviate this problem by addressing materials from an application perspective. The books do not assume prior knowledge of the applications and require from its readers the equivalence of only one introductory course in materials science. Due to the tutorial nature of the books in the series, the books are expected to be useful as textbooks in undergraduate and graduate levels. In addition, due to the timeliness and importance of the topics and to the inclusion of an up-to-date list of references, these books are expected to be used by professionals as reference books as well.

This book (Vol. 1 of the series), titled "High Performance Construction Materials: Science and Applications" and edited by Dr. Caijun Shi and Professor Y. L. Mo, is intended to provide scientific and practical information on a wide range of construction materials. In U.S.A., a large fraction of construction cost is for the repair and restoration of the existing infrastructure. The main reasons for the limited durability include the use of unsuitable materials and the improper use of materials. The latter usually relates to a flawed design. An example is the troubled Big Dig construction project in Boston. The development and application of high performance construction materials can enhance safety, extend the service life of the infrastructure and result in substantial cost savings and conservation of natural resources.

This book describes a number of high performance construction materials, including concrete, steel, fiber reinforced cement, fiber reinforced plastics, polymeric materials, geosynthetics, masonry materials and coatings. It discusses the scientific bases for the manufacture and use of these high performance materials. Testing and application examples are also included.

The collection of high performance construction materials covered in this book is covered in related books to limited degrees only. Books dealing with construction materials typically address traditional materials only and, as a consequence, they do not satisfy the increasing demands of today's society. On the other hand, books dealing with materials science are not engineering oriented, with limited coverage of the application to engineering practice. This book reflects the great advances made on high performance construction materials in recent years. It is intended to apply relatively new high performance construction materials to design practice.

This book is appropriate for use as a textbook for courses in engineering materials, structural materials and civil engineering materials at the senior undergraduate and graduate levels. In addition, it is suitable for use by practice engineers, including construction, materials, mechanical and civil engineers.

D.D.L. Chung
Book Series Editor
Buffalo, NY, U.S.A.
August 29, 2007

Contents

Biographical Sketch for each Author

Chapter 1

Dr. Caijun Shi is a Professor in the College of Civil Engineering, Hunan University and President of CJS Technology USA Ltd., USA. His research interests include cement and concrete technology and waste management. Dr. Shi has authored/ co-authored more than 120 technical papers and is a co-author/ co-editors of 6 books. He is a fellow of American Concrete Institute and a senior member of RILEM.

Y. L. Mo is a professor in the Civil and Environmental Engineering Department at the University of Houston. He has written five books and book chapters and has had more than 100 technical papers published in national and international journals, 90 conference papers, and 45 technical reports. Based on his research performance Dr. Mo has received the Alexander von Humboldt Research Fellow Award from Germany in 1995 and the Distinguished Research Award from the National Science Council of Taiwan in 1999.

Chapter 2

Caijun Shi is a Professor in the College of Civil Engineering, Hunan University and President of CJS Technology USA Ltd., USA. His research interests include cement and concrete technology and waste management. Dr. Shi has authored/ co-authored more than 120 technical papers and is a co-author/ co-editors of 6 books. He is a fellow of American Concrete Institute and a senior member of RILEM.

Y. L. Mo is a professor in the Civil and Environmental Engineering Department at the University of Houston. He has written five books and book chapters and has had more than 100 technical papers published in national and international journals, 90 conference papers, and 45 technical reports. Based on his research performance dr. Mo has received the Alexander von Humboldt Research Fellow Award from Germany in 1995 and the Distinguished Research Award from the National Science Council of Taiwan in 1999.

Dr. Hemant B. Dhonde is a structural design engineer with Bechtel Corporation, Houston, Texas, USA. He received his BE in civil engineering from University of Pune, India in 1997; his ME in geotechnical engineering in 2000 from University of Pune, India; and his Ph.D. in structural engineering from the University of Houston in 2006. His research interests include development of self-consolidating concrete and self-consolidating fiber reinforced concrete. Dr. Dhonde has several international publications and inventions to his credit.

Chapter 3

Antoine E. Naaman is Professor Emeritus of Civil Engineering at the University of Michigan, Ann Arbor, USA. He obtained his Ph.D. degree in Civil Engineering from the Massachusetts Institute of Technology, in 1972. He is a Fellow of the American Concrete Institute, the Prestressed Concrete Institute, and the Amercian Society of Civil Engineers. Dr. Naaman's research studies have led to more than 300 technical publications, including two textbooks, and eleven co-edited books. His research interests include prestressed concrete, high performance fiber reinforced cement composites, ferrocement, and the integration-tailoring of advanced construction materials to improve structural performance.

Chapter 4

Keh-Chyuan Tsai completed his Bachelor (1977), Master (1980) and Ph.D. (1988) degrees in National Taiwan University (NTU), Stanford University and University of California at Berkeley respectively. He has been a professor of civil engineering at NTU since 1993. He has served as the director of the National Center for Research on Earthquake Engineering since 2003. His research interests include steel structures.

Ying-Cheng Lin holds a B.S. (National Taiwan University 2002) in civil engineering and a M.S. (National Taiwan University 2004) in structural engineering. He has been a Ph.D. Student and assisting in conducting a NEES project at Lehigh University since 2006. His research interests include seismic responses of buildings, large-scale experimental testing, and structural applications of high-performance materials.

Jui-Liang Lin is an associate research fellow at National Center for Research on Earthquake Engineering, Taiwan. He holds a Ph.D. (National Taiwan University 2007) in structural engineering. He was a visiting scholar at Stanford University during 2006 to 2007. He has several-years industrial experiences before starting his Ph.D. study. His research interests include structural dynamics and earthquake engineering.

Po-Chien Hsiao holds a B.S. (National Taiwan University 2002) in civil engineering and a M.S. (National Taiwan University 2004) in structural engineering. He has been a Ph.D. student at University of Washington since 2007. He was a research assistant at National Center for Research on Earthquake Engineering during 2006 to 2007. His research interests include seismic responses of buildings and experimental testing.

Sheng-Lin Lin now is a Ph.D. student at Civil and Environmental Engineering Department at the University of Illinois at Urbana and Champaign (UIUC). He got his Bachelor of Science (2001) and

Master of Science (2003) at the National Taiwan University. His research interest includes seismic analysis and design of steel, concrete buildings, and earthquake simulation tests. He is a licensed Civil Engineer in Taiwan.

Chapter 5

Len Hollaway is Professor Emeritus of Composite Structures in the Faculty of Engineering and Physical Sciences, University of Surrey and was formerly a visiting Research Professor at the University of Southampton. He is a Fellow of the Institution of Civil Engineers, UK, Member of the Institution of Structural Engineers, UK, a Euro Engineer and a Fellow of the International Institute for FRP in Construction. He has published widely; has over 200 refereed technical papers and is author, co-author or editor of 10 books on various aspects of composites in the civil engineering industry.

Chapter 6

Richard E. Klingner is the L. P. Gilvin Professor of Civil Engineering at The University of Texas at Austin, where he teaches and conducts research on the dynamic response of structures, earthquake-resistant design of masonry and concrete structures, and anchorage to concrete. For the period 2002-2008, Dr. Klingner was Chair of the Masonry Standards Joint Committee, sponsored by the American Concrete Institute, The Masonry Society, and the American Society of Civil Engineers.

Chapter 7

Peter Stevenson has been active in geosynthetics for nearly three decades working with woven and knit reinforcements (founder and president of Xtex Inc, geogrid producer) as well as needled filters and separators (founder and president of Acme

STW). A founding member of the International Geosynthetic Society (IGS) he currently serves as Secretary of the Society.

Chapter 8

Cumaraswamy Vipulanandan is the chairman and professor of civil and environmental engineering department at the University of Houston, Texas. He is the Director for the Center for Innovative Grouting Materials and Technology (CIGMAT) and the Texas Hurricane Center for Innovative Technology (THC-IT) at the University of Houston. He received his MS and PhD in Civil Engineering from Northwestern University, Evanston, Illinois and BSc in Civil Engineering from University of Moratuwa, Sri Lanka. He has been Principal Investigator or Co-Principal Investigator for 60 funded projects amounting to over $6 million. His work has resulted in more than 150 refereed papers and over 100 presentations at national and international conferences. He has edited eight books/proceedings related to fracture of concrete, grouting, recycled materials, pipelines, soft and expansive soils, and piles.

Marko Isaac received his Master of Science in Civil Engineering from the University of Houston, Houston, Texas. Marko has been investigating different mixes of self-consolidating concrete and the effect of various coating materials on various types of concrete materials. Marko Isaac is a structural engineer at Bechtel Corporation at Houston, Texas, where he has been designing concrete and steel structures.

Chapter 9

Deborah Chung is a National Grid Endowed Chair Professor, State University of New York (Buffalo). She received her Ph.D. degree in Materials Science from Massachusetts Institute of Technology and B.S. degree from California Institute of Technology. She is Fellow of American Carbon Society and ASM International and she received the Pettinos Award, a triennial international award to recognize one person for outstanding research accomplishments in carbon. Her publications include over 400 journal papers and 6 books.

G. Song is the founding Director of the Smart Materials and Structures Laboratory and an Associate Professor of Mechanical Engineering at the University of Houston. Dr. Song is a recipient of the prestigious NSF CAREER grant in 2001. Dr. Song received his Ph.D. and MS degrees from the Department of Mechanical Engineering at Columbia University in the City of New York in 1995 and 1991, respectively. Dr. Song received his B.S. degree in 1989 from Zhejiang University, P.R.China. He has expertise in smart materials and structures, structural vibration control, and structural health monitoring and damage detection. He has developed two new courses in smart materials and published more than 170 papers, including 66 peer reviewed journal articles.

Ning Ma is currently a senior engineer at OptiSolar, a California based solar energy company. Dr. Ma obtained his Ph.D. degree and M.S. degree, all in Mechanical Engineering, from the University of Houston and the University of Akron in 2006 and 2002, respectively. His research interests include structural design, wind induced vibration, vibration control, and shape memory alloys. He is a member of ASME (American Society of Mechanical Engineer). His publication includes more than 20 conference papers and 8 journal papers.

Haichang Gu is a Postdoctoral Fellow in the Smart Materials and Structures Laboratory at the University of Houston. He obtained his Ph.D in Mechanical Engineering at the University of Houston in 2007. His research interests include smart materials and structures, structural health monitoring, vibration control and smart composites. Dr. Gu has more than thirty publications, including 9 journal papers. He is a member of ASME (American Society of Mechanical Engineer) and SPIE (International Society for Optical Engineering).

Chapter 1

Introduction

Caijun Shi

College of Civil Engineering, Hunan University
Changsha, China
CJS Technology USA Ltd, USA

Y. L. Mo

Department of Civil and Environmental Engineeriing
University of Houston, Houston, TX

1.1 Historical Development of Construction and Uses of Construction Materials

1.1.1 *Stone age habitats*

Mankind's earliest ancestors used natural shelters such as caves and rock overhangs for protection purposes. Gradually, they learned to improve their caves with inlaid stone floors, walls at the entrances and fireplaces. Ultimately they began to build entirely new habitats in locations where there had no natural shelter.

Some of the earliest structures built by man were very simple dolmens: by placing two stones vertically and one flat stone spanning them across. Other stone buildings include granaries and temples. The earliest evidence of a man-made habitat was found Olduvai Gorge in Central Africa, which date to about 2,000,000 BCE [1-2]. A small circle of stones was found stacked to hold branches in position. This was clearly the work of Homo Habilis, our tool-making ancestor. This precursor to Homo Erectus used fire as early as 3,000,000 BCE. They also used tools (more than a tool-maker) and had campsites. They used wooden poles and/or animal bones to erect a framework over which animal hides or leaves were draped. Hides and bones were from hunting. The tent was erected in its crude form by driving a pole in the ground, and slinging animal hides over it. The hide was then held down with stones.

Clay and wood were often used for permanent dwellings, in the so-called wattle-and-daub construction [1]. The walls were made of small saplings or reeds driven into the ground, and tied laterally with vegetable fibers, then plastered over with wet clay for rigidity.

1.1.2 *River valley civilizations — The first steps in permanence*

Once mankind began to settle, the agglomerations of people formed villages, towns and cities, mainly along the riverbanks of the Nile (The Egyptian Civilization), the Tigris and Euphrates (the Sumerian or Mesopotamian Civilization), the Indus (the Harappan Civilization) and the Yellow River (the Yellow River Civilization). Houses were constructed using sun-dried bricks [3]. The bricks were made from mud and straw, which were abundantly found in the river flood plains and were mixed together. This wet mud and straw mixture was formed in a wooden frame. After the mixture had hardened sufficiently through evaporation, the form was removed. The bricks were then left in the sun to be dried completely. The straw acted as reinforcement to hold the mud together when evaporation caused inevitable shrinkage of the mud. Fired bricks were produced and used later for heavy use areas, like pavements and sewers in Mesopotamia.

1.1.3 *Construction in ancient Egypt*

The main building material used in Ancient Egypt was stone – mainly limestone. Rough stones were used in building interior walls and foundations, while fine stones, cut out with special care, were used in decorating main walls or erecting colossal temples. Yellow limestone was brought from Al-Silsila Mount, white limestone from Tura, and gray or red granite from Aswan and alabaster from central Egypt. Basalt was often used in paving roads and laying lower courses of buildings. Generally, the above-mentioned stones, in addition to diorite, marble and porphyries were used in making statues and utensils. Convertible diorite was used in making the famous Chephren statue. Many scarabaeuses and other objets d'arte were made of soft steatite. Ancient Egyptians actually reached unmatched high levels in architectural design and construction engineering. Even at present it is still hard to conceive how they could have all these buildings and structures erected with such high level of perfection and precision, using only primitive and naive tools far incomparable with modern machinery and equipment [4]. Some good examples include pyramids and temples. The pyramids have the deepest impression on the whole world's imagination. Edfu Temple is 137-m-long, 79-m-wide and 36-m-high temple and is still exceptionally almost intact, with its hall, columns, stairs and ceiling still maintaining their original state. In addition, its colors and decorations still look fresh.

1.1.4 *Construction in the Greek Era*

In the Greek Era, Post 1800 B.C., the use of the stone construction continued. The focus, however, was on the ornamentation of the buildings [5]. The heavy stones, once ornamented, were lifted into place by simple machinery. The concealed faces of the columns had grooves and holes that held the ropes to lift the stones. Metal dowels and clamps were employed to fasten the stone pieces together. No mortar was ever used.

The stone surface was carved by master masons to enhance the visual appeal of the buildings. Moldings were used extensively to give each building a profile. "Orders" were developed to create standardized, symmetrical and precise buildings. The buildings were usually painted in brilliant colors. Majority of Greek temples were built with stone frame. The trabeated building was constructed with stone columns and timber beams. The maximum stone beam span was 5-6 m. Usually wooden beams were supported on the stone columns. Wood was the predominant material for roofs.

1.1.5 *Construction in the Romans times*

The Romans made great contributions to construction and developed three different construction methods [6]: (1) stone and masonry, (2) concrete construction, and (3) timber and metal.

Stone and Masonry Construction: Stone construction in the Roman Times was a carryover of the techniques used by the Greeks. The Romans understood the concept of the true arch and used it to its full extent. Stone arches of large spans were used extensively in the construction of aqueducts and buildings. Many of the Roman stone buildings became models for medieval European construction. Stone construction soon gave way to brick construction. Lime mortar, constituted of lime, sand and water, used until the 2nd Century B.C. The best-known example of brick technology developed by the Romans is the Hagia Sophia, with its brick dome spanning 32.6 m.

Concrete Construction: The Romans started to mix pozzolana with ime to make strong lime-pozzolana mortar and concrete in the 3rd Century B.C.   The use of lime-pozzolan cement mortar and concrete in construction was a landmark as it altered the interior space of buildings. The crowning example of concrete technology is the Pantheon, a massive structure with brick-faced concrete walls 6 m thick, spanned by a concrete dome 1.5 m thick, 43.2 m above the floor level.

Timber and Metal Construction: Timber technology was also highly developed, evident from the advanced timber form required for the brick and concrete construction. The Romans also developed the concept of the truss. A truss is a beam with some portions of it hollowed out. The areas, which are hollowed out are redundant in supporting loads but only increase the dead weight of the structure supporting these beams. The Romans used it extensively in their buildings and bridges. Bronze trusses were used where durability and longevity was desired.

Metals were also used extensively in Roman buildings for non-structural purposes. Roman introduced lead and gilded bronze tiles as roofing materials [6]. Lead was very popular as it was waterproof and could be used on low pitch roofs as well. Lead also made pipes to supply fresh water to buildings and to discharge wastewater from buildings. Another material used was glass. Although the Egyptians had discovered it, the Romans were the first to use it in their buildings. Various kinds of colored glass were used as mosaics for interior decorations. The Romans also made the first recorded clear window glass. Glazed sun porches became very popular in villas. It was also used in public buildings and middle-class housing. Thin slabs of colored stone were employed for decoration of interior surfaces in buildings. These stone slabs were fastened to concrete walls with metal fasteners.

1.1.6 *The early industrial age (18th–19th Century)*

The construction activity in the initial stages of the Industrial Age, that is, in the latter half of the 18th century, can be summarized by one single term – mass production [7]. Unforseen innovation in technology made new building prototypes possible during this period. The methods used by these prototypes were Iron members as the structural frame, and the balloon frame.

With the advent of steam engine and the knowledge of the smelting and puddling process, iron could be made easily for construction purposes at a large scale. Many different forms and members, such as hollow tubular columns, wrought iron trusses, tension beams, and rivets were standardized and mass-produced in factories. Although initially iron beams were supported by the masonry, soon iron was being used for the entire structure.

Brick was also being mass-produced in factories using the mechanical extrusion method and the fired beehive kilns. This reduced the cost of bricks and led to extensive use in buildings as infill with the iron members as the structure.

Glass was used extensively with the iron frame - the best and most known example being the Crystal Palace in London. The Crystal Palace was made entirely of prefabricated materials, which were simply assembled together on the site. The plate glass panels let in light like never before. The concept of construction underwent a major change. The same prototype was employed for train stations all across western Europe. The biggest one is the St. Pancras Station in London with the iron trusses spanning 74 m.

In the 18th century a big efforts started in Europe to understand why some limes possess hydraulic properties. In 1756, John Smeaton discovered that cement made from limestone containing a considerable proportion of clay would harden under water [8]. Based on his discovery, he rebuilt the Eddystone Lighthouse in 1759, which stood for 126 years before replacement was necessary.

Several other people, including L. J. Vicat and Lesage in France, Joseph Parker and James Frost in England, investigated hydraulic cement during the period from 1756 to 1830. In 1824, Joseph Aspdin, a bricklayer and mason in Leeds, England, took out a patent on a hydraulic cement that he called portland cement because its color resembled the stone quarried on the Isle of Portland off the British coast [8]. Aspdin's method involved the careful proportioning of limestone and clay, pulverizing them, and burning the mixture into clinker, which was then ground into finished cement.

Joseph Monier invented Reinforced concrete in 1849 and received a patent in 1867 [9]. Reinforced concrete combines the tensile or bendable strength of metal and the compressive strength of concrete to withstand heavy loads. Joseph Monier exhibited his invention at the Paris Exposition of 1867. Besides his pots and tubs, Joseph Monier promoted reinforced concrete for use in railway ties, pipes, floors, arches, and bridges, and four years later, he registered another patent to use it in beams and columns [9]. An American mechanical engineer, W. E. Ward, built the first landmark building in reinforced concrete near Port Chester, New York in 1875 [10]. It used reinforced concrete for walls, beams, slabs and staircase. A reinforced concrete system was patented in the United States by Thaddeus Hyatt in 1878 [11]. Later on, there were parallel developments of reinforced concrete frame construction by Ernest L. Ransome in the United States, by Francois Hennebique in France, and by G. A. Wayss in Germany/Austria. Ernest L. Ransome first used reinforcing in 1877 and patented a system using twisted square rods to help the development of bond between the concrete and reinforcing in 1884 [12]. His largest work of the time was the Leland Stanford, Jr. Museum at Stanford University, the first building to use exposed aggregate. He was also responsible for several industrial buildings in New Jersey and Pennsylvania, such as the 1903-1904 construction of the Kelly and Jones Machine Shop in Greensburg, Pennsylvania.

On the other side of the Atlantic, Francois Hennebique, a successful mason turned contractor in Paris, started to build reinforced concrete houses in the late 1870s. He took out patents in France and Belgium for the Hennebique system of construction and proceeded to establish an empire of franchises in major cities. He promoted the material by holding conferences and developing standards within his own company network. Most of his buildings (like Ransome's) were industrial. In 1879 G. A. Wayss, a German builder, bought the patent rights to Monier's system and pioneered reinforced concrete construction in Germany and Austria, promoting the Wayss-Monier system [12].

In 1903 Perret, designed and built a multi-storey building in Paris using reinforced concrete: this structure deeply influenced architecture and concrete constructions for decades, since it was built without load-bearing walls, replaced by columns, beams and slabs. Perret also built museums, churches, garages and theatres, such as the Theatre Champs Elysées.

1.1.7 *Constructions in the 20th century — High rise steel structures/buildings*

The construction during the late 19th up to the mid-20th Century can be identified by one predominant structural form - the High Rise [13]. This type of structure was usually constructed with either concrete or steel.

1.1.7.1   High rise steel structures/buildings

The first tall structure using steel as its principal building material was the Eiffel Tower in Paris. It was 300 m tall. Its height was not challenged until 1929. But the major innovation was the development of the steel frame, as a structural element. The home of the high rise is Chicago, where the first metal structure was built - the 10-story Home Insurance Company Building in 1885. The metal framing used in this building was completely encased in brick so as to render it fireproof. The Manhattan Building was the first to use vertical truss bracing to resist wind forces. The first all-steel building was the Ludington Building in 1891 [13].

Soon standard construction practice for high rise structures included the steel frame of rolled steel I-beams, riveted and/or bolted connections, diagonal wind bracing, brick or clay tile fireproofing and the caisson foundation. Vertical transportation was by electric-powered hydraulic elevators. But climate control within the structures was still through natural ventilation.

In the years after World War II, glass was used extensively in high-rise structures, culminating in the curtain-walled skyscraper. But the efficiency of such high-rise structures was possible only after the development of rubber as a sealant, artificial climate control, and through the use of aluminum as a building material. One of the major landmarks in curtain walled structures was the United Nations Secretariat Building (1949) in New York City

1.1.7.2 High rise concrete buildings

The major developments of reinforced concrete have taken place since the year 1900. The Ingalls Building, built in 1903 in Cincinnati, Ohio, was the world's first reinforced concrete skyscraper, as shown in Fig. 1.1 [14]. The 16 story building was designed by the Cincinnati architectural firm Elzner & Anderson and was named for its primary financial investor, Melville Ingalls. The building was considered a daring engineering feat at the time,

Fig. 1.1 The Ingalls Building built in 1903 in Cincinnati, Ohio [15].

But its success contributed to the acceptance of concrete construction in high-rise buildings in the United States. Notre Dame du Raincy, built in 1922, was an important breakthrough (especially compared with previous concrete buildings) and is regarded as a masterpiece of architectural design: the lofty arched ceiling and the slender columns bear witness to the exceptional features of this new building material.

The giants and midgiants of the 1930s were all of steel construction. The Johnson Wax Tower, however, provided the impetus for Bertrand Goldberg's twin towers of Marina City, though on a vastly different scale. The Chicago 60story high-rise, erected in 1962, heralded the beginning of the use of reinforced concrete in modern skyscrapers and with it, competition for the steel frame. Place Victoria in Montreal, constructed in 1964, reached height of 190 m utilizing 42 MPa concrete in the columns. Concretes of higher strength proved to be the key to increased height, permitting as they do a reasonable column size on the floors below. One Shell Plaza in Houston topped out at 218 ft in 1970 using 42 MPa concrete. The Chicago area, with its plentiful supply of high quality fly ash (which helps to achieve a more workable concrete at lower water/cement ratios), has spawned the greatest concentration of tall reinforced concrete buildings. The 70-story Lake Point Towers used 52 MPa concrete to reach 197 m in 1968. Water Tower Place reached 859 feet in 1973 with concrete strengths as high as 63 MPa [10].

The developments during this period have led to construction practices that have become standard for buildings and continue to be so even today. Skyscrapers are the expression of architectural and construction expertise of this century.

1.2   Recent Construction — High Performance Construction Materials

Since the 1980's, the design and construction use more and more high performance materials. High performance construction materials provide far greater strength, ductility, durability, and resistance to external elements than traditional construction materials, and can significantly increase the longevity of structures in the built environment and can also reduce maintenance costs for these structures considerably. These most significant high performance construction materials include high performance concrete, high performance steel, fiber reinforced cement composites, FRP composites, etc.

The United States, the Strategic Highway Research Programe (SHRP) sponsored a project on High Performance Concrete in 1987. In an effort to improve and extend the service life of bridges, the Federal Highway Administration (FHWA) initiated a national program in 1993 to implement high-performance concrete (HPC) in bridges [16]. Also, FHWA initiated a program to develop high-performance steels (HPS) for bridges. A 485 MPa grade of HPS was the first developed, and is specified in ASTM A709 as grade HPS-70W (HPS 485W) (the designation "W" stands for "weathering"). High Performance Steel, grade 70 (HPS-70W) became available for use in early 1996 for fabrication and testing in bridges. Fiber reinforced cement composites and FRP composites are also becoming more and more popular because their unique mechanical properties and corrosion resistance (see Chapters 3 and 5 for detailed discussions).

The other trend is the combined use of several different types of high performance materials in one structure. The combination or composite structures made of several high performance materials could maximize the advantages of these components. High strength steels and high strength concretes have been tested or used in composite construction [17,18]. It has been found that in addition to strength and serviceability, stability, local buckling and ductility are also important effects in the design of composite members incorporating high strength materials [17].

In addition to these technical advantages, the use of high performance materials can have very significant economic advantages as well. The materials cost of high performance materials is usually higher than conventional materials due to the special requirements for raw materials and manufacturing processes. However, these materials maybe only one component in construction, and the total cost of the finished construction is more important than the cost of an individual material. According to a study by Moreno [19], the use of 41 MPa compressive strength concrete in the lower columns of a 23-story commercial building requires a 865-mm square column at a cost of $9.90/m^2. The use of 83 MPa concrete allows a reduction in column size to 610 mm square at a cost of $5.60/m^2. In addition to the reduction in initial cost, a smaller column size results in less intrusion in the lower stories of commercial space and, thereby, more rentable floor space.

In addition to increasing the duration of structures, high-performance materials are also valuable because they can improve the efficiency of design and construction practices. For instance, sustainable design and construction, an innovative building approach that incorporates high standards of environmental protection with an emphasis on life-cycle cost considerations, often uses high performance materials because these substances are more environmentally friendly and possess greater recyclable capability than conventional construction materials. The use of high performance materials to increase service life of a structure from 50 years to 100 years will save far more than the amount of money for the original construction cost of the structure. Also, it will conserve natural resources and reduce negative impacts on the environment. Moreover, many of these materials are often much easier and faster to install than conventional construction materials, a key advantage for the many fast-track projects delivered using design-build.

More and more people have been harnessing the advancement in smart materials technology for structural engineering applications. Specifically, the use of smart sensors and actuators as well as advanced signal processing and computational techniques are explored and adapted for structural health monitoring and control of structures. One good example is the Confederation Bridge, which was completed in 1997 and connects the Provinces of Prince Edward Island (PEI) and New Brunswick (NB) on the east coast of Canada (Fig. 1.2) [20]. The 12.9 km long Confederation Bridge is the world's longest prestressed concrete box girder bridge built over salt water. With 45 main spans of 250 m each and a 100-year design life, the design criteria of the Confederation Bridge are not covered by any code or standard in the world. With a design life of 100 years, the use of high performance concrete and careful attention to production and construction practices were imperative. Over 400,000 cubic meters of concrete was used for the structure. The proposed high-performance concretes were extensively tested for durability, especially through freeze-thaw cycles, sulfate resistivity and chloride diffusivity testing, checking of alkali content and alkali/aggregate reactivity, evaluation of curing regimes for the huge components, etc. Precasting was chosen for improved quality, as well as reduced construction time. A comprehensive monitoring and research program is being carried out to monitor and study the behavior and performance of the bridge under ice forces, short-and long-term deformations, thermal stresses, traffic load and load combinations, dynamic response due to wind and earthquake, and corrosion, and to obtain critical information that engineers now lack in these areas [21].

Fig. 1.2 Confederation Bridge built in Canada, 1997 [20].

1.3   Design Codes and Specifications for Use of High Performance Construction Materials

While the benefits of using high performance construction materials are numerous, many building practitioners are unfamiliar with the behavior and characteristics of these materials. The design and construction community often lacks information about the use of these materials, limiting their capability of achieving the highest possible standards in quality assurance and control construction projects.

High performance construction materials may be derived in two ways: modified existing materials or complete newly developed materials. As for these high performance materials from modification of existing materials, the current design code or regulations may be still applicable. However, for these newly developed high performance materials, new codes or specifications are usually required to be developed for the purpose of design, construction and testing purpose.

For different applications or different construction procedures, different professional associations or committees may need to get involved in the development of specifications or guidelines for design, testing and construction. For example, a number of committee activities from professional organizations are addressing the recommended use and specification of FRP composites. Many organizations have published codes, standards, test methods and specifications for FRP composites and their products for the respective products. Table 1.1 lists some American professional associations and technical committees working on FRP. American Society of Civil Engineers (ASCE) has a technical committee called Structural Composites and Plastics (SCAP) to address the design and implementation of composites. This committee published a design manual in the early 1980's and is currently working to update this manual to address the many FRP composite products developed over the years.

Table 1.1 American professional associations and committees working on FRP.

Organization	Committee
American Society of Civil Engineers (ASCE)	Structural Composites and Plastics
American Concrete Institute (ACI)	440 – Composites for Concrete 440C – State-of-the-art-Report 440D – Research 410E – Professional Educations 440F – Repair 440G – Student Education 440H – Reinforced Concrete (rebar) 440I – Prestressed Concrete (tendons) 440J – Structural Stay-in-Place Formwork 440K – Material Characterization 440L – Durability 440M - Repair & Strengthening of Masonry Structures
American Society of Testing and Materials (ASTM)	ASTM D20.18.01 – FRP Materials for Concrete ASTM D20.18.02 – Pultruded Profiles ASTM D30.30.01 – Composites for Civil Engineering
AASHTO Bridge Subcommittee	T-21 - FRP Composites
Transportation Research Board	A2C07 – FRP Composites

The American Concrete Institute has a Committee 440 – Fiber Reinforced Polymer Reinforcement to develop and report information on fiber reinforced polymer for internal and external reinforcement of concrete. The committee consists of 13 subcommittees and has published several state-of-the-art review and guides for the use of FRP.

Several ASTM committees are currently working on consensus test methods for the use of rebars, repair materials, and pultruded structural profiles. ASTM D20.18.01 (FRP Materials for Concrete) committee is developing standard test methods for FRP rebar and repair materials. ASTM D20.18.02 focuses on the development of test methods for FRP pultruded profiles and shapes. ASTM D30.30.01 addresses FRP composites products used construction.

The American Association of State Highway and Transportation Officials (AASHTO) Bridge Committee established a subcommittee in 1997 called "T-21 Composites". This committee has an ongoing effort to develop design guidelines for of the use of composites in bridge applications including FRP reinforced concrete, concrete repair, and vehicular bridge deck panels.

1.4 Organization of This Book

This book is intended to provide some recent progresses and applications of several most commonly used high performance construction materials. Chapter 2 discusses high performance concrete; Chapter 3 discusses high performance fiber reinforced cement composites; Chapter 4 discusses high performance steel; Chapter 5 discusses fiber reinforced Polymer; Chapter 6 discusses masonry materials. Chapter 7 discusses geosynthetics; Chapter 8 discusses coatings and Chapter 9 discusses several smart materials and structures and presents several applications of these materials and structures.

This book can be used as a textbook for advanced courses in civil engineering, or a reference book for students, laboratory workers, engineers and scientists.

References

1. Bordes, F. (1968). The Old Stone Age. New York, McGraw-Hill.
2. Burkitt, M. C. (1972). Our Early Ancestors. New York, Benjamin Blom.
3. Hawkes, J., The First Great Civilizations. Alfred A. Knopf, Inc., New York, 1973.
4. The Egyptian Government, Construction in Ancient Egypt, http://www.touregypt.net/featurestories/construct.htm.
5. Construction in The Greek Era, http://www.suite101.com/ article.cfm/ building_construction/40459.
6. Adam, J.-P., *Roman Building: Materials and Techniques*, B. T. Batsford, London, 1994.
7. "Building Construction." Encyclopedia Britannica, 2007.
8. Straub, H., A History of Civil Engineering, London, Leonard Hill, 1952.
9. Ra'afat, A., The art of Artechitecture and Reinforced Concrete, Cairo, Halabi, 1970.
10. Shaeffer, R. E., Reinforced Concrete: Preliminary Design for Architects and Builders, McGraw-Hill, 1992.
11. Hassoun, N., Structural Concrete – Theory and Design, 2nd edition, Prentice Hall, 2002.
12. Collins, P. (1959). Concrete, The Vision of a New Architecture, Faber and Faber, London.
13. Dunn, A., *Structures: Skyscrapers.* New York: Thomson Learning, 1993.
14. Ali, M. M., Evolution Of Concrete Skyscrapers: From Ingalls To Jinmao, *Electronic Journal of Structural Engineering,* Vol. 1, No. 1, pp. 2-14, 2001.
15. http://en.wikipedia.org/wiki/Ingalls_Building.
16. Moore, J. A., High-Performance Concrete for Bridge Decks, Vol. 21, No. 2, pp. 58, 1999.

17.   Bridger, R., Patrick, M. and Webb, J., High strength materials in composite construction, Internation Conference on Composite constructive – conventional and innovative, September 16-18, 1997, pp. 29-40.

18.   Hoffmeister, B., Sedlacek, G., Muller, C., and Kuhn, B, High Strength Materials in Composite Structures, Composite Construction in Steel and Concrete IV 2002, American Society of Civil Engineers, Feb 25, 2002.

19.   Moreno, J. "High-Performance Concrete: Economic Considerations," *Concrete International*, Vol. 20, No. 3, pp. 68-70, 1998.

20.   www.**confederationbridge**.com.

21.   Ghali, A., *et al.*, Field monitoring and research on performance of the Confederation Bridge, Journal of Canadian Civil Engineering, Vol. 24, No. 6, pp. 951-962, 1997.

Chapter 2

High Performance Concrete

Caijun Shi
College of Civil Engineering, Hunan University
Changsha, China
CJS Technology USA Ltd, USA

Y. L. Mo and H. B. Dhonde
Department of Civil and Environmental Engineering
University of Houston

2.1 Introduction

2.1.1 *Historical development*

Use of High Performance Concrete (HPC) truly began in 1927 when engineers building a tunnel through the Rocky Mountains near Denver needed a quick way of supporting the loads on the tunnel [1]. At that time HPC was in the research stages and was not yet ready to enter the market. The engineers prevailed upon scientists to allow its use. Eventually, the tunnel was built using this material. The reason for the use of HPC lies in its ability to reach an adequate maturity in 24 hours rather than 7 days for regular concrete.

In the 80's, the development and use of HPC attract world-wide attention due to the poor durability and short service life of these existing structures. Many organizations in the world started research programs on high performance concrete. In the United States, the Strategic Highway Research Program (SHRP) sponsored a project on High Performance Concrete in 1987 [2]. Federal Highway Administration (FHWA) initiated a national program in 1993 to implement HPC in bridges [3]. In 1996, AASHTO established the "Lead States Team" to

"Promote implementation of high performance concrete technology for use in pavements and bridges and share knowledge, benefits and challenges with states and their customers". The first meeting of the Lead States was held in September 1996, in St. Louis, Missouri. Initially focused on bridges, the HPC team consisted of the states of New Hampshire, Texas, Missouri, Nebraska, Virginia and Washington. At the 1998 meeting, Iowa and Arizona were added along with another Washington representative to add HPC Pavement to the HPC Team.

During the past two decades, significant progresses have been made on the design, testing and use of HPC. Zia et al [4] summarized some of the early research activities and applications of HPC. Many structural design codes, specifications and guidelines have also been developed correspondingly. Many books and conference proceedings have been published even by American Concrete Institute (see www.concrete.org). This chapter discusses the selection of raw materials, mix design, performance and applications of some HPCs.

2.1.2 *Definitions of HPC*

Ever since the introduction of HPC, numerous definitions have been proposed or published. Russell summarized many of them in a paper published in 1999 [5]. The following sections main discuss the definitions proposed FHWA programs and ACI.

In the Strategic Highway Research Program (SHRP), HPC was initially defined by three requirements [6]:

- Maximum water-cementitious material ratio of 0.35
- Minimum durability factor of 80%, as determined by ASTM C 666 Method A, and
- Minimum compressive strength of either:

   (a) 21 MPa (3000 psi) within 4 hours after placement;
   (b) 34 MPa (5000 psi) within 24 hours;
   (c) 69 MPa (10,000 psi) within 28 days.

In response to FHWA's call for a clear definition of HPC based on long-term performance criteria, Goodspeed et al. [7] proposed a definition, consisting of four durability and four strength parameters, each one being supported by performance criteria, performance testing procedures, and recommendations to allow performance to be accurately related to specific adverse field conditions. The eight performance criteria are freeze/thaw durability, scaling resistance, abrasion, chloride penetration, strength, elasticity, shrinkage, and creep. Users of this definition can indicate the level of performance that they require for each performance characteristic, based on their own field and weather conditions, in order to determine the HPC mixture best suited to their specific need.

In a FHWA's publication [8], it states that "HPC is concrete that has been designed to be more durable and, if necessary, stronger than conventional concrete. HPC mixes are composed of essentially the same materials as conventional concrete mixes. But the proportions are designed, or engineered, to provide the strength and durability needed for the structural and environmental requirements of the project." HPC may be regarded as any concrete that has properties needed for some objectives that are not possessed by ordinary concrete. According to the American Association of State Highway and Transportation Officials (AASHTO) Technology Implementation Group, HPC, which is concrete that has been designed to be more durable, and, if necessary, stronger than conventional concrete, can help highway agencies to build bridges that are better able to hold up to traffic and environmental demands. These bridges must also be economical to build and maintain.

In 1998, ACI published the following definition: "HPC is defined as concrete which meets special performance and uniformity requirements that cannot always be achieved by using only the conventional materials and mixing, placing and curing practices. The performance requirements may involve enhancements of placement and compaction without segregation, long-term mechanical properties, early-age strength, toughness, volume stability, or service life in severe environments". Based on the ACI definition, HPC includes:

- High workability concrete
- Self-consolidating concrete (SCC)
- Foamed concrete
- High strength concrete
- Lightweight concrete
- No-fines concrete
- Pumped concrete
- Sprayed concrete
- Waterproof concrete

- Autoclaved aerated concrete
- Roller compacted concrete

These different types of HPC are used for different purposes and environments. ACI has published committee documents on all these types of concrete [9].

As the performance requirements are different for different applications, the following sections mainly discusses high performance concrete as specified by FHWA. SCC is also discussed in detail due to its fast and wide adoption by the industry.

2.2 Constituents and Mixture Proportions of HPC

2.2.1 *Constituents of HPC*

Portland cement is the most widely used cement because of its commercial availability and low cost. It is still the most used binder for HPC. However, supplementary cementing materials such as silica fume, ground granulated blast furnace slag, coal fly ash have predominantly used for HPC. The use of superplasticizer is essential to achieve high strength, good workability and good durability. The selection of proper gradation and quality aggregates is the other requirement to produce high HPC. The ingredients of HPC are to be chosen to result in the desired performance and yet ensure economy [10]. In this process, the materials are stretched to their limits of performance [11].

2.2.1.1 Cement

Proper selection of the type and source of cement is one of the most important steps in the production of high-strength concrete. Variation in the chemical composition and physical properties of the cement affect the concrete compressive strength more than variations in any other single material. In US and some countries or regions, Portland cement is the commercial cement. It is much easier to select cement for HPC. Freyne et al. [12] tested eight cements – four ASTM C 150 Type Is, two Type I/IIs, one Type II, and one Type III, and found that mixtures containing Type III cement achieved the highest compressive strength at all ages tested, most significantly at early ages. Compressive strength differences among the mixtures were most pronounced at one day, but diminished over time through 56 days. Type III cement is often selected for high early strength. The use of Type II cement and high dosages of either Class F fly ash or GGBF slag are often the best options to control heat of hydration.

In some other countries and regions, cements are produced and marketed based on strength grades and their ingredients. In that case, it is more difficult to select the right grade and type of cement to be used. Previous experience and trial batches will be very important during the selection process.

There is an optimum cement content beyond which little or no additional increase in strength is achieved by increasing the cement content [13]. However, if cement content is increased there will be a remarkable influence on the consistency of concrete for the same water-cement ratio [14].

2.2.1.2 Supplementary cementitious materials

Supplementary cementing materials such as ground blast furnace slag, silica fume, metakaoline, coal fly ash and natural pozzolan are being widely used due to both technical and economical advantages. They can be divided into two categories based on the type of reaction they undergo: hydraulic or pozzolanic. Hydraulic materials react directly with water to form cementitious compounds, while pozzolanic materials chemically react with calcium hydroxide, a soluble hydration product, in the presence of moisture to form compounds possessing cementitious properties. Coal fly ash, ground granulated blast furnace slag (GGBFS) and silica fume are the most commonly used SCMs. The production and characteristics of these materials can be found in several publications [15,16].

The typical cement replacement for coal fly ash and slag is between 20-50%, and 5 to 10% for silica fume by mass. The use of silica fume usually requires the use of a high range water reducer. There are numerous benefits to incorporate SCMs into HPC mix design, include improving the workability of fresh concrete, reducing/ eliminating the free lime content, decreasing the C/S ratio of C-S-H in hardened cement pastes, mitigating alkali-aggregate reactions, etc. The products resulting from the reactions between lime and SCMs refine the pore structure of hardened pastes and reduce the permeability of hardened pastes. In many cases, it is necessary to use SCMs to achieve low permeability.

There are numerous advantages to using multiple SCMs in blended cements or as a separate additive in ternary and quaternary systems. Silica fume can be used to offset the early strength gain associated with the use of fly ash, while fly ash and slag can also be used to increase the long-term strength gain of silica fume concrete. Silica fume can also be used to reduce the levels of fly ash or slag required for sulfate resistance and alkali silica reaction mitigation. Fly ash, and to a lesser extent slag, can be used to offset the increased water demand associated with the use of silica fume. Overall, the use of multiple SCMs will greatly improve the resistance of concrete-to-chloride ion penetration and reduce the potential alkali-aggregate reaction expansion.

2.2.1.3 Aggregate

For each concrete strength level, there is an optimum size for the coarse aggregate that will yield the greatest compressive strength per unit mass of cement. In general, a smaller size aggregate will result in a higher compressive strength concrete. On the other hand, the use of the largest possible coarse aggregate size is important in increasing the modulus of elasticity or reducing creep and shrinkage.

According to ACI 211.4R [9] fine aggregates with a fineness modulus in the range of 2.5 to 3.2 are preferable for high-strength concrete. Concretes with a fineness modulus less than 2.5 may be sticky and result in poor workability and high water requirement. The gradation and packing of aggregates can have significant effects on properties of HPC. Chang et al. [17] studied effects of various fineness moduli (FM) of fine aggregate on the engineering properties of HPC, and found that the coarsest fine aggregate (FM = 3.24) has better positive effects on the properties of the fresh and hardened HPC.

For conventional concrete (strength < 40 MPa), strength of aggregate seldom becomes the restraint of the strength of concrete. However, for high performance concrete with strength > 60 MPa, especially when strength >80 MPa, the strength of aggregate controls the strength of aggregate. Aitcin and Metha [18] investigated the influence of four coarse-aggregate types available in Northern California on the compressive strength and elastic behavior of a very high strength concrete mixture. Using identical materials and similar mix proportions, the diabase and limestone aggregates were found to produce concretes with significantly higher strength and elastic modulus than did the granite and river gravel. The mineralogical differences in the aggregate types are considered to be responsible for this behavior. For a given mortar, the modulus of the concrete increases as the modulus of coarse aggregate increases [19].

The packing of aggregate also affects the properties of HPC. Tasi et al [20] found that the denser the aggregate packing the better the workability and engineering properties are under sufficient paste content. The application of the densified mixture design algorithm (DMDA) on designing HPC for every aggregate packing type can obtain high flowability and suitable strength growth. The strength efficiency of HPC designed by DMDA is much higher than that by the traditional one.

2.2.1.4   Chemical admixtures

Many chemical admixtures such as water reducing admixtures, retarding admixtures, accelerating admixtures, air entrainment agents, shrinkage reducing agents, steel corrosion inhibitors, antiwashout admixtures and alkali-aggregate reaction inhibitors, are used for the production of concrete [21,22]. The compatibility between cement and chemical admixtures and the optimum dosage of an admixture or combination of admixtures should be determined by laboratory experiments [21,23].

Water-reducing admixtures and high-range, water-reducing admixtures in HPC minimize the quantity of water necessary to produce a concrete with the required workability. They are often used in HPC specified for durability and are almost always used in high-strength concrete. Retarding admixtures and accelerating admixtures are used to delay or accelerate the setting of concrete. Air-entraining admixtures are used in HPC primarily to increase the resistance of the concrete to freeze/thaw damage when exposed to water and deicing chemicals. Shrinkage reducing admixtures are designed to decrease the effects of drying shrinkage. They function by reducing capillary tension in pore water that develops within concrete as it dries. This reduction in capillary tension reduces drying shrinkage, attendant cracking, microcracking and compressive creep [24]. Corrosion inhibitors are used in concrete to raise the chloride threshold level at which corrosion starts and to reduce the rate of corrosion after it begins. They are used primarily in producing structures that are exposed to chloride salts, such as bridge decks, parking garages, and marine facilities. For casting under water, antiwashout admixtures may be used to increase the cohesiveness of concrete, thereby reducing the loss of cement and increasing resistance to segregation. Alkali-aggregate reaction inhibitors are mainly these lithium compounds used as either an admixture in new concrete or as a treatment of existing structures.

2.2.1.5 Water

Water for mixing concrete shall be free from objectionable quantities of oil, acid, alkali, salt, organic matter, or other deleterious materials and shall not be used. The quality of hardened concrete is greatly influenced by the amount of water used in relation to the amount of cement, or water to cementitious materials ratio. Higher water to cement ratio is detrimental to the mechanical properties, deformation and durability of concrete. ASTM C1602/C 1602M [25] specifies the requirements for water to be used for mixing concrete.

2.2.2 *Mixture proportions for HPC*

The selection of mix proportions is the process of choosing suitable ingredients of concrete and determining their relative quantities with the object of producing as economically as possible concrete of certain minimum properties, notably strength, durability, and a required consistency [26]. Mix proportions for HPC are influenced by many factors, including specified performance properties, locally available materials, local experience, personal preferences, and cost. HPC is usually designed based on the requirements for specific applications and environments, which decide the constitute materials and mixture proportions for the HPC to be produced. For example, for precast, prestressed concrete bridge components such as beams and piles, engineers generally specify both a minimum strength at release of the prestressing strands and a design strength. For conventional strength concretes, mix proportions are then selected to achieve the release strength while the specified design strength at 28 days is easily exceeded. With high strength concrete, the design strength is higher and the release strength is correspondingly higher. To achieve the higher strength, it is necessary to increase the cementitious material content. For cast-in-place high performance concrete, as used in bridge decks or substructures, durability criteria rather than strength often control the selection of concrete mix proportions. In many cases, multiple engineering performance requirements (in addition to strength) need to be met simultaneously. The production of HPC is more of an art than a science, although the basic principle is simple [27].

There are various methods for proportioning conventional concrete [28]. Generally, these methods are based on fundamental functions: water-to-cement ratio, the constancy of water demand and theory of optimum aggregates proportioning, all of which determine mixtures with the required properties. However, most high-strength concretes contain SCMs other than cement. Consequently, the water-cementitious materials ratio must be considered instead of the water-cement ratio where the cementitious materials include cement, fly ash, silica fume, and GGBFS as appropriate. The effect of chemical admixtures, such as plasticizers or superplasticizers, can also be incorporated into these existing methods. However, with the new generation of HPC the problem of designing the concrete mixture becomes more sophisticated. Several procedures for proportioning HPC have been proposed or developed [29-33]. Most of them are semi-analytical. They usually provide the proportioning of aggregates and the calculation of W/C using an equation of compressive strength. Lim et al. [33] described a method for design of high-performance concrete mixtures using genetic algorithm. Sobolev [34] investigated strength properties and the rheological behavior of a cement–silica fume–superplasticizer system and proposed a method for proportioning HPC based on how the constituents affect the properties of the concrete.

Recently, Zain et al. [35] developed an expert system called HPCMIX that provides proportion of trial mix of HPC and recommendations on mix adjustment. The system was developed using hybrid knowledge representation technique. It is capable of selecting proportions of mixing water, cement, SCMs, aggregates and superplasticizer, considering the effects of air content as well as water contributed by superplasticizer and moisture conditions of aggregates.

Magee and Olek [36] collected and analyzed approximately 260 HPC mixtures from more than 200 publications. The statistic analyses on these mostly commonly used water, binder, air content, fine and coarse aggregates are plotted in Fig. 2.1. It is obvious that the use of relatively low water content (150-175 kg/m3) and high binder content (350-500 kg/m3) is very common for HPC. Two air content ranges of 1-2% and 5-6% indicate that the mixtures without and with air-entrainers respectively. Like conventional concrete, the most commonly used fine and coarse aggregate ranges for HPC are 700-800 kg/m3 and

C. Shi, Y. L. Mo & H. B. Dhonde

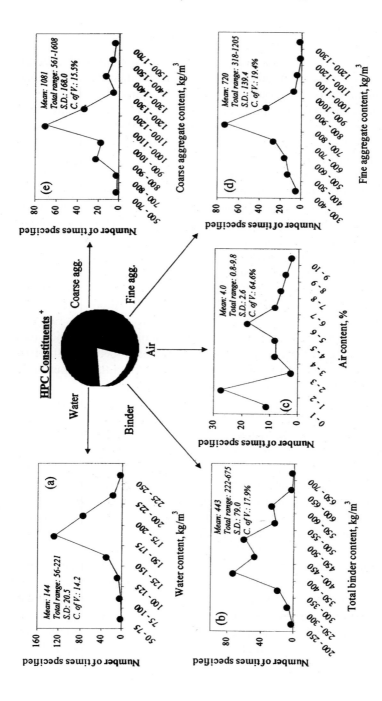

Fig. 2.1 Frequently used constituent quantities for HPC [36].

Table 2.1 Percentage of binder combinations for HPC [36].

No.	Combination	Percentage (%)
1	Portland Cement + Silica Fume	46
2	Portland Cement + Silica Fume + Fly Ash	17
3	Portland Cement	15
4	Portland Cement + Fly Ash	8
5	Portland Cement + Slag	7
6	Portland Cement + Silica Fume + Slag	7

1000-1100 kg/m3 respectively. The most commonly used admixtures include water reducers, air entrainers and retarders. High range water reducers or superplasticizers are the most frequently used admixtures, being specified for more than 50% of the mixtures reviewed.

As for the binders in HPC, six different combinations have been used, as summarized in Table 2.1. Combinations of portland cement and silica Fume are the most commonly used, 46% of the all mixtures reviewed. Silica fume is also often used in ternary combinations with portland cement with fl ash or slag. Figure 2.2 shows the detailed statistical analyses on the binders used for HPC.

2.3 Properties of High Performance Concrete

2.3.1 *Workability*

Properties of fresh concrete are very important since they affect the choice of equipment needed for handling and consolidation. Many terms such as consistency, flowability, mobility, pumpability, compactibility, finishability, and harshness, have been used to describe the properties of fresh concrete. Workability is often used to represent all those properties of fresh concrete. It is defined as the amount of mechanical work, or energy required to produce fully compacted concrete without segregation. A large number of tests have been proposed for the measurement of workability. The common ones include: (1) slump test, (2) compaction test, (3) flow test, and (4) Vebe test. Slump test is the most widely specified workability testing. The literature information

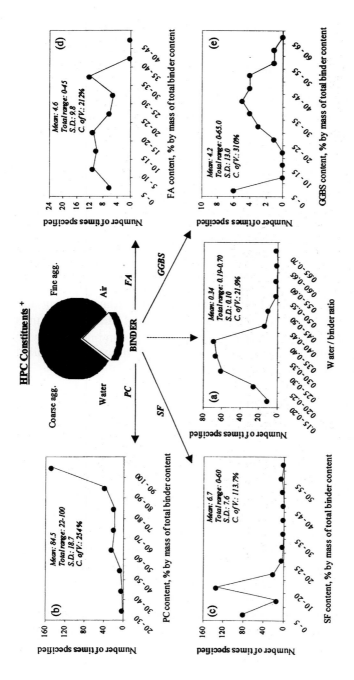

Fig. 2.2 Statistical analyses on the binders used for HPC [36].

analyses [37] indicated that 51% of HPC mixtures uses slump as the workability criteria and slump values for HPC are generally high, with 150-200 mm being the most common range, despite a slump as low as 6 mm being reported.

2.3.2 *Strength*

The strength of concrete is probably the most important overall measure of quality although other properties may also be critical since strength is directly related to the structural characteristics of the concrete. Many factors affect the strength of concrete. A simplified view of the factors affecting the strength of concrete is shown in Fig. 2.3. In the early 1970s, it was predicted that the practical limit of ready mixed concrete would be unlikely to exceed a compressive strength greater than 43 MPa. Over the past two decades, the development of high-strength concrete has enabled to produce very high strength concrete. Two buildings in Seattle, Washington contain concrete with a compressive strength of 130 MPa.

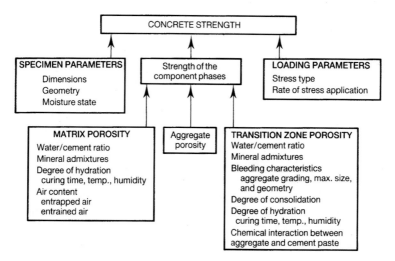

Fig. 2.3 Various factors affecting the strength of concrete.

The most common 28-day compressive strength range for HPC is 50-100 MPa, with strength as low as 30 MPa and as high as 150 MPa. ACI defines high-strength concrete as concrete with a compressive strength greater than 41 MPa. Manufacture of high strength concrete

(HSC) involves making optimal use of the basic ingredients that constitute normal-strength concrete. It needs to know what factors affect compressive strength and know how to manipulate those factors to achieve the required strength. In addition to selecting a high-quality portland cement, producers optimize aggregates, it is very important to optimize the combination of materials by varying the proportions of cement, water, aggregates, and admixtures. Since water to cement ratio determines the strength of cement paste, it has to use low water to cementitious materials ratio in order to achieve high strength. As mentioned above, for high performance concrete with strength > 60 MPa, especially when strength >80 MPa, the strength of aggregate controls the strength of aggregate. Thus, the selection of high quality aggregate is critical to produce very high strength concrete.

HSC is usually specified where reduced weight is important or where architectural considerations call for small support elements. By carrying loads more efficiently than normal-strength concrete, high-strength concrete also reduces the total amount of material placed and lowers the overall cost of the structure.

2.3.3 *Stress-strain relationship and modulus of elasticity*

Stress-strain relationship is very important in structural design. Concrete, like many other construction materials, shows elastic behaviour only to a certain degree. Strictly speaking, the Young's modulus of elasticity is only applied to the straight part of stress-strain curve. The tangent to the curve at the origin is called initial tangent modulus. The deformation occurring during loading is elastic and the sequent increase in strain is regarded as creep. The secant modulus is a static modulus. For comparative purpose, the maximum stress applied for determination of the secant modulus is specified as 40% of ultimate strength in ASTM C469 (2006) [38]. To eliminate creep, at least two cycles of preloading are required in ASTM C469.

The stress-strain behavior is dependent on a number of parameters which include material variables such as aggregate type and testing variables such as age at testing, loading rate, strain gradient and others. Many researchers have measured the complete stress-strain relationships of concrete with compressive strength upto 100 MPa. HSC demonstrates different stress-strain curves from conventional concrete in the following aspects [39]: (1) the matrix stiffness of HSC is larger than that of conventional concrete and approaches the stiffness of aggregate; (2) the bond strength between matrix and aggregate is higher in HSC; (3) the matrix tensile strength of HSC is higher, and (4) internatl cracking is reduced in terms of the number and size of cracking before loading. All these aspects mean that HSC behaves more elastic and brittle than conventional concrete. Figure 2.4 shows the schematic stress-strain curve and cracking patters of conventional concrete and HSC under uniaxial loading.

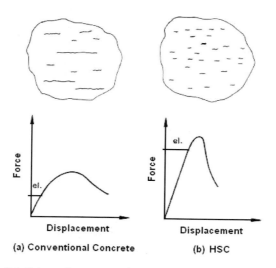

Fig. 2.4 Schematic stress-strain curve and cracking patterns of
conventional concrete and HSC [39].

Wee et al. [40] tested 169 cylinders with strength from 50 to 120 Mpa and found that the strain at stress peak increased with the strength of the concrete, however, for concretes with the same strength, their stress-strain curves demonstrated the same shapes regardless of different water to cement ratios, dosage and type of supplementary cementitious materials, and testing ages.

Cheng et al. [41] measured stress-strain curves of high strength concrete up to 800°C and found that HSC exhibited brittle properties below 600°C, and ductility above 600°C. HSC with steel fibers exhibits ductility for temperatures over 400°C. The compressive strength of HSC decreases by about a quarter of its room temperature strength within the range of 100–400°C. The strength further decreases with the increase of temperature and reaches about a quarter of its initial strength at 800°C. The strain at peak loading increases with temperature, from 0.003 at room temperature to 0.02 at 800°C.

In a recent FHWA project report [42], review of several published equations that relate the modulus of elasticity, E_c, to the concrete compressive strength, f'_c, and the unit weight of the concrete, w_c was presented. One such relationship is recommended in the ACI 318 Committee-ACI MCP06, 2006 [43] for concrete strengths in the range of 28 to 41 MPa:

$$E_C = 33(W_c)^{1.5}\sqrt{f'_c} \tag{2.1}$$

And also used in Ahmad's equation [44] for concrete strengths of 55, 69, and 83 MPa:

$$E_C = (W_c)^{2.5}(f'_c)^{0.325} \tag{2.2}$$

Collins et al. [45] have indicated that the strain at peak stress, ε'_c, can be calculated from the following equation:

$$\varepsilon_C' = \frac{f'_c}{E_c} \cdot \frac{n}{n-1} \tag{2.3}$$

where

$$k = 0.8 + \frac{f'_c}{17} \text{ (in MPa unit)} \tag{2.4}$$

These equations were used to calculate the strain at peak stress from the modulus of elasticity and concrete compressive strength.

Thorenfeldt et al. [46] proposed the following equation to increase the slope of the descending portion of the stress-strain curve as follows:

$$\frac{f_C}{f_c'} = \frac{\varepsilon_C}{\varepsilon_C'} \cdot \frac{n}{(n-1)+(\frac{\varepsilon_C}{\varepsilon_C'})^{nK}} \tag{2.5}$$

where, $k = 1$ on the ascending portion of the curve, and Collins et al. [45] suggested the following for the descending portion of the curve.

$$k = 0.67 + \frac{f'_c}{62} \text{ (in MPa unit)} \tag{2.6}$$

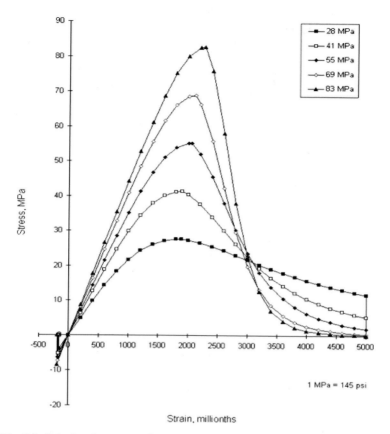

Fig. 2.5 Calculated stress–strain curves for concrete with different strengths [42].

Based on these equations, the calculated stress-strain relationships are plotted in Fig. 2.5.

The direct tensile stress-strain curve is difficult to obtain. Due to difficulties in testing concrete in direct tension, only limited and often conflicting data are available. Virtually no data is available regarding the strain capacity in flexural tension.

2.3.4 *Shrinkage*

Shrinkage is the reduction in volume at constant temperature without external loading. It is an important material property that has significant effects on the long-term performance of designed structures. Shrinkage can be classified into autogenous shrinkage, drying shrinkage and carbonation shrinkage. Autogenous shrinkage refers to volume changes caused by the hydration of cement. Drying shrinkage results from the drying of cement and concrete materials. Carbonation shrinkage occurs when the hydration products of cement react with CO_2 in the environment.

The decline of pore water pressure or international relative humidity within concrete due to the hydration of cement is the driving force for the autogenous shrinkage to occur. Autogenous shrinkage of conventional concrete is much smaller than that of HPC such as HSC and SCC with a low water-to-cement ratio. HPC might crack at early age as a consequence of restrained autogenous deformation, which is also confirmed by numerical analysis [47].

Nawa and Horita [48] measured autogenous shrinkage of HPC and divided it into four stages (as shown in Fig.2.6): (1) the initial period, (2) the induction period, (3) the acceleration period, in which the main shrinkage first begins to occur rapidly, and (4) the deceleration period, in which low shrinkage continues slowly.

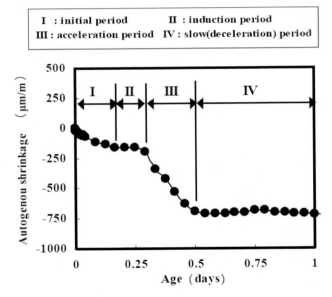

Fig. 2.6 Typical pattern of autogenous shrinkage at early age [48].

The autogenous shrinkage was dependent on w/c, type of cement used, and type and content of SCMs, dosage and type of superplasticizer, temperature, age, etc. The type of cement strongly affects autogenous shrinkage in all stages. Alumina cement and high-early-strength cement exhibit large early autogenous shrinkage and lead to large ultimate shrinkage. On the other hand, moderate-heat cement and belite-type low-heat cement show small autogenous shrinkage. Autogenous shrinkage increase as the cement fineness increases [49]. The W/C greatly influences autogenous shrinkage within the age of one day, particularly autogenous shrinkage in the acceleratory period, but has less influence in the deceleration period after the age of one day. On the other hand, what the superplasticizer influenced was mainly the length of the induction period of autogenous shrinkage [48].

The replacement of cement with fly ash reduces autogenous shrinkage [47,48]. The higher the replacement percentage, the lower the autogenous shrinkage. While the autogenous shrinkage increases as silica fume [50-52] or GGBFS content increases [53,54]. Several studies [55-58] have confirmed that the use of porous aggregate could reduce or eliminate the autogenous shrinkage. It is reported that replacement of 25% normal weight aggregate with water saturated lightweight concrete could eliminate the autogenous shrinkage [56].

The key factors affecting the magnitude of shrinkage are: aggregate, water-cementitious material ratio, member size, medium ambient conditions, admixtures, pozzolans, chemical admixtures, cement type. Drying shrinkage is a function of the paste, but is significantly influenced by the stiffness of the coarse aggregate. Pozzolans typically increase the drying shrinkage; chemical admixtures will tend to increase shrinkage unless they are used in such a way as to reduce the evaporable water content of the mix, in which case the shrinkage will be reduced.

Alfes [59] examined how shrinkage was affected by the aggregate content, the aggregate modulus of elasticity, and the silica fume content. Using W/C ratio in the range of 0.25 to 0.3 with 20% silica fume by weight of cement and varying amount and type of aggregates (basalt, LD-slag, and iron granulate), he produced concretes with 28-day strength in the range of 102 to 182 MPa. The test results showed that there is a direct and linear relationship between the shrinkage value and the modulus of elasticity of the concrete.

Hindy et al. [60] measured dry shrinkage of concrete specimens as well as on instrumented reference columns made with two different ready-mixed high-performance concretes. They found that the longer the curing time the lower the dry shrinkage, and that the lower the W/CM ratio the lower the dry shrinkage`. Dry shrinkage of small specimens measured by the conventional laboratory test seemed to over-estimate shrinkage of the concrete in the real structure. The ACI 209 predictive equation [61] could be used for the high performance concretes only if new values for the parameters were introduced. Mokhtarzadeh and French [62] found that the drying shrinkage strains of HPC observed in a laboratory study ranged between 63 to 83% of values predicted by ACI 209 equations. Huo et al [63] noticed that the prediction equations for shrinkage strains and creep coefficients in the ACI 209 Committee Report overestimated the actual shrinkage strains and creep coefficients of HPC. They proposed some new equations of shrinkage strains and creep coefficients, modified from the ACI 209-92 equations with strength correction factors, are proposed throughout the study. The proposed equations take into account the effects of concrete strength and can be used for both conventional concrete and HPC.

The effect of supplementary cementitious materials on drying shrinkage of concrete is dependent upon the nature, replacement level and concrete mixture proportions. Drying shrinkage in practical high-strength concrete is either equal to or somewhat lower than that for concrete without silica fume [64]. However, Li and Yao [65] found that the use of ultrafine GGBS and/or silica fume significantly decreases the drying shrinkage of HPC. Clearly, the water-reducing property of fly ash can be advantageously used for achieving a considerable reduction in the drying shrinkage of concrete mixtures [66].

Carbonation shrinkage is caused by the reaction between carbon dioxide (CO_2) present in the atmosphere and calcium hydroxide (($CaOH)_2$) present in the cement paste. The amount of combined shrinkage varies according to the sequence of occurrence of carbonation and drying process. If both phenomena take place simultaneously, less shrinkage develops. The process of carbonation, however, is dramatically reduced at relative humidities below 50%. Persson [67] suggested that the carbonation rate of HPC was related to age and w/c. The conditions for carbonation shrinkage of HPC were settled related to w/c and content of silica fume. At low w/c and high content of silica fume all the calcium hydroxide was consumed in HPC, which more or less eliminated the carbonation and thus also the carbonation shrinkage. Figure 2.7 indicated that there is a good relationship between 28-day compressive strength and carbonation depth of concrete containing PFA and SF [68]. The higher the compressive strength, the lower the carbonated depth. A recent study also indicated that when compared at an equal strength, the effect of the type of fly ash on carbonation becomes insignificant [69].

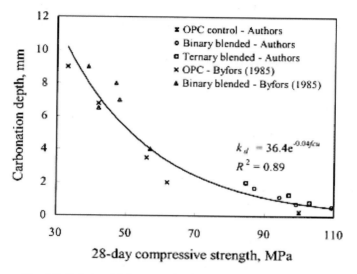

Fig. 2.7 Relationship between 28-day compressive strength and 2-year carbonation depth of concrete [68].

2.3.5 *Creep*

Creep refers to the deformation of concrete under loads at a constant temperature. There are two types of creep: basic creep and drying creep. Basic creep refers to the creep generated under constant humidity conditions, and drying creep is the creep during drying. On unloading, deformation decreases immediately due to elastic recovery. This instantaneous recovery is followed by a more gradual decrease in deformation due to creep recovery. The remaining residual deformation, under equilibrium conditions, is called "irreversible creep. Creep of concrete is normally evaluated using unsealed loaded and unloaded companion specimens exposed at a constant drying environment. Thus, the total deformation may be separated into the elastic compression, basic creep, and drying creep (moisture loss, autogeneous and carbonation shrinkage). Creep coefficient, specific creep, or creep compliance are generally used to describe creep strain by different mathematical prediction models. The creep coefficient is defined as the ratio of creep strain (basic plus drying creep) at a given time to the initial elastic strain. The specific creep is defined as the creep strain per unit stress. The creep compliance is defined as the creep strain plus elastic strain per unit stress, whereas the elastic strain is defined as the instantaneous recoverable deformation per unit length of a concrete specimen during the initial stage of loading.

Creep tests are usually required for concrete used in both high-rise buildings and long-span bridges. When doing creep test, it is important to specify specimen size; curing conditions before the test begins; curing conditions during the test; age of loading of the test specimens; and duration of load. Size of specimens is important because creep is size dependent. Smaller specimens creep faster than larger ones. Curing conditions before and during the test influence the properties of the concrete and the amount of creep. The amount of creep is also dependent on the age at which specimens are loaded. Specimens loaded at later ages undergo less creep than specimens loaded at early ages.

Müller et al. [70] compared the shrinkage and creep characteristics of high-performance concrete and normal strength and found that the prediction of creep and shrinkage of normal and high-performance concrete may be covered by one single approach being closely related to the model given by the CEB-FIP Model Code 1990, if some modifications and extensions of the original prediction formulae are introduced. Mokhtarzadeh and French [62] found that the recommended form of ACI 209 equation: $V_t = [t^{0.60}/(10 + t^{0.60})]*Vu$, was determined to be suitable for predicting the creep coefficient of high-strength concrete at any time t. The range of ultimate creep coefficients Vu predicted in this study varied between 0.92 to 2.46, as compared with the 1.30 to 4.15 range reported by ACI 209 for normal-strength concrete. The specific creep of high-strength concrete followed the general trend of more conventional concrete, i.e., concrete specific creep decreased as compressive strength increased. Because HPC and HSC exhibit less creep and shrinkage than conventional concrete, the majority of traditional creep and shrinkage models and methods for estimating prestress losses, over-predict the prestress losses of HPC and HSC girders [71].

In the SHRP C-205 studies [72], it was found that the creep behavior of very high strength concretes with different aggregates (crushed granite, marine marl, and rounded gravel) was evaluated. Creep strain measurements were made for 90 days in each case. The observed creep strains of the different groups of very high strength concrete ranged from 20% to 50% of that of conventional concrete. The creep strains were especially low for concretes with a 28-day strength in excess of 70 MPa (10,000 psi). Ma and Orgass [73] tested the creep of a self-compacting ultra-high performance concrete (UHPC) with a compressive strength between 150 and 170 MPa. Compared to conventional high strength concrete, the creep coefficient of UHPC is in principle lower than that for conventional high strength concrete and the drying creep of UHPC can be neglected.

The use of ultrafine ground GGBS and/or silica fume significantly decreases the creep and drying shrinkage of HPC [74]. In another study [54], it was found that the basic creep of HPC decreases as the silica fume replacement level increases. They also proposed some alternative prediction models because the existing models for predicting creep and shrinkage were inaccurate for high-strength concrete containing silica fume.

2.3.6 *Durability of HPC*

2.3.6.1 Introduction

Durability is one of the most important desired properties of concrete. Concrete is inherently a durable material. However, concrete is potentially vulnerable to attack in a variety of different exposures unless some cautions are taken. HPC that has a water/binder ratio between 0.30 and 0.40 is usually more durable than ordinary concrete not only because it is less porous, but also because its capillary and pore networks are somewhat disconnected due to the development of self-desiccation [75]. However, self-desiccation can be very harmful if it is not controlled during the early phase of the development of hydration reaction, therefore, HPC must be cured with more acres in order to achieve the designed properties. This section discusses the different durability aspects of HPC.

2.3.6.2 Permeability

Permeability of concrete plays an important role in durability of Portland cement concrete because it controls the rate of entry of moisture that may contain aggressive chemicals and the movement of water during heating or freezing. For a given cement, water-to-cement ratio (W/C) has the largest effect on permeability of concrete. A decreased W/C increases the strength of concrete and hence improves its resistance to cracking from the internal stresses that may be generated by adverse reactions.

The permeability of mature cement paste is very low even though it has a high total porosity since water cannot move through very small discontinuous gel pores. For a given W/C ratio, the water permeability coefficient of concrete is about two orders higher than that of the cement paste due to the effects from the internal cracking and continuous pores in the cement paste-aggregate interfacial zone. Also, the water permeability of concrete increases with the increase of maximum aggregate size.

There is no recognized standard test method to measure the permeability of concrete although many methods have been proposed in publication. Hearn et al. [76] gave an excellent review on the permeability of concrete and test methods. Actually, chloride permeability, instead of water permeability, is often specified for HPC, as described in the next section.

2.3.6.3 Transport of chloride ion in HPC

Chloride ions have the special ability to destroy the passive oxide film of steel even at high alkalinities. The penetration of chlorides in sufficient amounts to the depth of reinforcing steel will cause corrosion of the reinforcing steel and damage of reinforced concrete structures. Chloride-induced corrosion is a very common cause of concrete deterioration along sea coasts and in cold areas where deicing salts are used. The repair of damaged reinforced concrete structure caused by chloride-induced corrosion is very expensive. Thus, selection of a quality concrete material with low chloride permeability can be critical in construction of a durable steel reinforced concrete structure where chloride-induced corrosion is a concern.

Many test methods have been proposed or developed to measure the transport of chloride ion in hardened cement pastes and concretes.

However, each of them has certain advantages or limitations [77]. Based on the principle of transport of chloride ion during the testing, all these tests can be classified into non-steady state and steady state transport tests directly or indirectly. Steady state diffusion coefficients cannot be compared directly with non-steady state diffusion coefficients. A comparison of different test methods indicates that NT Build 492 [78] is the most appropriate one for the accelerated chloride migration test under non-steady conditions, and NT Build 355 [79] is the most appropriate one for testing under steady state conditions [77]. ASTM C1202 [80] or AASHTO T 277-1980 [81] is virtually a measurement of electrical conductivity of concrete and can be used for quality control purpose, not for evaluating the chloride permeability of concrete without correlations with other test results [82,83]. However, the passed charge from ASTM C1202 or AASHTO T 277 test, or so-called "rapid chloride permeability", is the main specified criteria for concrete in North America. A charge value of less than 700 to 1000 coulombs is typically specified for HPC, which is characterized as very low chloride permeability based on the rating in ASTM C1202.

It is generally believed that the use of low water to cement ratio and SCMS such as silica fume, slag and fly ash is critical to achieve low permeability or high resistance to the penetration of chloride ion into concrete. Analyses of 254 HPC mixes from publications indicated that 33% of them had durability performance attached in the form of chloride permeability, and only 4% of them specified coefficients of diffusion [84]. Further analyses indicated that relationships were evident between the "rapid chloride permeability" or chloride ion diffusion of coefficient and water to binder ratio, total binder content or silica fume content, while fly ash, slag and Portland cement contents have no evident relationships with the "rapid chloride permeability" or diffusion coefficient [84]. Bajorski et al. [85] plotted a contour map for concrete containing Portland cement, silica fume (both at varying quantities) and fly ash at a constant dosage of 30% by mass of total binder based on a limited number of experimental mixes, as shown in Fig. 2.8, which can which can be helpful to the explanation of experimental results. However, it should be keep in mind that the low "rapid chloride permeability" of concrete containing SCMS, especially silica fume, is partially due to the change of OH⁻ concentration in the pore solution, which has little to do with the permeability of concrete [82,83].

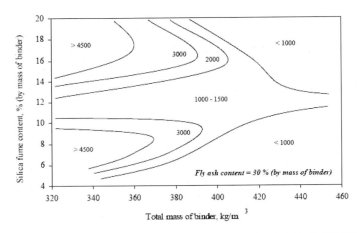

Fig. 2.8 Contour map of chloride permeability (Coulombs) for HPC [85].

2.3.6.4 Chemical resistance

- Acid corrosion

The acid corrosion of hardened cementing materials has drawn more and more attention recently due to the corrosion of concrete sewer pipes and concrete structures in municipal wastewater treatment plants [86], the impact of aggressive substances from animal feed and manure [87,88] and concerns regarding the acid corrosion resistance of cement-solidified wastes [89]. As discussed in published papers [90-92], many factors such as such as the nature of the cement, water to cement ratio, characteristics of aggregates, pore structure of hardened materials and curing conditions can affect the acid corrosion resistance of hardened cement and concrete. Common hydration products from conventional cementing materials include C-S-H, $Ca(OH)_2$, AFt, AFm, C_4AH_{13} , C_3AH_6 and C_2ASH_8. Except C-S-H, all other hydration products start to decompose between pH 11 and 12. C-S-H will also undergo a series of changes as pH drops [93]. There are two plateaus for the equilibrium pH. One corresponds to around pH 12, where the C/S ratio is greater than approximately 1.0. The other plateau corresponds to pH 10, where the C/S ratio of C-S-H varies from 0.05 to 0.6. Thus, as pH of the solution is lower than 10, C-S-H starts to decalcify very quickly and silica gel remains. The solubility of amorphous silica at room temperature is almost constant, varying from 100 to 150 ppm, when pH is below 9 [94]. Remained silica gel after the decomposition or decalcification of hydration products has a very low solubility and can provide certain protection on uncorroded cement pastes from further corrosion [91]. Thus, the remained products on the uncorroded cement pastes after acid corrosion play a very important role in determining the acid corrosion resistance of the cement materials, even more import than some other factors such as water to cement ratio, pore structure of hardened materials or curing conditions. This means that the use of siliceous pozzolanic materials can greatly improve the acid corrosion resistance of hardened cement and concrete, which is in agreement with published results [95,96].

- Sulphate attack resistance

Sulfate attack on concrete is a relatively rare but complex damage phenomenon caused by exposure of concrete products or structures to an excessive amount of sulfate from internal or external sources [97]. External sulphate attack is due to penetration of sulfates in solution, in groundwater for example, into the concrete from outside; while internal sulphate attack is due to a soluble source being incorporated into the concrete at the time of mixing, gypsum in the aggregate, for example. External sulfate attack is the more common type and typically occurs where water containing dissolved sulfate penetrates the concrete. A fairly well-defined reaction front can often be seen in polished sections; ahead of the front the concrete is normal, or near normal. Behind the reaction front, the composition and microstructure of the concrete will have changed. Concrete quality is the most important issue in sulfate attack. The negative effects of environment, sulfate concentration, and composition are most noticeable in concrete of lower quality.

De Almeida [98] invested the sulfate resistance of concretes with compressive strengths between 60 and 110 MPa was evaluated. The test comprises several soaking/drying cycles of samples in a $Na_2SO_4.10H_2O$ solution, followed by measurement of mass variation and residual compressive strength. It was found that the resistance to sulfate attack depends on concrete porosity and capillary absorption, and not on permeability, because pozzolanic reactions seem to interrupt pore continuity. The reduced w/c ratio obtained with the aid of the super-plasticizer was much more effective than the chemical characteristics related to the presence of mineral admixtures in concrete, as regards its resistance to sulfates.

The use of aggregates contaminated with sulfate salts cause the strength loss due to internal sulphate attack, but HPC is still more resistant than normal concrete [99]. Shannag and Shaia [100] suggested that the high strength concrete mix made using ordinary portland cement only did not perform satisfactorily in sulfate solutions and sea waters, but high-performance concretes that contain various combinations of silica fume and natural pozzolan, can provide good balance between strength and durability.

Properly designed (low water cementitious ratio, adequate cement content), produced (properly mixed and cured to reach high density and low permeability), placed, finished and cured concrete, will not suffer from sulfate attack to the same degree as an inadequately cured, higher water-to-cement ratio concrete. It might not suffer any damage even in the harshest sulfate environment [97]. Thus, it can be expected that HPC has much better sulphate resistance than conventional concrete. A good quality control program during the construction and curing will be very important to ensure the HPC to achieve designed performance targets.

- Alkali-aggregate reaction

The expansion and cracking of concrete caused by alkali-aggregate reaction was first studied in the early 1930's. There are two types of alkali-aggregate reaction, which can lead to cracking of concrete: (a) alkali-silica reaction (ASR) and (b) alkali-carbonate reaction (ACR). ASC is a chemical reaction within concrete between specific siliceous constituents, which sometimes occur in the aggregate and the alkalis in the concrete mixture. The reaction product is alkali silicate gel that swells after absorbing moisture and cause expansion and cracking of concrete. ACR is an expansive de-dolmitisation process resulting from the reaction between dolomite in carbonate aggregate and alkalis in the concrete mixture. ASR is a very common durability problem in the world, while ACR is rare. The following paragraphs will focus ASR.

It is widely accepted that there are three essential components for ASR-induced damage to happen in concrete: (1) reactive silica (from aggregates); (2) sufficient alkalies (mainly from portland cement, but also from other constituent materials); and (3) sufficient moisture. Eliminating any one of the above components effectively will prevent damage due to ASR. The most common methods of minimizing the risk of expansion due to ASR include [101]:

- Using nonreactive aggregates.
- Limiting the alkali content of concrete.
- Using supplementary cementing materials.
- Using lithium compounds.

The use of nonreactive aggregates is certainly a viable method to prevent ASR to happen. Limiting the alkali content of concrete mixtures below some threshold value is generally effective in preventing ASR-induced damage, but this approach is not always effective by itself since unanticipated high concentrations of alkalies may result from exposure to deicing salts, alkali release from aggregates, drying gradients (resulting in alkali migration), and other field effects. Using lithium compounds, especially $LiNO_3$, can be a viable approach to controlling ASR-induced damage [102]. The use of SCMs to control ASR in concrete is the most common mitigation measure used in concrete construction. The benefits of properly using SCMs include not only ASR mitigation, but also improved resistance to other durability problems, including sulfate attack, corrosion of reinforcing steel, and freezing and thawing.

Fly ash is one of the most commonly used SCMs in the world for several reasons, including economic and technical benefits. Perhaps the most important parameter affecting the ability of fly ash to control ASR expansion is the CaO content of the ash [103]. Generally, lower-lime ashes are more effective than higher-lime ashes in controlling ASR, mainly due to the higher alkali-binding capacity of concretes containing lower-lime ashes. Slag also is commonly used to mitigate ASR and typically is used in higher dosages than fly ash, typically 35 to 50% (by mass of cement) and in some cases, in even higher dosages [104]. The specific dosage needed to mitigate ASR in a given concrete mixture depends on the reactivity of the aggregate and the total alkali content of the concrete. Silica fume has not been used as frequently as fly ash and slag to control ASR. Although the efficacy of silica fume in minimizing the risk of ASR-induced damage depends on the reactivity of the aggregate, it appears to depend more on the total amount of alkalies available within the concrete. Actually, the use of combinations of fly ash, slag and silica fume, known as ternary blends, either through the use of a blended cement in conjunction with another SCM or through the use of plain cement with two SCMs, may reduce the quantities that would otherwise be used individually as combinations are often synergistic in their ability to control ASR [105,106]. Such combinations may improve the economic situation, workability, early strength development, and durability properties. Thus, the use of proper type and amount of SCMs in HPC can have advantages in several aspects.

- Corrosion in Chloride Solutions

It is well known that chlorides cause the corrosion of steels in concrete, while high concentration of calcium chloride also deteriorates concrete very quickly. Hydrated monochloroaluminate ($C_3A.CaCl_2.xH_2O$) is always identified in calcium chloride attacked cement concrete. However, it cannot be regarded as the main cause of deterioration because this salt forms in cement pastes, which do not deteriorate. At the same time, a complex chloride-containing salt was also noted but could not be fully characterized [107]. Thus, the mechanism of the $CaCl_2$ attack on Portland cement concrete was not clearly stated until the elucidation of $3CaO.CaCl_2.12H_2O$ in calcium-chloride-attacked cement concrete [108-110]. However, $CaO.3CaCl_2.12H_2O$ is a very unstable compound and can even decompose in a dry air [111]. The use of SCMs can reduce or eliminate free $Ca(OH)_2$, or the formation of $CaO.3CaCl_2.12H_2O$ so improve the resistance to $CaCl_2$ attack [110].

- Corrosion of steel reinforcement in HPC

The pore solution of a fully hydrated Portland cement usually has a pH over 13 and consists mainly of alkali hydroxides. The high pH results in the formation of a passive layer on the surface of steel reinforcement in the concrete, which is very dense, impenetrable film preventing the further corrosion of the steel. However, the passive layer can be destroyed due to the transport of chlorides to the surface of the steel, or due to the carbonation of the concrete, which will drop the pH down to around 8. Once the passive layer is destroyed, the steel can be corroded very quickly. The corrosion of the reinforcing steel in concrete by de-icing salts is one of the major issues concerning the durability of reinforced concrete.

The chloride corrosion threshold can vary between concrete in different bridges, depending on the type of cement and mix design used, which can vary the concentrations of tricalcium aluminate (C_3A) and hydroxide ion (OH^-) in the concrete. In fact, it has been suggested that because of the role that hydroxide ions play in protecting steel from corrosion, it is more appropriate to express corrosion threshold in terms of the ratio of chloride content to hydroxide content, [Cl⁻]/[OH⁻], which was recently established to be between 2.5 to 6 [112,113].

For most corrosion protection measures, the basic principles are to prevent chloride ion from react with or extend the time to reach the surface of steel [114]. The enhanced durability of HPC helps it resist penetration of chloride-laden snow and ice melt water, so to protect reinforcement against corrosion, resist chemical and physical deterioration. This results in longer life for the reinforcing steel within, and a reduction in spalling, cracking and associated repairs. HPC with low water/cement ratio of 0.28 to 0.32 is notably more effective for improving durability against chloride attack than conventional concrete. In order to realize durable concrete bridges with a long service life in severe environment such as ocean splash zones, both HPC and additional cover depth are essential to protect the chloride ions from rebars and tendons at reasonable cost [114].

Recently, many models have been proposed to predict the rate of steel corrosion in concrete structures [115-117]. Usually, these models consider initiation and propagation stages and include the effects of changes in exposure conditions during the propagation stage on corrosion and the effects of the corrosion reactions on the properties of concrete. However, the boundary conditions for solving the Laplace's equation for electrochemical potential may be different in different models. The use of these models may enable designers to carry out comprehensive sensitivity analyses and to gauge the significance of variations in the values of certain parameters on the rate of corrosion in concrete structures.

2.3.6.5 Frost resistance

The frost resistance of concrete is of considerable importance in areas where freezing happens during the winter. Unprotected cement pastes dilate as they are frozen, which results in internal tensile stresses and cracking. Several mechanisms have been proposed to explain the paste behaviour during freezing: generation of hydraulic pressure by ice formation, volume increase (9%) due to the conversion of water to ice, desorption of water from C-S-H, and segregation of ice.

For the normal strength concrete, entrained air of 4 to 8% by volume of concrete provides an effective defense against frost damage and the exact amount is dependent on the maximum size of the coarse aggregate, provided that the coarse aggregate itself is frost resistant. Some argue that with low W/CM ratio and mineral admixtures, the amount of freezable water in high-strength concrete would be low and its pore size would be decreased to the extent that water in the pore cannot freeze. Li et al. [118] investigated the freezing and thawing (F/T) durability of non-air-entrained cement pastes, mortars, and concrete. They used four different W/C ratios of 0.24, 0.27, 0.30, and 0.33 and cured the specimens in a moist environment until testing at 14, 28 or 90 days. Freezing and thawing was performed according to ASTM C 666 [119], Procedure A. The F/T durability of non-air-entrained pastes and mortars was evaluated by measuring the decrease in compressive strength, but the F/T durability of non-air-entrained concrete was determined relative dynamic modulus. For comparison purposes, the relative dynamic modulus of the mortar was also measured. At the W/C ratio of 0.24, both the paste and mortar showed excellent F/T resistance at 0, 5, and 10% silica fume levels. When the W/C ratio was higher than 0.24, the paste and mortar durability was significantly reduced. Similarly, at the W/C ratio of 0.24, the non-air-entrained concretes were F/T durable regardless of the silica fume and total cementitious content but the durability was decreased for concretes with higher W/C ratios. The results indicated that factors other than the W/C ratio had little influence on the F/T durability and the critical W/C value was 0.24. The damage in the paste was characterized by surface scaling while in the mortar and concrete a few large cracks led to final failure. Another laboratory study [120] found that it was possible to obtain concrete with good frost resistance based on the modified ASTM C 666 at w/c less than 0.40 without air entrainment, and no air entrainment was necessary for mixtures with w/c of 0.25 or less as far as scaling resistance is concerned.

Analyses of 254 HPC mixture specifications indicated that 37% of them still specify the air content as a criteria for freezing-thawing resistance, only next to strength and slump specifications, which are 89% and 51% respectively [84]. Thus, it is still important to introduce a proper amount of air with reasonable bubble size distribution to ensure the freezing-thawing resistance of the concrete, especially when water to cement ratio is relatively high [121].

2.3.6.6   Wear resistance

Wear can be classified into three catalogues: abrasion, erosion and cavitation. Wear refers to repeated rubbing or frictional process, which is usually in connection with traffic wear on pavements and industrial floors. Erosion is the abrasive action of fluids and suspended solids. It is a special case of abrasion and occurs in water-supply installations such as canals, conduits, pipes and spillways. Cavitation is the impact damage caused by high velocity disturbed liquid flow, which happens at spillways and sluiceways in dams and irrigation installation. Mechanical abrasion is the dominant abrasion for pavements and bridge decks exposed to studded tires.

Laplante et al. [122] found that coarse aggregate was the most important factor, followed by W/CM ratio in rank, affecting the abrasion resistance of concrete. The abrasion resistance of concrete is strongly influenced by the relative abrasion resistance of its mortar and coarse aggregate. When the coarse aggregate and mortar have nearly the same abrasion resistance, the surface wear of the concrete would be fairly uniform and the concrete can present serious skidding and slipping problems when wet. When the W/CM is very low, it can make the concrete almost as abrasion resistant as high-performance rocks.

De Almeida [123] investigated the abrasion resistance of high-strength concretes containing chemical and mineral admixtures following a Portuguese Standard, which is similar to the Brazilian Standard [124] and the German Standard DIN 52108 [125]. The concrete mixtures contained silica fume, fly ash or natural pozzolan, with or without a superplasticizer, with workability being kept constant. They had W/CM varied from 0.24 to 0.42 and compressive strength from 60 to 110 MPa at 28 days. It was found that the abrasion resistance of concrete generally varies inversely with the W/CM ratio, the porosity, and the cement paste volume in the concrete. The least abrasion resistant concrete produced in the study resulted in surface wear that was only 17% of ordinary concrete. Therefore, by using superplasticizer to reduce substantilly the W/CM ratio, the abrasion resistance of concrete would be improved considerably. Introducing mineral admixture without using superplasticizer would reduce the abrasion resistance of concrete since more water would be needed to maintain a constant workability. Naik et al. [126] found that the abrasion of fly concrete with strength greater than 40 MPa is very minimal. The abrasion resistance of high volume fly ash concrete with 70% replacement with cement was found to be higher than that of counterpart control conventional Portland cement concrete and concrete made with 50% fly ash [127].

There is no doubt that the wearing resistance of concrete is a direct function of its strength, and thus its water-cement ratio and constituent materials. High quality paste and strong aggregates are essential to produce an abrasion resistant concrete. Atis [128] suggested that both strength and porosity are important to abrasion resistance. A model was proposed to relate the compressive strength and porosity of concrete to its abrasion are offered as the models that could be used to aid mixture proportioning for performance and durability, and to assess concrete pavement where signs of poor performance have been detected even when the structural properties of the concrete are adequate.

2.3.6.7   Fire resistance

Concrete is a non-combustible material and has good fire resistant property, and is often used to protect steel from the effects of fire. However, concrete is damaged by exposure to high temperatures and will suffer loss of strength, cracking and spalling. Hydrated Portland cement contains a considerable portion of free calcium hydroxide, which decomposes above 400-500°C, leaving calcium oxide (quick lime). This will result in significant strength loss. If the calcium oxide is cooled down and becomes wetted after cooling, it hydrates again to calcium hydroxide, accompanied by an expansion in volume and may disrupt a concrete, which does not disintegrated during the fire. In hydrated alkali-activated cements, free $Ca(OH)_2$ usually does not exist. Thus, it can be expected that disruption of concrete due to the rehydration of CaO will not happen.

HSC provides a high level of structural performance, especially in strength and durability, compared to conventional concrete. However, the performance of HSC differs generally from that of conventional concrete and may not exhibit good fire performance. Spalling under fire conditions is one of the major concerns with HSC due to its low water/cement ratio [129,130]. HSC is believed to be more susceptible to the pressure build-up during exposure to fire because of its low permeability, compared to that of conventional concrete. The extremely high water vapor pressure, generated during exposure to fire, cannot escape because of the high density (and low permeability) of HSC. This pressure often reaches the saturation vapor pressure, which at 300°C is about 8 MPa. Such internal pressures are often too high to be resisted by the HSC, which has a tensile strength of about 5 MPa [130]. Since the spalling occurs in the initial stages of a fire, it may pose a risk to evacuating occupants and firefighters. Recently, Kodur et al. [131] proposed a numerical model accounts for spalling of HSC under fire conditions. Using the model, the fire resistance of HSC columns can be evaluated for any value of the significant parameters, such as load, section dimensions, column length, concrete strength, aggregate type and fiber reinforcement, without the necessity of testing.

Several studies have confirmed that the use of organic fibers may be able to eliminate the spalling problem [132-134]. Actually, polypropylene fiber is the most efficient one among various organic fibers. The addition of about 2 kg of polypropylene fiber in each cubic meter of concrete can solve the spalling problem [133,134]. Detailed microstructure investigation indicated that polypropylene fiber will be molten at around 200°C which results in the formation of a continuous pore network within the concrete and to accumulate the high pressure due to the water vapor from the decomposition of hydration products and enclosed air.

2.4 Self-Consolidating Concrete

2.4.1 *Introduction*

Self-consolidating concrete (SCC) is highly flowable, nonsegregating concrete that can spread into place, fill the formwork, and encapsulate the reinforcement without any mechanical consolidation. It has also been described as self-compacting concrete, self-placing concrete, and self-leveling concrete, which are subsets of SCC [135]. SCC was initially developed to ensure proper consolidation in applications where concrete durability and service life were of concern. It was later used to facilitate construction operations and reduce construction time and cost, and has many advantages over conventional concrete [136, 137]:

(1) eliminating the need for vibration;
(2) decreasing the construction time and labor cost;
(3) reducing the noise pollution;
(4) improving the filling capacity of highly congested structural members;
(5) improving the interfacial transitional zone between cement paste and aggregate or reinforcement;

(6) decreasing the permeability and improving durability of concrete, and

(7) facilitating constructibility and ensuring good structural performance.

However, SCC also has some disadvantages:

(1) Considerable experience and skill is essential to manufacture quality SCC.

(2) Strict quality control is important during the production, testing and placing of SCC.

(3) SCC has a relatively high cement paste content generating more heat of hydration, which may result in excess drying shrinkage. The fact that SCC has a lower aggregate content makes it even more susceptible to drying shrinkage.

(4) Forms are required to be relatively tight and stronger for SCC [138].

Many of the above mentioned drawbacks of SCC could be reduced or eliminated. SCC has attracted more and more attention world-wide since its discovery, and new applications for SCC are being explored because of its advantages. SCC has gained widespread attention in the world in the last few years for its obvious advantages of savings in labor costs, shortened construction time, better finish and improved work environment [139,140].

2.4.2 *Constituents and mixing proportions*

The basic ingredients used to make SCC, such as cement, supplementary cementing materials, aggregates, chemical admixtures, are the same as those used in making conventional concrete. However, SCC mix constituents and their proportions are to be carefully selected so as to achieve a concrete with lower rheological shear stress and viscosity that would remain homogenous during its use. Thus, rheological properties, i.e. properties dealing with the deformation and flow characteristics of fresh concrete, are important for successful production and use of SCC. Advancement in SCC technology was primarily possible due to the introduction of a new generation of chemical admixtures that improved and controlled the rheological properties of SCC.

There is no standard method for SCC mix design at the moment. Several documents suggested an empirical method for proportioning SCC [135,138,141]. Many academic institutions, admixture, ready-mixed, precast and contracting companies have developed their own mix proportioning methods.

Mix designs of SCC often use volume as a key parameter because of the importance of the need to over fill the voids between the aggregate particles. Some methods try to fit available constituents to an optimised grading envelope. Another method is to evaluate and optimise the flow and stability of first the paste and then the mortar fractions before the coarse aggregate is added and the whole SCC mix tested. In general, these procedures fall into the following three categories:

(1) combination of superplasticizer and high content of mineral powders,
(2) combination of superplasticizer and viscosity-modifying admixture (VMA) with or without defoaming agent; and
(3) combination of superplasticizer, powder and VMA with or without defoaming agent.

Powder based SCC is the first generation of SCC which has generally high strength and good durability. Chemical based SCC was an extension of the anti-washout underwater concrete. The advantage of this type of SCC is that it develops high flowability even with low powder contents and offers a better quality control of the mix. The new generation of combination types of SCC benefits from the advantages of the previous two types. Combination types of SCC have a wide range of flowability along with a better control on the stability of the mix [142].

SCC requires high flowability and a low yield value of the rheological characteristics. High flowability can be obtained by decreasing the yield value of the mortar paste and increasing the plastic viscosity of the concrete to resist segregation. Superplasticizers reduce the yield value of the mortar paste by a pronounced dispersion effect. Thus, the addition of HRWR is required to impart flowability and passing ability to the mix. But excessive HRWR may result in segregation and bleeding. To alienate segregation and bleeding, VMA can be added to the concrete [143]. Numerous research findings have shown that it is no longer a dream to make a flowable yet stable SCC, tailored for any application.

The flowability of SCC is affected by the degree of dispersion of the cement particles due to the physiochemical effect of the admixtures. Belite (C_2S) rich cement brings out the dispersing action of superplasticizers quite effectively [144]. High early strength cement and ordinary Portland cement produce high yield values in SCC whereas moderate heat cement and belite (C2S) cement are best for SCC. Mechanically stabilized cements using fly ash, silica fume, slag and limestone powder having high-fineness are best suited for SCC [144,145].

SCC uses a higher proportion of powder, if the powder content is increased, the viscosity increases while the yield value decreases, resulting in a highly flowable SCC [142]. In SCC, cement could be partially replaced by various supplementary powders such as fly ash, silica fume, lime fines, finely ground slag, etc [142]. This would reduce the cost as well as the problem of drying shrinkage. The use of expansive additives such as belite (C2S) in cement is very effective in compensating the shrinkage of SCC [142].

Aggregates suitable to prepare satisfactory medium-high slump traditional concrete can be utilized to make SCC. Ideally, aggregates should be well-graded and meet the requirements of ASTM C33 [146] or equivalent. It may be acceptable to use an aggregate source that does not meet the requirements of ASTM C33 if it produces a mixture that meets the performance targets for the fresh and hardened concrete. The effects of coarse and fine aggregates on the rheological, static and kinetic properties of SCC are critical. When the amount of coarse aggregate is increased, the flowability and compactibility of SCC decreases as the internal friction between aggregates increase. Flowability falls as the maximum size of coarse aggregate increases. Fine aggregates help in lowering down the yield value of concrete up to a certain point. Large amounts of fine aggregates lower the compactibility of SCC as the thickness of the mortar paste covering the fines decreases [142].

Unlike the normal vibrated concrete, SCC has a ratio of fine aggregate to coarse aggregates slightly above unity. Normal vibrated concrete has a coarse to fine aggregate ratio generally in the range of 1.6 to 2.5 [147]. Moreover, SCC with high aggregate contents, almost approaching that used in conventional concretes, has been successfully produced [148]. The difficulties encountered with challenging aggregates (gap graded, excessively harsh, and so on) may be improved by: adding material from other sources (improving gradation); using fine powders, such as limestone fines; or using a VMA. A field trial should be mandatory for determining the suitability of an aggregate for a project.

Polycarboxylate based HRWRA are the most typical admixture materials used for developing and proportioning SCC. VMA are also beneficial materials for adjusting the viscosity and improving the stability of SCC. Not all HRWRA or VMA products have the same properties. Some HRWRA impart the characteristics of stability and cohesiveness, others do not. VMA used with HRWRA improve concrete viscosity and increase water tolerance of the mixture. The use of VMA is not always necessary, but VMA can be advantageous when using lower powder contents gap graded or demanding aggregates. VMA also lessen the sensitivity of the mixture to changes in the mixture water. In addition to HRWRA and VMA, other admixtures and additives such as air-entraining admixtures, normal and mid-range water-reducing admixtures, liquid and dry color admixtures, accelerating admixtures, retarding admixtures, extended set-control admixtures, corrosion-inhibiting admixtures, shrinkage-reducing admixtures, and fibers can be specified for use in SCC.

Table 2.2 gives an indication of the typical range of constituents in SCC by weight and by volume. These proportions are in no way restrictive and many SCC mixes will fall outside this range for one or more constituents.

Table 2.2 Typical range of SCC mix composition [141].

Constituent	Typical range by mass (kg/m^3)	Typical range by volume (liters/m^3)
Powder		380–600
Paste		300–380
Water	150–210	150–210
Coarse aggregate	750–1000	270–360
Fine aggregate (sand)	Content balances the volume of the other constituents, typically 48–55% of total aggregate weight.	
Water/Powder ratio by Vol.		0.85–1.10

2.4.3 *Testing of SCC*

2.4.3.1 Introduction

The properties that differentiate SCC from conventional concrete are those of the fresh concrete. The required workability for casting concrete depends on the type of construction, selected placement and consolidation methods, complexity of the formwork, and structural design details that affect the degree of congestion of the reinforcement.

Table 2.3 Test properties and methods for evaluating SCC [141].

Characteristic	Test method	Measured value
Flowability/filling ability	Slump-flow	total spread
	Kajima box	visual filling
Viscosity/flowability	T_{500}	flow time
	V-funnel	flow time
	O-funnel	flow time
	Orimet	flow time
Passing ability	L-box	passing ratio
	U-box	height difference
	J-ring	step height, total flow
	Kajima box	visual passing ability
Stability (Segregation resistance)	penetration	depth
	sieve segregation	percent laitance
	settlement column	segregation ratio

In outlining and defining the fresh properties, two points of view can be used. The first is in evaluating the fundamental rheological properties of the SCC mixture. The second method of defining the fresh properties of SCC is to evaluate them strictly based upon practical field related requirements. These practical characteristics are the properties of stability, filling ability, and passing ability [135,141].

- Filling ability – The property which determines how fast SCC flows under its own weight and completely fills intricate spaces with obstacles, such as reinforcement, without loosing its stability.
- Passing ability – the ability of SCC to pass through congested reinforcement and adhere to it without application of external energy.
- Stability – the ability of SCC to remain homogenous by resisting segregation, bleeding and air popping during transport, placing and after placement.

These properties provide SCC with a unique rheology that distinguishes it from conventional concrete. The required levels of stability, filling ability, and passing ability of SCC are determined by the application. Producing such a special concrete requires an improved work environment and strict quality control measures. A wide range of test methods have been developed to measure and assess the fresh properties of SCC, as shown in Table 2.3. No single test is capable of assessing all of the key parameters, and a combination of tests is required to fully characterize an SCC mixture.

ASTM has standardized several of them, including such as slump flow/VSI (ASTM C1611 [149]), J-Ring (ASTM C1621/C1621M [150]), L-Box (being standardized) and column segregation (ASTM C1610/ C1610M [151]) to determine the stability, filling ability, and passing ability of SCC. These methods are briefly described as follows:

2.4.3.2 Slump flow/VSI (filling ability/deformability and stability)

The slump flow test is a measure of mixture filling ability, as shown in Fig. 2.9. This test is performed similar to the conventional slump test using the standard ASTM C143-2006 [152] slump cone. Instead of measuring the slumping distance vertically, however, the mean spread of the resulting concrete patty is measured horizontally. This number is recorded as the slump flow. The VSI is determined through rating the apparent stability of the slump flow paddy [153].

2.4.3.3 J-Ring (passing ability)

The J-Ring consists of a ring of reinforcing bar that will fit around the base of a standard ASTM C 143 slump cone. The slump cone is filled with concrete and then lifted in the same fashion as if one were conducting the slump flow test, as shown in Fig.2.10. The final spread of the concrete is measured, and the difference between the conventional slump flow value and the J-ring slump flow value is calculated.

Fig. 2.9 Slump flow testing of SCC.

Fig. 2.10   J-Ring testing of SCC.

2.4.3.4 L-Box (passing ability)

The L-Box test consists of an L-shaped container divided into a vertical and horizontal section. A sliding door separates the vertical and horizontal sections. An obstacle of three reinforcing bars can be positioned in the horizontal section adjacent to the sliding door. The vertical section of the container is filled with concrete and the sliding door is immediately removed, allowing the concrete to flow through the obstacle into the horizontal section, as shown in Fig. 2.11. The height of the concrete left in the vertical section (h_1) and at the end of the horizontal section (h_2) is measured. The ratio of h_2/h_1 is calculated as the blocking ratio.

Fig. 2.11 L-Box testing of SCC.

2.4.3.5 Column segregation (stability)

This test evaluates the static stability of a concrete mixture by quantifying aggregate segregation. This test consists of filling a 26 in. (610 mm) high column with concrete. The column is sectioned into three pieces, as shown in Fig. 2.12. The concrete is allowed to sit for 15 minutes after placement. Each section is removed individually and the concrete from that section is washed over a No. 4 (4.75 mm) sieve and the retained aggregate is weighed. A non-segregating mixture will

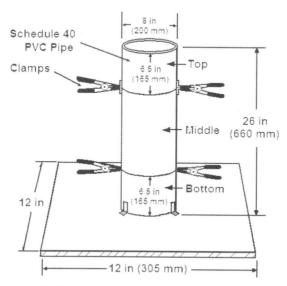

Fig. 2.12 Column segregation testing apparatus for SCC.

have a consistent aggregate mass distribution between the top and bottom sections. A segregating mixture will have a higher concentration of aggregate in the lower section.

2.4.4 *Self-Consolidating Fiber Reinforced Concrete (SCFRC)*

Fiber Reinforced Concrete (FRC) requires a high degree of vibration to get good compactness. This increases the labor costs and noise pollution at the work site. Moreover, if the reinforcement is dense or the form is intricate in shape, it becomes even more difficult to place and vibrate the concrete. Unfortunately, when one tries to enhance the workability of FRC by adding more superplasticizers or intensifying the degree of vibration, segregation invariably occurs. Hence, the development of a self-consolidating fiber reinforced concrete (SCFRC) should make for easier placement of concrete, save labor and avoid noise pollution. SCC offers several economic and technical benefits; the use of steel fibers extends its possibilities [148].

When steel fibers are added to the concrete mix, the tensile and shear resistance of the composite material is enhanced. However, fibers are also known to impede the workability of plain concrete. Moreover, the end zones are densely reinforced making it necessary to use a highly workable concrete with steel fibers that would not only reduce or completely eliminate the conventional reinforcement but also make it easier to place concrete. The use of SCFRC would guarantee the following advantages:

- Fibers in SCFRC are expected to partially or fully replace the dense reinforcement and also control concrete cracking.
- SCFRC would be easier to place and finish than the conventional FRC.
- The shear, flexural and tensile strengths along with ductility and toughness are expected to be improved with the use of SCFRC mixtures.
- SCFRC might prove to be more economical in the long run, owing to the fact that labor and time is saved with the use of SCFRC.

Some research experiments as well as field applications have been successfully carried out on SCFRC. The mix design of SCFRC could be based on the mix design of an existing SCC mix [154]. The workability of SCFRC is affected by fibers as they posses high surface area. The degree to which workability decreases depends on the type and content of fibers, the matrix composition and the properties of the constituents of the matrix on their own. The higher the fiber content in SCFRC, more difficult it becomes to uniformly distribute the fibers in the matrix [148]. Concrete with satisfactory workability could be made self-consolidating even with a large fiber content of up to 1.3% by volume [155].

2.4.5 *Properties of hardened SCC*

2.4.5.1 Mechanical properties

Properties of hardened SCC are expected to be similar to, or better than, those of a comparable conventional concrete mixture. Changes in mixture proportions and in fluidity can influence the hardened properties, which can diverge from what is commonly expected from conventional concrete. If specific key properties are important in a particular application, these need to be considered when developing SCC mixtures.

2.4.5.2 Deformation

Autogenous shrinkage can be particularly high in mixtures made with relatively low *w/cm*, high content of cement, and supplementary cementitious materials exhibiting a high rate of pozzolanic reactivity at early age. Special attention should be given to protect the surface of SCC at early ages to minimize any desiccation. In very low *w/cm* mixtures, wet curing may be required. Drying shrinkage is related to the water and paste contents, as well as aggregate volume, size, and stiffness. The increased paste volume in SCC and reduction in aggregate content can lead to greater potential for drying shrinkage, which should be managed in the mixture proportioning process. The drying shrinkage increases with the increase in the content of powder materials, which can be particularly high in SCC mixtures. As in the case of drying shrinkage, creep of SCC is highly dependent on the mixture composition, paste volume, and aggregate content. For the same mixture proportions as that for conventional concrete, creep of SCC is expected to be similar to that of conventional concrete. When the SCC is proportioned with greater paste volume, however, it can exhibit higher creep than conventional concrete of similar compressive strength.

2.4.5.3   Bonding with aggregate or reinforcements

Several studies have reported that the interface between cement paste and aggregate or reinforcements is less porous and the bonding characteristics between them are better than conventional concrete. SCC flows easily around the reinforcement and bonds well. Up to 40% higher bonding strengths were measured for SCC compared to conventional concrete [156, 157 & 158]. This may be due to the lower water content and the higher powder volume in the SCC mixtures relative to the reference mixtures, which reduces the accumulation of bleed water under horizontally embedded reinforcement bars. In normal concrete, bleed water accumulation can increase the local *w/cm* under the bar and weaken the strength of the bond [159].

2.4.5.4   Long-term durability

When a proper air-void system is provided, SCC can exhibit excellent resistance to freezing and thawing and to deicing salt scaling [160, 161 & 162]. It is difficult to stabilize air voids in segregating concrete. In such cases, increasing the concrete viscosity by use of a VMA or by changing the mixture proportions (through the addition of more powder and/or reduction in water) may be necessary to ensure adequate air-void distribution [163].

2.4.5.5   Aesthetics

One benefit of SCC is that it provides improved surface appearances and aesthetics in finished concrete. Pour lines, bugholes, honeycombs, and other surface imperfections are largely reduced [164]. The fluidity of SCC as well as the elimination of vibration will result in improved aesthetics. Selection of form release agents is very important in achieving the desired smooth finish.

2.5 Specifications and Guidelines for HPC

2.5.1 *Structural design of HSC*

ACI Innovation Task Group ITG-4 is also working on the document to allow normal weight concrete with a specified compressive strength of

40 MPa or greater to be used in cast-in-place buildings constructed in reg ions where the seismic hazard is moderate to high [165].

At its 2004 Annual Meeting, the AASHTO Subcommittee on Bridges and Structures approved revisions to the AASHTO LRFD Bridge Design Specifications for the 2005 edition to permit the use of several articles with concrete compressive strengths up to 103 MPa. A project has been completed to develop recommended revisions to the AASHTO LRFD Bridge Design Specifications to extend the applicability of shear design provisions for reinforced and prestressed concrete structures to concrete compressive strengths greater than 70 MPa [166].

2.5.2 *Performance specifications for HPC*

During the past decade, many States developed specifications and guidelines for the raw materials, mixing, placement, curing and performance of HPC based on applications of project [84,167,168, 169].

In November 1999, FHWA awarded a three-year HPC pooled-fund study entitled "Compilation and Evaluation of Results from High Performance Concrete Bridge Projects". The project collects and compiles information from each of the joint State-FHWA HPC bridge projects and other HPC bridge projects; analyzes and evaluates the compiled information in comparison with existing AASHTO specifications and guidelines for materials, testing, structural design, and construction; and recommend equations, specifications with commentary, and guidelines for material and structural properties where sufficient research results exist [170]. The following AASHTO specifications were reviewed to identify provisions that directly impact the use of HPC:

- AASHTO Standard Specifications for Transportation Materials and Methods of Sampling and Testing, Parts I and II.
- AASHTO Standard Specifications for Highway Bridges.
- AASHTO LRFD Bridge Design Specifications.
- AASHTO LRFD Bridge Construction Specifications.

The proposed revisions include 15 material specifications, 14 test methods, 30 articles of the standard design specifications, 17 articles of the Load and Resistance Factor Design (LRFD) bridge design specifications, and 16 articles of the LRFD bridge construction specifications were developed.

2.5.3 *Guidelines and specifications for SCC*

Since its 1989 introduction in Japan, self-consolidating concrete (SCC) has been used worldwide for precast and cast-in-place applications. In 1998, Japan published guidelines for the design, testing and use of SCC [138]. Europe and China published similar guidelines in 2005 [141,171].

In US, SCC is being increasingly used during the past years. Precast/Prestressed Concrete Institute published an internal guideline in 2003 [135] and so did the Portland Cement Association in 2005 [172]. Further advancing the practice, a National Cooperative Highway Research Program (NCHRP) Project (No. 18-12) is underway to develop SCC mixes, structural design parameters, and construction specifications for precast/prestressed concrete elements. Initiated in 2004 and scheduled for completion in 2007, NCHRP Project 18-12, "Self-Consolidating Concrete for Precast, Prestressed Concrete Bridge Elements," aims to increase the acceptance and use of SCC in highway bridge construction. The project will develop guidelines for SCC use in precast, prestressed concrete bridge elements and recommend relevant changes to the Load and Resistance Factor Design Bridge Design and Construction Specifications issued by the AASHTO. Also addressed during the course of the study will be issues of workability, strength development, creep and shrinkage properties, durability, and other factors influencing constructibility and performance.

2.6 **Applications of HPC**

2.6.1 *Introduction*

HPC has been widely used for all kinds of concrete structures in the world, including buildings with beautiful aesthetic or complicated architecture. The owner of the project considers not only the beauty of the structure but also the time and money that they need to save for the project. This section introduces several examples of the use of high performance concrete.

2.6.2 *High rise buildings*

The development of high-flowing and high strength concrete with strengths greater than 420 kgf/cm^2 was used in 1990 for Taipei Far-Eastern Plaza. However, in this stage pozzolanic material is not allowed due to the possibility of degradation after the addition of pozzolanic material.

In 1993 HPC with strengths greater 560 kgf/cm^2 was first developed to apply the thinking of Prof. Mehta, that the lesser the cement and water in concrete the better the durability, to achieve the needs of pumping concrete up directly from the bottom to 85 story height column. The HPC design is allowed to add suitable amount of pozzolanic materials and superplasticizer otherwise the strength cannot be easily achieved. The HPC is needed to be high flowing and self- consolidating to aid the pump-ability. However, the rheology of HPC should be less viscous but not too sticky. After the success of producing HPC, a lot of remarkable concrete structures were built with such excellent material. It is believed that HPC is a green material with only half amount of cement for the same strength grade.

Until 1999, 921 Earthquake has punished the poor quality concrete structure to collapse, especially the structures with heavy reinforced steel. Concrete with superior workability and strength is aimed to solve the problem. SCC was then developed and is supposed to be a kind of HPC that can fulfill all of those wishes. But then it was found that because SCC usually uses large amount of cement and superplastizer, segregation would occur. The surface cracking is liable to occur as well. Normally, the workability is correlated to water content and the strength is directly related to cement content.

As mentioned before, HPC was made with reduced water and cement content in order to minimize the possibility of segregation and bleeding, designed with proper water/binder ratio (W/B) to control the design strength. The specification of HPC is to limit the water content to be less than 150 kg/m^3 and the cement content to be less than given by the formula: characteristic strength of (HPC / 0.14).

The design specifications for flowing HPC or SCC in Taiwan take into consideration not only the chemical point of view, but also the physical point of view of the concrete. Fly ash and other fine particle are used to fill the void of aggregate to physical pack the concrete matrix and react chemically with alkali from cement to form low-density gel and develop chemical bond between the concrete ingredients. This type of HPC has five salient features [173-175]:

- *Durability*:   Resistivity > 20 kΩ; Chloride penetration < 2000 Coulomb; low volume stability; low autogenous shrinkage (w/c $>$ 0.42); low drying shrinkage (water content < 150 kg/m^3).
- *Workability*: Slump 0~270 mm; slump flow 203~700 mm without bleeding and segregation i.e. maintaining stability; flow like honey.
- *Economy*: Low cement content, large amount of pozzolans, easy and rapid construction.
- *Safety*: Strength growth is sustainable through cement hydration and pozzolanic reaction.
- *Ecology*: Low CO_2 emission (low cement content) and more recycling materials (i.e., fly ash, slag, silica fume, rice husk ash, recycling aggregate, fine sediment and mud, steel slag); sustainable design.

Some projects that have used HPC or SCC for construction are shown in Table 2.4 (Courtesy of Prof. C.L. Hwang).

Qinghai-Tibet Railway

The Qinghai–Tibet railway was completed by July 1 2006 and is a high-altitude railway that connects Xining, Qinghai Province, to Lhasa, Tibet Autonomous Region, China. The section between Xining and Golmud is 815 km long and was built in the 1950s. Owing to a lack of technologies, it was unable to construct the railroad beyond Golmud. The construction for the section between Golmud and Lhasa was started in 2001. Of the 1,142 km section, more than 960 km, or over 80% of the section is built at an altitude of more than 4,000 m, the world's highest railway; over half section lays on permafrost. The 1,338 m Fenghuoshan tunnel is 4,905 m above sea level and the highest rail tunnel in the world. There are 675 bridges, a total length of 160 km in this section.

Table 2.4 shows the conditions and requirements for concrete for tunnels and bridges on the Qinghai–Tibet Railway.

2.6.3 *Bridges*

HPC with its beneficial strength and durability properties is an ideal material for prestressed concrete highway bridges. The use of HPC in these structures can lead to both short- and long-term cost savings. Since 1993, the Texas Department of Transportation (TxDOT, USA) has been working on the use of HPC for bridge. The two Texas HPC bridges, the Louetta Road Overpass and the North Concho River/U.S.87/S.O.R.R., are direct applications of HPC to highway bridge structures. These bridge structures have been successfully constructed using HPC and are being carefully monitored to learn more about their long-term performance and to help develop criteria for future designs. The design, construction, and instrumentation processes to date verify that HPC can be used successfully for innovative and efficient highway bridge structures.

Table 2.4 Projects in Taiwan that Utilized HPC/SCC for construction.

Building	Location	Year	Concrete Grade	Remarks
Far Eastern Plaza	Taipei, Taiwan	1989~ 1990	420 kgf/cm^2	Pumping up to 23rd floor.
T & C Tower (85 floors)	Kaohsiung, Taiwan	1995~ 2000	560 kgf/cm^2	Pumping up to 85th floor.
Museum of Marine Biology & Aquarium	Pingtung, Taiwan	1996~ 2003	140 kgf/cm^2; 280 kgf/cm^2; 350 kgf/cm^2	High flowing and corrosion resistance
Automobile Research &Testing Center	Changhwa, Taiwan	1999~ 2000	140 kgf/cm^2; 280 kg/cm^2 (56 days)	Corrosion resistance
Taiwan High Speed Railway	Taiwan	1999~ 2001	560 kgf/cm^2	High flowing
Rainbow Gate at Ceramic Museum	Yingge, Taiwan	1999	700 kgf/cm^2	Aesthetics and flexible.
Taipei 101 Financial Center (Fig. 2.13)	Taipei, Taiwan	1999~ 2001	280 kgf/cm^2; 420 kgf/cm^2; 700 kgf/cm^2	Pumping up to 88th floor.

Fig. 2.13 Taipei 101, Taipei, Taiwan – Tallest building in the world until now, 508 m
(Courtesy of Prof. C.L. Hwang).

Fig. 2.14 Qinghai–Tibet Railway (www.qh.xinhuanet.com).

The Louetta Road Overpass, in Houston, Texas, is composed of two three-span highway overpasses each 119 m in total length. The design incorporated the recently developed 1372 mm deep Texas U54 section and HPC with compressive strengths of up to 90.3 MPa. As is common in Texas, the composite decks are composed of a layer of precast pretensioned concrete panel units and a layer of cast-in-place concrete, both of which are HPC. Although designed as simply-supported structures, some degree of continuity is assumed to be provided for live load by the cast-in-place decks, which were placed continuously along the entire length of each bridge in a single pour. Piers were designed to support each girder individually, resulting in an aesthetically pleasing substructure. These piers also utilized HPC, and were constructed using precast post-tensioned segments. A photograph of the Louetta Road Overpass is shown in Fig. 2.15.

Fig. 2.15 Louetta Road Overpass, Houston, Texas.

The North Concho River/U.S. 87/S.O.R.R. Overpass in San Angelo, Texas is made up of an Eastbound eight-span HPC bridge 292 m (958 ft.) in total length and a Westbound nine-span normal strength concrete bridge 290 m (951 ft.) in total length. Both structures utilize the standard 1372 mm (54 in.) deep AASHTO Type IV cross-section for the main spans, with concrete compressive strengths of up to 55 MPa (8000 psi) for the Westbound normal strength concrete bridge and 95 MPa (13,800 psi) for the Eastbound HPC bridge. The companion bridge structures provide an excellent opportunity to compare an HPC structure and a structure constructed using standard materials and typical concrete strengths. Precast prestressed concrete panels were used in the construction of the composite bridge decks. Cast-in-place deck segments were cast in single span units, resulting in simply-supported spans. As can be seen in Fig. 2.16, the piers are single-column cast-in-place elements, designed with aesthetics in mind.

The use of HPC can be very beneficial for the design and construction of highway bridges. Longer spans, larger girder spacing, and shallower cross-sections are short-term benefits, which can result from utilizing the strength properties of HPC. Reduced maintenance and life-cycle costs are expected to be long-term benefits resulting from the superior durability properties of HPC.

Fig. 2.16 North Concho River/U.S. 87/S.O.R.R. Overpass, San Angelo, Texas.

2.6.4 *Application of SCC*

Numerous applications of SCC are successfully carried out in the construction industry throughout the world in the past decade. High-rise commercial and industrial buildings, precast products, bridges, highway pavements etc. are some of the common applications of SCC. Precast prestressed concrete I-beams are used extensively as the primary superstructure elements in Texas highway bridges. A commonly observed problem in these beams is the appearance of end zone cracking due to the prestress forces, hydration of concrete, shrinkage and temperature variation, as shown in Fig. 2.17 although this end is heavily reinforced in the lateral direction, as shown in Fig. 2.18. Workable Self-Consolidating Fiber Reinforced Concrete (SCFRC) mixes were developed that would be capable of not only replacing partially/ completely the dense reinforcement but also to eliminate cracking [176]. Four Traditional Fiber Reinforced Concrete (TTFRC), two SCC and

Fig. 2.17 End zone cracking in a prestressed concrete beam.

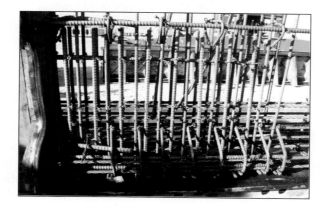

Fig. 2.18 End zone reinforcement in a prestressed concrete beam.

three SCFRC mixes with two different types and variable amounts of hook-ended steel fibers were tested for their workability and hardened properties. The two fibers RC80/60BN (L = 2.4 in.) and ZP305 (L = 1.2 in.) had an aspect ratio of 80 and 55, respectively.

Figure 2.19 shows various tests such as slump flow, VSI rating, V-funnel and J-ring conducted at the precast plant, to obtain the flowability, stability, filling ability and passing ability of SCFRC mixes, respectively.

Based on their performance in the workability and hardened properties tests, suitable TTFRC and SCFRC mixes with optimum fiber contents were selected to cast full-scale beams. Seven, 25 ft. long, prestressed concrete beams were cast in a precast plant using TTFRC and

SCFRC mixes. Conventionally used equipments and techniques were applied for mixing, transporting, placing and steam curing the beams at the precast plant. The beams were load tested till failure.

Fig. 2.19 Workability tests of SCFRC mixes.

Major findings:

(1) It was possible to produce SCFRC mixes with Texas conventional materials, equipments and techniques. The mixes showed satisfactory workability and were deemed suitable to cast the end regions of prestressed I-beam. Optimum steel fiber content was found to be governed by the workability criteria for the SCFRC mixes. Tests of hardened properties confirmed the effectiveness of steel fibers in enhancing the tensile strength, flexural strength and ductility of concrete. On an average steel fiber increased the tensile strength of concrete by about 50% in case of SCFRC mixes.

(2) Casting SCFRC mixes in beams was relatively easy. The SCFRC mixes were highly workable and demonstrated self-compactability without any signs of fiber blocking when placed in the beam form. SCFRC was observed to flow from one end of the beam to the other without loosing its stability.

(3) The steel fibers had performed quite effectively in controlling or completely eliminating the end zone cracks. Temperature logger data revealed that the maximum thermal loading for the traditional beam was

about 84°F, much more than the value found in literature of 60°F. SCC produced more thermal loading (120°F) due to higher cement content. Hence, it is prudent to incorporate steel fibers in SCC mixes to counteract the enhanced thermal load in the matrix.

(4) Load tests of beams have shown that steel fibers were phenomenally effective in increasing the shear and flexural strength and ductility of the beams. These tests have proved the ability of steel fibers to partially or completely replace traditional transverse steel reinforcement in the end region of the beams.

2.7 Summary

This chapter discusses the definition, raw materials, mix design, performance, specification and applications of HPC. It can be summarized as follows:

(1) HPC may be regarded as any concrete that has properties needed for some objectives that are not possessed by ordinary concrete.

(2) Although the constituents for HPC are the same as those for ordinary concrete, they have to be properly selected so to achieve the designed properties.

(3) Many methods have been proposed/developed to design mixtures for HPC. Most of them are semi-analytical, and usually provide the proportioning of aggregates and the calculation of W/C using an equation of compressive strength.

(4) HPC can be designed to have all mechanical properties and durability superior to conventional concrete.

(5) SCC is very different from conventional concrete in fresh state and needs to be characterized differently. However, the properties of hardened SCC can be characterized like conventional concrete.

(6) Depending on the applications and requirements, some existing specifications/guidelines need to be revised or new specifications/guidelines need to be developed in order to use HPC.

(7) HPC has been widely used for a variety of projects in the world.

References

1. Ali, M. M., Evolution of Concrete Skyscrapers: From Ingalls to Jinmao, Electronic Journal of Structural Engineering, Vol. 1, No. 1, pp. 2-14, 2001.
2. Zia, P.; Leming, M. L.; and Ahmad, S. H., High-Performance Concretes, A State-of-the-Art Report, Report No. SHRP-C/FR-91-103, Strategic Highway Research Program, National Research Council, Washington, D.C., 1991.
3. Moore, J. A., et al., High-Performance Concrete for Bridge Decks, Concrete International, Vol. 21, No. 2, February, pp. 58-68, 1999.
4. Zia, P., Ahmad, S., and Leming, M., High-Performance Concretes: A State-of-Art Report (1989-1994), FHWA-RD-97-030, FHWA, 1997.
5. Russell, H.G., ACI Defines High-Performance Concrete, Concrete International, Vol. 21, No. 2, pp. 56-57, February, 1999.
6. Zia 1991 (reference not available).
7. Goodspeed, C. H, Vanikar, S., and Cook, R. A., High-Performance Concrete Defined for Highway Structures, Concrete International, Feb, Vol. 18, No. 2, pp. 62-67, 1996.
8. FHWA, High-Performance Concrete Bridges, Building Bridges for the 21st Century, U. S. Department of Transportation, Publication No. FHWA-SA-98-084, 1998.
9. ACI MCP06, ACI Manual of Concrete Practice 2006 (Six-part set), American Concrete Institute, 01-Jan-2006.
10. Aitcin, P. C., and Neville, A., High-Performance Concrete Demystified. Concrete International, January, Vol. 15, No. 1, pp. 21-26, 1993.
11. Basu, 2001 (reference not available).
12. Freyne et al 2004 (reference not available).
13. ACI Committee 363, State of the Art Report on High Strength Concrete, American Concrete Institute, Detroit, 2006.
14. Gutierrez and Canovas, 1996 (reference not available).
15. Malhotra, V. M., Supplementary Cementing Materials for Concrete, CANMET Publication, Ottawa, Canada, 1987.
16. Shi, C., Roy, D. M., and Krivenko, P., Alkali-Activated Cements and Concretes, Taylor & Francis, January 2006.
17. Chang, T., Lin, S., Lin, H., and Lin, P., Effects of Various Fineness Moduli of Fine Aggregate on Engineering Properties of High-

Performance Concrete, Journal of the Chinese Institute of Engineers, Vol. 24, No. 3, pp. 289-300, 2001.

18. Aitcin, P. C., and Mehta, P. K., Effect of Coarse Aggregate Characteristics on Mechanical Properties of High-Strength Concrete, ACI Materials Journal, Vol. 87, No. 2, pp. 103-107 1990.

19. Zhou, F. P., Lydon, F. D., and Barr, B. I. G., Effect of Coarse Aggregate on Elastic Modulus and Compressive Strength of High Performance Concrete, Cement and Concrete Research, Vol. 25, No. 1, pp. 177-186, 1995.

20. Tasi, C. T., and Hwang, C. L., The Effect of Aggregate Gradation on Engineering Properties of High Performance Concrete, Journal of ASTM International (JAI), Vol. 3, No. 3, 2006.

21. Ramachandran, V. S., Concrete Admixtures Handbook – Properties, Science and Technology, Second Edition, Noyes Publication, New Jersey, 1995.

22. Shi, C., Berke, N., Jeknavorian, A. and Zhong, H., Roles of Chemical Admixtures in Sustainable Concrete Materials and Structures, Proceedings of the 6th International Symposium on Cement and Concrete, Vol. 2, pp. 1631-1643, Xian, China, September 19-22, 2006.

23. Bedard, C., and Mailvaganam, N. P., The Use of Chemical Admixtures in Concrete. Part I: Admixture-Cement Compatibility, J. Perf. Constr. Fac., Vol. 19, No. 4, pp. 263-266, 2005.

24. Ai, H., and Young, J. F., Mechanisms of Shrinkage Reduction using a Chemical Admixture, Proceedings of the 10th International Congress on the Chemistry of Cement, Vol. 3, Gothenburg (Sweden), pp. 8., 1997.

25. ASTM C 1602/C 1602M, Standard Specification for Mixing Water Used in the Production of Hydraulic Cement Concrete, Annual Book of ASTM Standards, Vol. 04.02, Concrete and Aggregate, American Society for Testing & Materials (ASTM), Philadelphia, USA, 2006.

26. Neville, A. M., Properties of Concrete (4th Ed.), Longman Group Limited, Essex, 1995.

27. Atcin and Neville, 1998 (reference not available).

28. Kosmatka, S., Kerkhoff, B., and Panarese, W., Design & Control of Concrete Mixtures, 14th Edition, Portland Cement Association, Skokie, IL., 2002.

29. Gutierrez, P. A., and Canovas, M. F., High-Performance Concrete: Requirements for Constituent Materials and Mix Proportioning Method, ACI Mater. Jr., Vol. 3, pp. 233-241, 1996.

30. Wang, D., Chen, Z., and Qin, W., Computerized Mix Proportioning for HPC, Concrete International, Vol. 19, No. 9, pp. 42-45, September, 1997.

31. Aitcin, P.C., High-Performance Concrete, Taylor & Francis Ltd, UK 1998.

32. Larrard, D., and Sedran, T., Mixture-Proportioning of High-Performance Concrete, Cement Concrete Research, Vol. 32, pp. 1699-1704, 2002.

33. Lim, C. H., Yoon, Y. S., and Kim, J. H., Genetic Algorithm in Mix Proportioning of High-performance Concrete, Cement and Concrete Research, Vol. 34, No. 3 , pp. 409-420, 2004.

34. Sobolev, K., The Development of a New Method for the Proportioning of High-Performance Concrete Mixtures, Cement and Concrete Composites, Vol. 26, No. 7, pp. 901-907, 2004.

35. Zain, M. F. M., Islam, M. N., and Basri, I. H., An Expert System for Mix Design of High Performance Concrete, Advances in Engineering Software, Vol. 36, No. 5, pp. 325– 337, 2005.

36. Magee, B. J. and Olek, J., High-Performance Concrete for Highway Structures: A General Review of Definitions, Mixture Proportions and Performance Levels, Symposium Proceedings of CI/FHWA/FIB International Symposium on High Performance Concrete, Ed Johal, L. S., Orlando, Florida, USA, September 25-27, 2000.

37. Olek, et al., 2002 (reference not available).

38. ASTM C469 -02, Standard Test Method for Static Modulus of Elasticity and Poisson's Ratio of Concrete in Compression, Annual Book of ASTM Standards, Vol. 04.02., Concrete and Aggregate, American Society for Testing & Materials (ASTM), Philadelphia, USA, 2006.

39. Reinhardt, H. W., Structural Behaviour of High Performance Concrete, Otto-Graf-Journal, Vol. 11, pp. 9-18, 2000.

40. Wee, T. H., Chin, M. S. and Mansur, M. A., Stress-Strain Relationship of High-Strength Concrete in Compression, Journal of Materials in Civil Engineering, Vol. 8, No. 2, pp. 70-76, 1996.

41. Cheng, F. P, Kodur, V. K. R., and Wang, T. C., Stress-Strain Curves for High Strength Concrete at Elevated Temperatures,

Journal of Materials in Civil Engineering, Vol. 16, No. 1, pp. 84-90, January/February 2004.

42.	Turner-Fairbank Highway Research Center, Optimized Sections for High-Strength Concrete Bridge Girders-Effect of Deck Concrete Strength, Publication No. FHWA-HRT-05-058, October 2006.

43.	ACI Committee 318, ACI MCP06, 2006, Building Code Requirements for Reinforced Concrete, American Concrete Institute, Detroit, 2006.

44.	Ahmad, S. H., and Shah, S. P., Structural Properties of High Strength Concrete and its Implications for Precast Prestressed Concrete, Journal of the Prestressed Concrete Institute, Vol. 30, No. 6, pp. 92–119, November/December 1985.

45.	Collins, M. P., Mitchell, D., and Macgregor, J. G., Structural Design Considerations for High-Strength Concrete, Concrete International, American Concrete Institute, Vol. 15, No. 5, pp. 27-34, May 1993.

46.	Thorenfeldt, E., Tomaszewicz, A., and Jensen, J. J., Mechanical Properties of High-Strength Concrete and Application in Design, Proceedings of the Symposium on Utilization of High-Strength Concrete, Tapir, Trondheim, pp. 149-159, 1987.

47.	Lee, H. K., Lee, K. M., and Kimy, B. G., Autogenous Shrinkage of High-Performance Concrete Containing Fly Ash, Magazine of Concrete Research, Vol. 55, No. 6, pp. 507-515, 2003.

48.	Nawa, T., and Horita, T., Autogenous Shrinkage Of High-Performance Concrete, Proceedings of the International Workshop on Microstructure and Durability to Predict Service Life of Concrete Structure, Sapporo, Japan, February 2004.

49.	Tazawa, E., and Miyazawa, S., Influence of Cement and Admixture on Autogenous Shrinkage of Cement Paste, Cement and Concrete Research, Vol. 25, No. 2, pp. 281-287, 1995.

50.	Tazawa, E., and Miyazawa, S., Autogenous Shrinkage of Cement Paste with Condensed Silica Fume, 4th CANMET/ACI International Conference on Fly Ash, Silica Fume, Slag and Natural Pozzolans in Concrete, ACI, pp. 875-894, Istanbul, 1992.

51.	Brooks, J. J., Cabrera, J. G., Factors Affecting the Autogenous Shrinkage of Silica Fume High-Strength Concrete, Proceedings of International Workshop on Autogenous Shrinkage of Concrete, Japan Concrete Institute, pp. 185-192, Hiroshima, Japan, 1998.

52. Zang, M. H., Tam, C. T., Leow, M. P., Effect of Water-to-Cementitious Materials Ratio and Silica Fume on the Autogenous Shrinkage of Concrete, Cement and Concrete Research, Vol. 33, No.10, pp. 1687-1694, 2003.

53. Matsushita, H., Tsuruta, H., and Nakae, K., The Influence of Physical Properties of Aggregate on Shrinkage on High Strength Slag Cement Concrete, Fly Ash, Silica Fume, Slag, and Natural Pozzolans in Concrete. Sixth CANMET/ACI/JCI Conference, pp. 685-700, Bangkok, 1998.

54. Mazloom, M., Ramezanianpour, A. A., and Brooks, J. J., Effect of Silica Fume on Mechanical Properties of High-Strength Concrete, Cement and Concrete Composites, Vol. 26, No. 4, pp.347-357, May 2004.

55. Van Breugel, K., and De Vries, H., Potential of Mixtures with Blended Aggregates for Reducing Autogenous Deformation in Low Water/Cement Ratio Concrete, Proceedings of the 2nd International Symposium on Structural Lightweight Aggregate Concrete, Eds. Helland, S., Holand, I., and Smeplass, S., pp. 463-472, Kristiansand, Norway, 2000.

56. Bentur, A., Igarashi, S. and Kovler, K., Prevention of Autogenous Shrinkage In High Strength Concrete By Internal Curing Using Wet Lightweight Aggregates, Cement and Concrete Research, Vol. 31, No. 11, pp. 1587-1591, 2001.

57. Shi, C. and Wu, Y., Mix Design and Properties of Self-Consolidating Lightweight Concrete Containing Glass Powder, ACI Materials, Vol. 102, No. 5, pp. 355-363, 2005.

58. Bentz, D. P., Lura, P., and Roberts, J. W., Mixture Proportioning for Internal Curing, Concrete International, Vol. 27, No. 2, pp. 35-40, 2005.

59. Alfes, C., Modulus of Elasticity and Drying Shrinkage of High-Strength Concrete Containing Silica Fume, Fly Ash, Silica Fume, Slag, and Natural Pozzolans in Concrete, Proceedings of the Fourth International Conference, Ed. by Malhotra, V. M., ACI, Detroit, MI, Vol. 2, pp. 1651-1671. (ACI SP-132), Istanbul, Turkey, May 1992.

60. Hindy, E. E., Miao, B., Chaallal, O., and Aitcin, P.C., Drying Shrinkage of Ready-Mixed High-Performance Concrete, ACI Materials Journal, Vol. 91, No. 3, pp. 300-305, 1994.

61. ACI Committee 209, Prediction of Creep, Shrinkage and Temperature Effects in Concrete Structures, American Concrete Institute, Detroit, 2006.

62. Mokhtarzadeh, A., and French, C., Time-Dependent Properties of High-Strength Concrete with Consideration for Precast Applications, ACI Materials Journal, Vol. 97, No. 3., pp. 263-271, 2000.

63. Huo, X. S., Al-Omaishi, N., and Tadros, M. K., Creep, Shrinkage, and Modulus of Elasticity of High-Performance Concrete, ACI Materials Journal, Vol. 98, No. 6, pp. 440-449, 2001.

64. Luther, M. D., and Hansen, W., Comparison of Creep and Shrinkage of High-Strength Silica Fume Concretes with Fly Ash Concretes of Similar Strengths, Proceedings of the Fourth International Conference on Fly Ash, Silica Fume, Slag and Natural Pozzolans in Concrete, Ed. Malhotra, V. M., ACI SP-114, American Concrete Institute, Vol. 1, pp. 573-91, Detroit, MI, USA, 1989.

65. Li, J., and Yao, Y., A Study on Creep and Drying Shrinkage of High Performance Concrete, Cement and Concrete Research, Vol. 31, No. 8, pp. 1203-1206, 2001.

66. Mehta, P. K., High-Performance, High-Volume Fly Ash Concrete For Sustainable Development, Proceedings of the International Workshop on Sustainable Development and Concrete Technology, pp. 3-14, Beijing, China, May 20-21, 2004.

67. Persson, B., Experimental Studies on Shrinkage in High-Performance Concrete, Cement and Concrete Research, Vol. 28, No. 7, pp. 1023-1036, 1998.

68. Khan, M. I., Lynsdale, C. J., Strength, Permeability, and Carbonation of High-performance Concrete, Cement and Concrete Research, Vol. 32, No. 1, pp. 123-131, 2002.

69. Khunthongkeaw, J., Tangtermsirikul, S., and Leelawat, T., A Study on Carbonation Depth Prediction for Fly Ash Concrete, Construction and Building Materials, Vol. 20, No. 9, pp. 744-753, 2006.

70. Müller, H. S., Küttner, C. H., Kvitsel, V., Creep and Shrinkage Models of Normal and High-Performance Concrete – Concept for a Unified Code-Type Approach, Proceedings of ACI-RILEM Workshop on Creep and Shrinkage of Concrete Structures, Paris, France, Vol. 3, No. 3-4, pp. 113-132, 1999.

71. Cousins, T. E., Investigation of Long-Term Prestress Losses in Pretensioned High Performance Concrete Girders, VTRC 05-CR20, Virginia Transportation Research Council, Charlottesville, Virginia, June 2005.

72. Zia, P., Ahmad, S. H., Leming, M. L., Schemmel, J. J., and Elliott, R. P., Mechanical Behavior of High Performance Concretes, Vol. 5, (SHRP-C-365) Very High Strength Concrete. Strategic Highway Research Program, National Research Council, Washington, D. C., Vol. 11, pp. 101, 1993.

73. Ma, J., and Orgass, M., Creep of Ultra High Performance Concrete, LACER, No. 10, pp. 329-340, 2005.

74. Li, J., and Yao, Y., A study on Creep and Drying Shrinkage of High Performance Concrete, Cement and Concrete Research Vol. 31, No. 8, pp. 1203-1206, 2001.

75. Aïtcin, P. C., The Durability Characteristics of High Performance Concrete, A Review, Cement and Concrete Composites, Vol. 25, No. 4-5, pp. 409-420, 2003.

76. Hearn, N., Hooton, R. D., and Nokken, M. R., Pore Structure, Permeability and Penetration Resistance Characteristics of Concrete, Significance of Tests and Properties of Concrete and Concrete-Making Materials, Ed. by Lamond, J. and Pielert, J., STP169D, ASTM International, 2006.

77. Shi, C., Yuan, Q., Deng, Q., and Zheng, K., Test Methods For The Transport of Chloride Ion in Concrete, Journal of Chinese Ceramic Society, Vol. 35, No. 4, 2007.

78. NT BUILD 492, Chloride Migration Coefficient From Non-Steady-State Migration Experiments, Nordtest, Tekniikantie 12, FIN-02150, Espoo, Finland, 1999.

79. NT BUILD 355, Chloride Diffusion Coefficient From Migration Cell Experiments, Nordtest, Tekniikantie 12, FIN-02150, Espoo, Finland, 1997.

80. ASTM C1202, Electrical Indication of Concrete's Ability to Resist Chloride Ion Penetration, Annual Book of American Society for Testing Materials Standards, Vol. C04.02, 2006.

81. AASHTO T 277 [1980] (reference not available).

82. Shi et al, 1998 (reference not available).

83. Shi, 2004 (reference not available).

84. Olek, J., Lu, A., Feng, X., Magee, B., Performance-Related Specifications for Concrete Bridge Superstructures, Vol. 2, High-

Performance Concrete, FHWA/IN/JTRP-2001/8, SPR-2325, Purdue University, 2002.

85. Bajorski, P., Streeter, D. A., and Perry, R. J., Applying Statistical Methods for Further Improvement of High Performance Concrete for New York State Bridge Decks, Transportation Research Record, No. 1574, pp. 71-79, Nov., 1996.

86. Flemming, H. C., Eating Away at the Infrastructure – The Heavy Cost of Microbial Corrosion, Water Quality International, No. 4, pp. 16-19, 1995.

87. Belie, D., Verselder, H. J., Blaere, B. D., Nieuwenburg, D. V. and Verschoore, R., Influence of the Cement Type on the Resistance of Concrete to Feed Acids, Cement and Concrete Research, Vol. 26, No. 11, pp. 1717-1725, 1996.

88. Bertron, A., Duchesne, J., and Escadeillas, G., Attack of Cement Pastes Exposed to Organic Acids in Manure, Cement and Concrete Composites, Vol. 27, No. 9-10, pp. 898-909, 2005.

89. Stegemann, J. A., and Shi, C., Acid Resistance of Different Monolithic Binders and Solidified Wastes, in Waste Materials in Construction: Putting Theory into Practice, Elsevier Science, pp. 551-562, 1997.

90. Pavlik, V., Corrosion of Hardened Cement Paste by Acetic and Nitric Acids, Part I: Calculation of Corrosion Depth, Cement and Concrete Research, Vol. 24, No. 3, pp. 551-562, 1994.

91. Shi, C., and Stegemann, J. A., Acid Corrosion Resistance of Different Cementing Materials, Cement and Concrete Research, Vol. 30, No. 6, pp. 803-808, 2000.

92. Zivica, V., and Bajza, A., Acid Attack of Cement-Based Materials – A Review Part 2, Factors Of Rate Of Acidic Attack And Protective Measures, Construction and Building Materials, Vol. 16, No. 4, pp. 215-222, 2002.

93. Beaudoin, J. J., and Brown, P. W., The Structure of Hardened Cement Paste, Proceedings of 9th International Congress on the Chemistry of Cement, Vol. I, pp. 485-525, New Delhi, 1992.

94. Iler, R. K., The Chemistry of Silica-Solubility, Polymerization, Colloid and Surface Properties and Biochemistry, John Wiley & Sons, 1979.

95. Torii, K. and Kawamura, M., Effects of Fly Ash and Silica Fume on the Resistance of Mortar to Sulphric Acid and Sulphate

Corrosion, Cement and Concrete Research, Vol. 24, No. 2, pp. 361-370, 1994.

96.   Tamimi, A. K., High-Performance Concrete Mix for an Optimum Protection in Acidic Conditions, Materials and Structures, Vol. 30, pp. 188-191, 1997.

97.   Skalny, J. P., Odler, I., Marchand, J., Sulfate Attack on Concrete, Spon. Press, 2001.

98.   De Almeida, I. R., Resistance of High Strength Concrete To Sulfate Attack: Soaking And Drying Test, Proceedings of The Second International Conference on Concrete Durability, SP 126, Vol. II, pp. 1073-1092, American Concrete Institute, MI, 1991.

99.   Al-Attar, T. S., Al-Khateeb, A. M., and Bachai, A. H., Behavior of High Performance Concrete Exposed to Internal Sulfate Attack (Gypsum-Contaminated Aggregate), Proceedings of 10th Biennial International Conference on Engineering, Construction, and Operations in Challenging Environments (Earth & Space 2006) and 2nd NASA/ARO/ASCE Workshop on Granular Materials in Lunar and Martian Exploration held in League City/Houston, TX, March 5-8, 2006.

100.   Shannag, M. J., and Shaia, H. A., Sulfate Resistance of High-Performance Concrete, Cement and Concrete Composites Vol. 25, Issue 3, pp. 363-369, April 2003.

101.   Fournier, B., Berube, M. A., and Rogers, C. A., Proposed Guidelines for The Prevention of Alkali-Silica Reaction in New Concrete Structures, Transportation Research Record, No. 1668, pp. 48-53, 1999.

102.   Folliard, K. J., Thomas, M. D. A., and Kurtis, K. E., Guidelines for the Use of Lithium to Mitigate or Prevent Alkali-Silica Reaction (ASR), Publication No. FHWA-RD-03-047, FHWA, July 2003.

103.   Shehata, M. H., and Thomas, M. D. A., The Effect of Fly Ash Composition on the Expansion of Concrete due to Alkali Silica Reaction, Cement and Concrete Research, Vol. 30, pp. 1063-1072, 2000.

104.   Thomas, M. D. A., A Review of the Effect of Fly Ash and Slag on Alkali-Aggregate Reaction in Concrete, Building Research Establishment, Garston, Watford WD2 7JR, U.K., pp. 117, 1996.

105.   Bleszynski, R., Thomas, M. D. A, and Hooton, D., The Efficiency of Ternary Cementitious Systems for Controlling Expansion due to Alkali-Silica Reaction in Concrete, Proceedings of the

11th International Conference on Alkali-Aggregate Reaction (ICAAR), Quebec, Canada, pp. 583-592, June 11-16 2000.

106. Lane, D. S., and Ozyildirim, C., Preventive Measures for Alkali-Silica Reactions (Binary and Ternary Systems), Cement and Concrete Research, Vol. 29, No. 8, pp. 1281-1288, 1999.

107. Chatterji, S., Mechanism of CaCl2 Attack on Portland Cement Concrete, Cem. Concr. Res., Vol. 8, No. 4, pp. 461-468, 1978.

108. Monosi, S., Alver, I., and Collepardi, M., Chemical Attack of Calcium Chloride on the Portland Cement Paste, Cemento, Vol. 86, No. 2, pp. 97-104, 1989.

109. Collepardi, M., Coppola, L., Pistolesi, C., Durability of Concrete Structures Exposed to CaCl2. Based Deicing Salts, Proceedings of the Third CANMET/ACI International Conference on Durability of Concrete, pp. 107-120, American Concrete Institute, MI, 1994.

110. Torii, K., Sasatani, T., Kawamura, M., Effects of Fly Ash, Blast Furnace Slag and Silica Fume on Resistance of Mortar to Calcium Chloride Attack, Proceedings of 5th International Conference on Fly Ash, Silica Fume, Slag and Natural Pozzolans in Concrete, Vol. 2, pp. 931-950. American Concrete Institute, Detroit, MI, 1995.

111. Shi, C., Formation and Stability of $3CaO.CaCl_2.12H_2O$, Cement and Concrete Research, Vol. 31, No. 9, pp. 1273-1275, 2001.

112. Pettersson, K., Chloride Threshold Value and the Corrosion Rate in Reinforced Concrete, Cement and Concrete Research, Vol. 20, pp. 461-470, 1994.

113. Hussain, S. E., Rasheeduzzafar, A., Al-Mussallam, A., and Al-Gahtani, A. S., Factors Affecting Threshold Chloride for Reinforcement Corrosion in Concrete, Cement and Concrete Research, Vol. 25, No. 7, pp. 1543-1555, 1995.

114. Smith, L. J., and Virmani, Y. P., Materials and Methods for Corrosion Control of Reinforced and Prestressed Concrete Structures in New Construction, FHWA, Publication No. 00-081, 2000.

115. Balabanic, G., Bicanic, N., and Durekovic, A., Mathematical Modeling of Electrochemical Steel Corrosion in Concrete, Journal of Engineering Mechanics, Vol. 122, No. 12, pp. 1113-1122, 1996.

116. Liang, M. T., Jin, W. L., Yang, R. J., and Huang, N. M., Predeterminate Model of Corrosion Rate of Steel in Concrete,

Cement and Concrete Research, Vol. 35, No. 9, pp. 1827-1833, 2005.

117. Isgor, O. B., and Razaqpur, A. G., Modeling Steel Corrosion in Concrete Structures, Materials and Structures, Vol. 39, No. 3, pp. 259-270, 2006.

118. Li, Y., Langan, B. W., and Ward, M. A., Freezing and Thawing: Comparison between Non-Air-Entrained and Air-Entrained High-Strength Concrete, Proceedings of ACI International Conference, (ACI SP-149), Singapore, Ed. Malhotra, V. M., American Concrete Institute, Detroit, MI, pp. 545-560, 1994.

119. ASTM C666 (reference not available).

120. Pinto, R. A., and Hover, K. C., Frost and Scaling Resistance of High-Strength Concrete, RD122, Portland Cement Association. Skokie, IL, 2001.

121. Shi, C., Berke, N., Jeknavorian, A. and Zhong, H., Roles of Chemical Admixtures in Sustainable Concrete Materials and Structures, Proceedings of the 6th International Symposium on Cement and Concrete, Vol.2, pp. 1631-1643, Xian, China, September 19-22, 2006.

122. Laplante, P., Aitcin, P. C., and Vezina. D., Abrasion Resistance of Concrete, Journal of Materials in Civil Engineering, Vol. 3, No. 1, pp. 19-28, 1991.

123. De Almeida, I. R., Abrasion Resistance of High Strength Concrete with Chemical and Mineral Admixtures. Durability of Concrete, Proceedings of the Third International Conference, ACI SP-145, Nice, France, Ed. Malhotra, V. M., American Concrete Institute, Detroit, MI, pp. 1099-1113, 1994.

124. Brazilian Standard (reference not available).

125. German Standard DIN 52108 (reference not available).

126. Naik, T. R., Singh, S. S., and Hossain, M. M., Abrasion Resistance of High-Strength Concrete Made with Class C Fly Ash, ACI Materials Journal, Vol. 92, No. 6, pp. 649-659, 1995.

127. Atis, C. D., High Volume Fly Ash Abrasion Resistant Concrete, J. Mater. Civ. Eng., Vol. 14, No. 31, pp. 274–277, 2002.

128. Atis, C. D., Abrasion-Porosity-Strength Model for Fly Ash Concrete, Journal of Materials in Civil Engineering, Vol. 15, No. 4, pp. 408-410, 2003.

129. Phan, L. T., Fire Performance of High-Strength Concrete: A Report of the State-of-the-Art, National Institute of Standards and Technology, Gaithersburg, MD, 1996.

130. Kodur, V. K. R., and Sultan, M. A., Structural Behavior of High Strength Concrete Columns Exposed to Fire, International Symposium on High Performance and Reactive Powder Concretes, pp. 217-232, Sherbrooke, QC, 1998.

131. Kodur, V. K. R., Wang, T. C., and Cheng, F. P., Predicting the Fire Resistance Behavior of High Strength Concrete Columns, Cement and Concrete Composites, Vol. 26, No.2, pp. 141-153, 2004.

132. Sarvaranta, L., and Mikkola, E., Fiber Mortar Composites in Fire Conditions, J. Fire Materials, Vol. 18, pp. 45-50, 1994.

133. Kalifa, P., Chene, G., and Galle, C., High-Temperature Behaviour of HPC with Polypropylene Fibers from Spalling to Microstructure, Cement and Concrete Research, Vol. 31, No. 10, pp. 1487-1499, 2001.

134. Han, C. G., Hwang, Y. S., Yang, S. H., and Gowripalan, N., Performance of Spalling Resistance of High Performance Concrete with Polypropylene Fiber Contents and Lateral Confinement, Cement and Concrete Research, Vol. 35, No. 9, pp. 1747-1753, 2005.

135. PCI, TR-6-03, Interim Guidelines for the Use of Self-Consolidating Concrete in Precast/Prestressed Concrete Institute Member Plants, Precast/Prestressed Concrete Institute, Chicago, Illinois, 2003.

136. Shi, C., Wu, Y. and Riefler, M., Comparison of Self-Consolidating with Conventional Concretes, 5th International Symposium on Cement and Concrete, Vol. 2, pp. 1013-1019, Shanghai, China, October 28-November 31, 2002.

137. Shi, C. and Wu, Y. and Riefler, C., Use of Self-Consolidating Lightweight Concrete for Insulated Concrete Form System, Concrete International, Vol. 28, No. 2, pp. 40-43, February 2006.

138. Japan Society of Civil Engineers (JSCE), Recommendation for Construction of Self-Compacting Concrete, Japan Society of Civil Engineers, 1998.

139. Gaimster, R., and Foord, C., Self-compacting concrete, Concrete, Vol. 34, pp. 23-25, 2000.

140. Khayat, K.H., Hu, C., and Monty, H., Stability of Self-Consolidating Concrete, Advantages, and Potential Applications, Proceedings of First International RILEM Symposium on Self-Compacting Concrete, Stockholm, Sweden, pp. 143-152, 1999.

141. EFNARC, The European Guidelines for Self-Compacting Concrete Specification, Production and Use, May 2005.

142. Toyoharu, N., Tatsuo, I. and Yoshinobu, E., State-of-the-Art Report on Materials and Design of Self-Compacting Concrete, International Workshop on Self-Compacting Concrete, pp160-189, Japan, 1998.

143. Okamura, H. and Ozawa, K., Mix-Design for Self-Compacting Concrete, Concrete Library of the Japanese Society of Civil Engineers, pp. 107-120, 1995.

144. Nawa, T., Eguchi, H. and Fukaya, Y., Effect of Alkali Sulfate on the Rheological Behavior of Cement Paste containing a Superplasticizer, ACI SP-119, pp. 405-424, 1989.

145. Edamatsu, Y., Kobayashi, T. and Nagaoka, S., Influence of Powder Properties on Self-Compactibilty of Fresh Concrete, JCA Proceedings of Cement and Concrete, No. 51, pp. 394-399 (in Japanese), 1997.

146. ASTM C33 (reference not available).

147. Holschemacher, K. and Klug, Y., A Database for the Evaluation of Hardened Properties of SCC, LACER, No. 7, pp. 123-134, 2002.

148. Grünewald, S. and Walraven, J.C., Parameter-Study on the Influence of Steel Fibers and Coarse Aggregate Content on the Fresh Properties of Self-Compacting Concrete, Cement and Concrete Research, 31, pp. 1793-1798, 2001.

149. ASTM C1611, Test Method for Slump Flow of Self-Consolidating Concrete, Annual Book of ASTM Standards, Vol. 04.02., Concrete and Aggregate, ASTM International, West Conshohocken, PA, USA, 2006.

150. ASTM C-1621/C-1621M-06, Standard Test Method for Passing Ability of Self-Consolidating Concrete by J-Ring, Annual Book of ASTM Standards, Vol. 04.02, Concrete and Aggregate, ASTM International, West Conshohocken, PA, USA, 2006.

151. ASTM C1610/C1610M-06, Standard Test Method for Static Segregation of Self-Consolidating Concrete Using Column Technique, Annual Book of ASTM Standards, Vol. 04.02,

Concrete and Aggregate, ASTM International, West Conshohocken, PA, USA, 2006.

152. ASTM C143/C143M, Standard Test Method for Slump of Hydraulic Cement Concrete, Annual Book of ASTM Standards, Vol. 04.02, Concrete and Aggregate, ASTM International, West Conshohocken, PA, USA, 2006.

153. Daczko, J. A., and Kurtz, M. A., Development of High-Volume Coarse Aggregate Self-Compacting Concrete, Proceedings of the Second International Symposium on Self-Compacting Concrete, Tokyo, Japan, 2001.

154. Petersson, O., SCC Task 9 Reseach Report, Swedish Cement and Concrete Research Institute, Stockholm, Sweden, 1997.

155. Ambroise, J., Rols, S., Pera, J., Properties of Self-Levelling Concrete Reinforced by Steel Fibers, IBRACON, Brazilian Concrete Institute, 2001.

156. Zhu, W., Sonebi, M., and Bartos, P. J. M., Bond and Interfacial Properties of Reinforcement in Self-Compacting Concrete, Materials and Structures, Vol. 37, No. 7, pp. 442-448, 2004.

157. Sonebi, M., and Bartos, P. J. M., Hardened SCC and Its Bond with Reinforcement, Proceedings of the First International RILEM Symposium on Self-Compacting Concrete, Stockholm, September, pp. 275-289, 1999.

158. Chan Y. W., Chen Y. S., and Liu Y. S., Development of Bond Strength of Reinforcement Steel in Self-Consolidating Concrete, ACI Structural Journal, July-August, Vol. 100, No. 4, pp. 490-498, 2003.

159. Sonebi, M., Zhu, W., and Gibbs, J., Bond of Reinforcement in Self-Compacting Concrete, Concrete, Vol. 35, No. 7, July-August, pp. 26-28, 2001.

160. Persson B., Internal Frost Resistance of Salt Frost Scaling of Self Compacting Concrete, Cement and Concrete Research, Vol. 33, pp. 373-379, 2003.

161. Khayat, K. H., Optimization and Performance of Air-Entrained Self-Consolidating Concrete, ACI Materials Journal, Vol. 97, No. 5, pp. 526-535, 2000.

162. Beaupré, D., Lacombe, P., and Khayat, K. H., Laboratory Investigation of Rheological Properties and Scaling Resistance of SCC, RILEM Materials and Structures, Vol. 32, No. 217, pp. 235-240, 1999.

163. Khayat, K. H., and Assaad, J., Air-Void Stability of SCC, ACI Materials Journal, Vol. 99, No. 4, pp. 408-416, 2002.

164. Gaimster R., and Foord C., Self-compacting Concrete, Concrete, Vol. 34, No. 4, pp. 23-25, 2000.

165. Ghosh, S. K., and Ahmad, S. H., High-Strength Concrete for Seismic Applications, Concrete International, Vol. 28, No. 9, pp. 47-49, 2006.

166. Hawkins, N., Application of the LRFD Bridge Design Specifications to High-Strength Structural Concrete: Shear Provisions, NCHRP Report 579, FHWA, 2007.

167. Ozyildirim, C., Permeability Specifications for High-Performance Concrete Decks, Transportation Research Board, No. 1610, pp. 1-5, 1998.

168. Rangaraju, P. R., Development of Some Performance-Based Material Specifications For High-Performance Concrete Pavement, Transportation Research Record, No. 1834, pp. 69-76, 2003.

169. Sprinkel, M. M., Performance Specification For High Performance Concrete Overlays On Bridges, VTRC 05-R2, Virginia Transportation Research Council, Charlottesville, Virginia, 2004.

170. Russell, H. G., Compilation and Evaluation of Results from High-Performance Concrete Bridge Projects, Volume I: Final Report FHWA-HRT-05-056, FHWA, 2005.

171. Central South University (CSU), Design Guidelines for Self-Consolidating Concrete, CSCE002-2004, Chinese Construction Press, Beijing, 2004.

172. PCA, Guide Specification for High Performance Concrete for Bridges, EB233, Portland Cement Association, 2005.

173. Xie, Y. J., Zhong, X. H., Zhu, C. H., and Zhang, Y., Durability of HPC for Bridge and Tunnel Structure on Qinghai-Tibet Railway, China Railway Science, Vol. 24, No. 1, pp. 108-112, 2003.

174. 郁慕贤, 青藏铁路耐久性混凝土的配制与施工控制, Railway Construction, No. 9, pp. 78-80, 2004.

175. Luo, Y. G., Construction Technique for Durability Concrete for Qinghai - Tibet Railway, Journal of Railway Engineering Society, No.3, pp.51-55, 2006.

176. Dhonde, H. B., Mo, Y. L. and Hsu, T. T. C., Fiber Reinforcement in Prestressed Concrete Beams, Technical Report 0-4819-1, Texas Department of Transportation, Austin, Texas, USA, 2005.

Chapter 3

High Performance Fiber Reinforced Cement Composites

Antoine E. Naaman
Department of Civil and Environmental Engineering
University of Michigan, Ann Arbor, MI, 48109-2125, U.S.A.

3.1 Introduction

Fiber reinforced cement based composites have made striking advances and gained enormous momentum in recent years. This is due in particular to several developments involving the matrix, the fiber, the fiber-matrix interface, the composite production process, a better understanding of the fundamental mechanisms controlling their particular behavior, and a continually improving cost performance ratio. Examples include: (1) the commercial introduction of a new generation of additives (superplasticizers and viscous agents) which allow for high matrix strengths to be readily achieved with little loss in workability, (2) the increasing use of active or inactive micro-fillers such as silica fume and fly ash and a better understanding of their effect on matrix porosity, strength, and durability, (3) the increasing availability for use in concrete of fibers of different types and properties which can add significantly to the strength, ductility, and toughness of the resulting composite, (4) the use of polymer addition or impregnation of concrete which adds to its strength and durability but also enhances the bond between fibers and matrix thus increasing the efficiency of fiber reinforcement, and (5) some innovations in production processes (such as self-consolidation or self-compacting) to improve uniform mixing of high volumes of fiber with reduced effects on the porosity of the matrix. Substantial progress has also been made in modeling the behavior of these composites [1-20]. Reference [14] is of particular interest to the subject of high performance.

Generally the attribute "advanced" or "high performance" when applied to engineering materials is meant to differentiate them from the conventional materials used, given available technologies at the time and geographic location considered for the structure. It also implies an optimized combination of properties for a given application and should be generally viewed in its wider scope. Combined properties of interest to civil engineering applications include strength, toughness, energy absorption, stiffness, durability, freeze-thaw and corrosion resistance, fire resistance, tightness, appearance, stability, construct-ability, quality control, and last but not least, cost and user friendliness.

For the purpose of this chapter, the attribute "high performance" is limited to the particular class of FRC composites that shows strain-hardening behavior in tension after first cracking, accompanied by multiple cracking up to relatively high strain levels. Such an attribute applies to low and high strength, normal weight or lightweight cement composites.

3.2 Definitions

3.2.1 *Fiber reinforced cement (FRC) composites*

For practical purposes and mechanical modeling, fiber reinforced cement (FRC) or concrete composites are generally defined as composites with two main components, the fiber and the matrix (Fig. 3.1). While the cementitious matrix may itself be considered a composite with several components, it will be assumed to represent, in the context of this chapter, the first main component of the FRC composite. The fiber represents the second main component. The fiber is assumed to be discontinuous and, unless otherwise stated, randomly oriented and distributed within the volume of the composite. Both the fiber and the matrix are assumed to work together, through bond, and provide the synergism needed to make an effective composite. The matrix, whether it is a paste, mortar, or concrete is assumed to contain all the aggregates and additives specified. Air voids entrapped in the matrix during mixing are assumed to be part of the matrix.

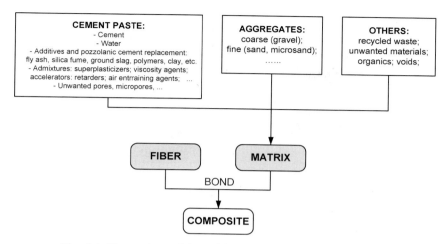

Fig. 3.1 Composite model considered as a two component system,
namely fiber and matrix.

3.2.2 *High Performance Fiber Reinforced Cement (HPFRC) Composites*

One approach followed by the author to define whether a fiber reinforced cement composite qualifies as "high performance", is based on the shape of its stress-strain curve in direct tension (Fig. 3.2).

If the stress-strain curve shows strain hardening (or pseudo-strain hardening) behavior after cracking then the qualification "high performance" is used (Figs. 3.3(b) and 3.4(b)); otherwise, for a conventional FRC composite, the stress-strain curve would show strain-softening response immediately after first cracking (Figs. 3.3(a) and 3.4(a)). This is equivalent to saying that, for HPFRC composites the shape of the stress-strain curve in tension is at least elastic-plastic or better. Thus the following definition is used for this chapter:

> *High performance fiber reinforced cement composites are a class of FRC composites characterized by a strain-hardening behavior in tension after first cracking, accompanied by multiple cracking up to relatively high strain levels.*

A. E. Naaman

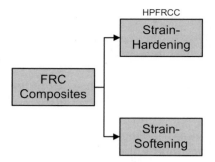

Fig. 3.2 Simple classification of FRC composites based on their tensile response.

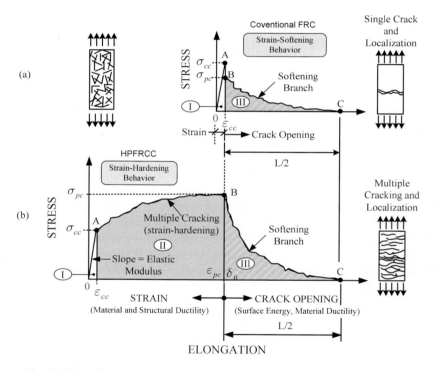

Fig. 3.3 Typical stress-strain or stress-elongation curve in tension up to complete separation. (a) Conventional strain-softening FRC composite. (b) Strain-hardening FRC composite or HPFRCC.

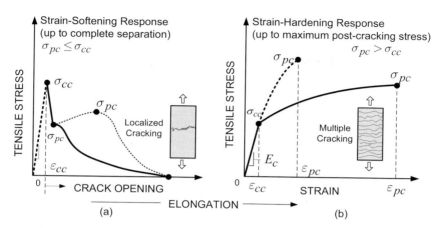

Fig. 3.4 Typical stress-elongation curves in tension of fiber reinforced cement composites. (a) Strain-softening behavior. (b) Strain-hardening behavior (HPFRCC).

Strain-hardening behavior is a very desirable property, generally accompanied by multiple cracking and related large energy absorption capacity. Five international symposia have taken place using such a definition. One simple way to express the condition to achieve strain-hardening behavior is to write that the post-cracking strength of the composite is larger than its cracking strength, that is, $\sigma_{pc} \geq \sigma_{cc}$, (Fig. 3.3(b)). Other approaches based on energy balance whereas the energy necessary to create a new crack is smaller than the energy needed to open an existing crack have also been developed but are not covered here for brevity [11,21]; they are discussed and compared in [15].

Typically the stress-strain curve of an HPFRC composite (Figs. 3.3(b) and 3.4(b)) starts with a steep initial ascending portion up to first structural cracking (part I), followed by a strain-hardening branch where multiple cracking develops (part II). The point where first structural cracking occurs is characterized by its stress and strain coordinates $(\sigma_{cc}, \varepsilon_{cc})$; the peak point at the end of the strain-hardening branch leads to the maximum post-cracking stress and strain $(\sigma_{pc}, \varepsilon_{pc})$. At the peak point, one crack becomes critical defining the onset of crack localization. The resistance drops thereafter. No more cracks can develop, and only the critical crack will open under increased deformation. Other cracks will gradually unload or become smaller (narrower in width). Following the peak point, there is generally a descending branch which corresponds mainly to the load versus opening of the critical crack (part III of Fig. 3.3(b)). Along that branch fibers can pull-out, fail, or a combination of these phenomena may occur. Also, the cement matrix may contribute some resistance along that part of the curve up to a certain crack opening.

Note that before crack localization, the elongation of the composite can be translated into tensile strain, whether strain-softening or a strain-hardening behavior prevails. However, after localization, the elongation is controlled by the opening of the critical crack. Along part III of the curves in Figs. 3.3(a) and 3.3(b), the strain, that is unit deformation, looses its meaning.

The stress-strain curve of a conventional strain-softening FRC composite (Figs. 3.3(a) and 3.4(a)) starts the same way as for a high performance FRC composite (part I); however it will have localization occurring immediately following first cracking; no strain-hardening and multiple cracking occur (that is, it does not have a part II). After localization the descending branch follows a pattern similar to that of a HPFRC composite (part III).

3.2.3 *Stress at first cracking*

Note that the term "first cracking" has been the subject of much discussion and interpretation in numerous prior studies. It may imply first visible cracking or deviation from linearity as detected along the initial ascending portion of the stress-strain curve. In this discussion it implies that a crack has "percolated" through the structural tensile member. While, during the ascending loading branch, many micro-cracks may develop either along the member or within a given cross-section of it, a percolation crack is defined here as a crack providing a complete separation between two parts of the tensile member. The corresponding crack opening may be very small and possibly invisible to the naked eye. For a typical tensile prism, a percolation crack would have its perimeter-boundary on the lateral surface of the prism. A percolation crack may have joined together a number of other smaller cracks or micro-cracks and is not necessarily normal to the direction of the tensile load, but can be considered normal for the convenience of modeling. The stress and strain coordinates of the first cracking point are termed σ_{cc} and ε_{cc}, respectively (Figs. 3.3 and 3.4).

3.2.4 *Maximum post-cracking stress or composite strength*

The stress in the composite after percolation cracking is termed the post-cracking stress; its maximum value is defined as σ_{pc}, the post-cracking strength. For strain-hardening composites, $\sigma_{pc} \geq \sigma_{cc}$, while for strain-softening composites $\sigma_{pc} < \sigma_{cc}$ (Figs. 3.3 and 3.4). The strain at maximum post-cracking stress is termed ε_{pc}. Note that after the maximum post-cracking point, localization occurs at the critical crack and the term "strain" becomes inappropriate; instead crack opening, crack width, or member elongation become more appropriate to describe the behavior. This is also true for strain-softening composites immediately after cracking (Figs. 3.3(a) and 3.4(a)).

3.2.5 *Strain-hardening and deflection-hardening FRC composites*

The discussion following the 2003 fourth international workshop on High Performance Fiber Reinforced Cement Composites (HPFRCC-4) led to suggesting a general classification of FRC composites such as shown in Fig. 3.2 [22,23]. One key distinguishing fundamental material characteristic is whether the response of the composite is strain-hardening or strain-softening in tension (see also Figs. 3.3 and 3.4). The implication of tensile behavior on the bending response of structural members is illustrated in Fig. 3.5, leading to deflection-hardening or deflection-softening response. Practically all fiber reinforced cement composites currently available are covered by the simple classification of Figs. 3.2 and 3.5. Everything else being equal, it is observed that: (1) all strain-hardening composites lead to deflection-hardening structural elements, (2) a tension strain-softening composite can lead to structural elements with either deflection-hardening or deflection-softening behavior, and (3) a strain-hardening composite provides a better mechanical performance than a strain-softening one. Deflection-hardening behavior is useful in structural applications where bending prevails, while deflection-softening composites cover a wide range of practical applications starting at the lower end by the control of plastic shrinkage cracking of concrete, to the higher end where they are used in concrete pavements and slabs on grade. Note that, as with other materials, scale and size effects can be significant [24]; thus the tensile response of very small specimens may not be indicative of the response of real scale structural elements in either tension or bending.

Figure 3.5 contains additional information on the critical volume fraction of fibers needed to achieve strain-hardening or deflection-hardening behavior. These equations are described in details in Sections 3.14 and 3.15.

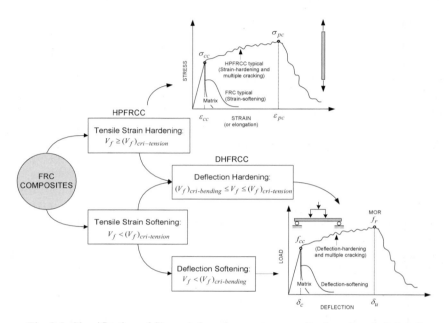

Fig. 3.5 Classification of fiber reinforced cement composites based on their tensile response and implication for bending response of structural elements.

3.3 Cement Matrices and Fibers for FRC Composites

3.3.1 *Cement or cementitious matrices*

Cement based matrices have evolved enormously since the 1960's due in particular to a better understanding their microstructure and the influence of packing and porosity, as well as the development of mineral and chemical additives to achieve a number of particular properties. Typically the matrix for fiber reinforced cement composites can be a cement paste, a mortar (that is, essentially a paste with sand), or a concrete (that is, essentially a mortar with coarse aggregates or gravel)

(Fig. 3.1). The cement paste is made of cement and water, and may contain additives which can be mineral (such as fly ash) or chemical such as air entraining admixtures. The paste itself can be made to have low viscosity, be stiff like clay, or to flow like a liquid; in the latter case, the term slurry or slurry paste is often used to describe the matrix.

The cement powder, which reacts with water to eventually leads to the binder and hardened matrix, may be blended, that is, containing supplementary cementitious materials, such as flash ash, ground furnace slag, micro-silica, and the like (Fig. 3.1). Indeed, mineral components such as silica fume and fly ash are now commonly used either as additives to, or as replacement of cement. They help provide a denser composite, reduce porosity, improve fresh properties, improve strength, corrosion resistance and durability, control the hydration reaction, etc. Chemical admixtures, such as water reducing agents, superplasticizers, and viscosity agents, help control and improve a host of other properties in the fresh state to facilitate the fabrication and manufacturing phase.

Today, self-consolidating and self-compacting cementitious mixtures allow us to rethink construction procedures for fiber reinforced cement composites. Such mixtures, for instance, allow us to use fiber reinforced concrete in congested areas of reinforced concrete structures such as in beam column connections of seismic frames, coupling beams and the like, without compromising the full penetrating and encapsulating function of the matrix.

The reader is referred to specialized texts where extensive in-depth information can be found on cement and concrete matrices and related compositions and mixtures [25,26].

Table 3.1 provides examples of mixtures for steel fiber reinforced mortar and concrete where no mineral additives were added, and in which conventional sand was used. Table 3.2 gives examples of mixtures for high performance fiber reinforced cement composites with high fiber volume contents in which special sand, mineral additives, and superplasticizers were used. This table helps understand the influence of various parameters including the type of sand and the water to binder ratio. Note that, whenever possible, the use of fly ash (FA) as a replacement for part of the cement is recommended.

The relative cost of the matrix in fiber reinforced cement composites, even when improved with relevant additives, decreases (in comparison to the cost of the fiber) with an increase in fiber content. This is

Table 3.1 Typical examples of mix proportions (by weight of cement)
of steel fiber reinforced mortar and concrete.

Mix ID	Cement	Sand	Coarse Agg.	Water	Target f'_c MPa
Mortar 1	1	3	----------	0.5	49
Mortar 2	1	2	----------	0.5	60
Mortar 3	1	1	----------	0.35	70
Concrete 1	1	2.5	2.1	0.6	25
Concrete 2	1	2	2.3	0.5	42
Concrete 3	1	1	1.6	0.35	67

Note: These mixtures were used for different matrix compressive strengths, and steel fiber contents ranging from 1% to 3% by volume of mortar in the mix or about 3% to 8% by weight of mortar; cement is ASTM Type I; aggregate is crushed limestone of 12 mm maximum size; fiber aspect ratio is less than 100; fiber length is less than 30 mm; superplasticizer was added as needed especially for the high fiber content.

particularly true for high performance fiber reinforced cement composites where the fiber content is relatively high. For instance, in the case of steel fiber reinforced concrete with conventional hooked steel fibers, the cost of the fiber at 2% fiber content by volume may range from about 2 to 10 times the cost of the concrete matrix. Therefore, there is advantage in utilizing the best possible matrix for the conditions of use of the final structure.

3.3.2 *Fibers for cement and concrete matrices*

Short discontinuous fibers used in concrete can be characterized in different ways (Fig. 3.6). First, according to the fiber material: natural organic (such as cellulose, sisal, jute, bamboo, horse hair, etc.); natural mineral (such as asbestos, rock-wool, etc.); man-made such as steel, titanium, glass, carbon, polymers or synthetic, etc. Second, according to their physical/chemical properties: density, surface roughness, chemical stability, non-reactivity with the cement matrix, fire resistance or flammability, etc. Third, according to their mechanical properties such as tensile strength, elastic modulus, stiffness, ductility, elongation to failure, surface adhesion property, etc. Fourth, according to the geometric

Table 3.2 Examples of mixtures by weight of cement of high performance fiber reinforced cement composites with up to 2% fibers by volume (mostly steel and Spectra fibers were used).

ID	C	FA	SF	W	S	A	SP	f'_c MPa
M1	0.3	0.7	--	0.9	5[c]	--	--	7
M2	0.3	0.7	--	0.575	3.5[c]	--	--	13
M3	0.7	0.3	--	0.65	3.5[a]	--	--	20
M4	0.8	0.2	--	0.45	1[a]	--	--	44
M5	0.8	0.2	--	0.45	1[b]	--	0.03	55
M6	0.8	0.2	--	0.27	1.1[d] +0.38[b]	--	0.02	63
M7	0.8	0.2	0.07	0.26	1[d]	--	0.04	76
M8	0.8	0.2	0.07	0.26	1[a]	--	0.04	84
M9	0.8	0.2		0.26	0.5[b] +0.5[d]	--	0.02	86
M10	1	--	0.24	0.27	1.1[d] +0.38[b]	--	0.10	90
M11	1	--	0.12	0.28	0.67[a]	1	0.05	101

Note: C = Cement; FA = Fly Ash; SF = micro-silica from Silica Fume; W = Water; S = Sand; A = coarse Aggregate, here crushed limestone with maximum size 10 mm; SP = Superplasticizer; f'_c = compressive strength from cylinders. The sand used is a silica sand with the following characteristics: (a) ASTM -50-70; (b) ASTM -270; (c) ASTM 30-70; silica sand passing ASTM sieve No. 16. These mixtures were selected from various investigations carried out by the author and his students at the University of Michigan.

properties of the fiber: length, diameter or perimeter, cross-sectional shape, and longitudinal profile.

Once a fiber material (such as steel) has been selected, an infinite combination of geometric properties related to its cross sectional shape, length, diameter or equivalent diameter, and surface deformation can be selected. In some fibers the surface is etched or plasma treated to improve bond at the microscopic level. For steel, the cross section of the fiber can be circular, rectangular, diamond, square, triangular, flat, polygonal, or any substantially polygonal shape. To develop better

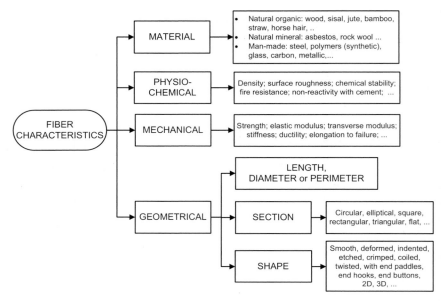

Fig. 3.6 Main fiber characteristics of interest in fiber reinforced cement composites.

mechanical bond between the fiber and the matrix, the fiber can be modified along its length by roughening its surface or by inducing mechanical deformations.   Thus fibers can be smooth, indented, deformed, crimped, coiled, twisted, with end hooks, end paddles, end buttons, or other anchorage system [27,28].   Typical examples of steel fibers are shown in Fig. 3.7.   Some other types of steel fibers such as ring, annulus, or clip type fibers (Fig. 3.7(b)) have also been used and shown to significantly enhance the toughness of concrete in compression; however, work on these fibers did not advance much beyond the research level.

Similarly to the geometric properties a wide variety of fiber materials and corresponding mechanical properties exists.   The fibers most commonly used in cement composites at time of this writing are listed in Table 3.3, and a range of values is given for their tensile strength and elastic modulus.   While fibers can have equal tensile strengths, the higher modulus fibers are generally more effective for normal weight cement matrices.

Table 3.3 Typical properties of some fibers used in FRC composites.

Material	Specific Gravity (1 for water)	Tensile Strength MPa	Tensile Modulus GPa	Remark
Steel	7.8	Up to 3500	200	The elastic modulus of steel is almost independent of its tensile strength. Yield strength of steel fibers varies widely depending on the fabrication process and alloying. The higher the strength the lower the strain capacity to failure.
E-Glass	2.6	Up to 3500	76	Modulus of glass can vary from 33 GPa for A-Glass to 98 GPa for S-Glass. While filament strength remains high from 3300 to 4800 MPa, the strength of a yarn or fiber bundle, which is made from a large number of filaments, will have much lower equivalent tensile strength.
Carbon	1.8	Up to 4500	100 to 300	Depending on grade and fabrication, such as pitch carbon or PAN carbon, significantly different properties can be achieved.
Kevlar (Aramid)	1.44	2800	124	Kevlar is a trade name of Dupont. The fiber material source is aramid. Aramid fibers have properties similar to Kevlar.
Spectra (HPPE)	0.97	2585	117	Spectra is a trade-name. The material is an ultra-high molecular weight polyethylene, also termed high performance polyethylene (HPPE). Other similar trade-named fiber include: Dyneema.
PVA (PolyVinyl Alcohol)	1.31	880-1600	25 to 40	A large variety of fibers exists with a wide range of tensile strengths and moduli. The higher the diameter, the lower the properties.
PP (Poly-Propylene)	0.91	Up to 800	Up to 10	Strength and modulus depend on the manufacturing process, and heat stretching leading to highly oriented long-chain molecules.

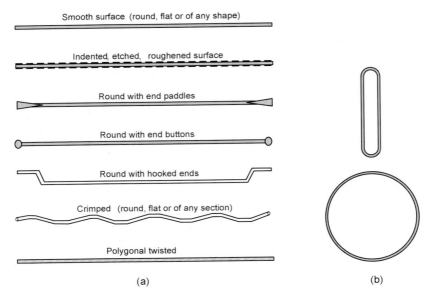

Fig. 3.7 (a) Typical profiles of steel fibers commonly used in concrete (twisted fiber is
new). (b) Closed loop fibers tried in some research studies.

3.3.3 *Micro-fibers*

For the purpose of this chapter, micro-fibers are defined here as having a
diameter or equivalent diameter of 100 µm (0.1 mm) or less. Their
length to diameter ratio should be preferably high, say above 100, and
may exceed 1000. Asbestos fibers qualify as micro-fibers. Micro-fibers
of carbon, glass, Kevlar, PVA or other polymeric materials are becoming
increasingly available for use in cement composites. Micro-fibers offer
very large numbers of fibers per unit volume; they induce micro-
toughness, reduce the size (width and length) of microcracks, and, if
properly bonded, generally improve the tensile strength at cracking (or
proportionality limit) of the composite. Because fine grain matrices with
micro-fibers can be used to produce manufactured products, the
machine-ability of the product is improved. The use of very fine grain
matrices reinforced with short microfibers or combinations of micro- and
macro-fibers, in stand alone applications such as small manufactured
products is likely to expand. Figure 3.8 shows typical examples of
macro- and micro-fibers.

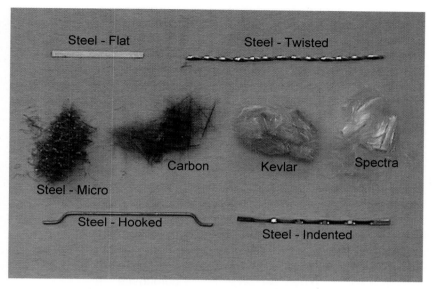

Fig. 3.8 Photograph illustrating typical fibers and micro-fibers used in cement composites.

3.3.4 *Current range of fiber geometric properties*

Most common steel fibers are round in cross-section, have a diameter ranging from 0.4 to 0.8 mm, and a length ranging from 25 to 80 mm. Their aspect ratio, that is, the ratio of length over diameter or equivalent diameter, is generally less that 100, with a common range from 40 to 80. Steel fiber diameters as low as 0.1 mm are also available but are much more costly than larger diameter fibers. The length and diameter of synthetic fibers vary greatly. Single filament fibers can be as little as 10 micrometers in diameter such as for Kevlar or carbon fibers, and as large as 0.8 mm such as with some polypropylene or poly-vinyl-alcohol (PVA) fibers. Generally in concrete applications, the aspect ratio of very fine fibers exceeds several hundred while that of coarser fibers is less than 100. Most synthetic fibers (glass, carbon, Kevlar) are round or substantially round in cross section; flat synthetic fibers cut from plastic sheets and fibrillated are suitable wzhen very low volume content is used such as for the control of plastic shrinkage cracking of concrete.

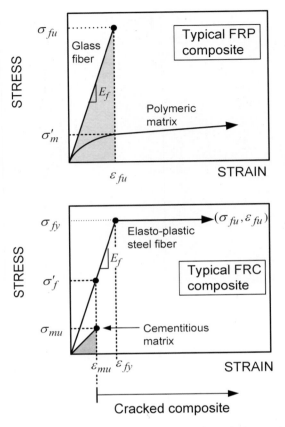

Fig. 3.9 Typical stress-strain diagrams of fiber and matrix in a typical: (a). Fiber reinforced polymeric composite, and (b). Fiber reinforced cement or concrete composite.

3.4 Key Difference Between Fiber Reinforced Cement and Fiber Reinforced Polymeric Composites for Mechanical Modeling

Generally, cementitious matrices have mechanical characteristics that distinguish them from metallic and polymeric matrices, namely they have relatively high compressive strength, poor tensile strength, and brittleness at failure, particularly under tension. They are also noted for their very small tensile strain at cracking and failure. Adding fibers to such matrices is meant to enhance their mechanical properties, particularly toughness, control cracking, and increase their range of use in stand-alone applications or in combination with conventional reinforced and prestressed concrete. Fiber reinforced polymeric and metallic composites typically use fibers with tensile strain at failure smaller than that of the matrix (Fig. 3.9(a)). Hence, failure of the composite implies either failure of the fibers, or their complete debonding while the matrix may be in a yielding state.

The main common characteristic of fiber reinforced cementitious composites, as compared to fiber reinforced polymeric and metallic composites, is that the ultimate tensile strain of the fiber is significantly larger than that of the matrix (Fig. 3.9(b)). This implies that at some level of tensile loading, the matrix will crack and the resistance to full separation is entirely born by the fibers bridging the cracked surfaces.

This also generally implies that the contribution of the fibers can be fully utilized only after matrix cracking.

3.5 Notation

The following notation is used:

A_c = cross-sectional area of composite

A_f = cross-sectional area of one fiber

A_m = cross-sectional area of matrix

E_f = elastic modulus of fiber; if the elastic modulus in the longitudinal direction is different from that in the transverse direction, the terms $E_{f//}$ and $E_{c\perp}$ are used.

E_m = elastic modulus of the matrix

d = diameter of fiber

L = length of fiber

N_v = average number of fibers per unit volume of composite

N_s = average number of fibers crossing a unit area of composite, or number of fiber intersections per unit area

V_f = volume fraction of fibers

V_m = volume fraction of matrix

ε_{fu} = tensile strain at failure of the fiber

ε_{fy} = tensile strain at yield of the fiber

ε_{mu} = tensile strain at failure of the matrix

σ_{fy} = yield strength of the fiber

σ_{fu} = tensile strength of the fiber

σ_{mu} = tensile strength of the matrix

τ = bond strength at the fiber matrix interface

Note that $V_f + V_m = 1$.

3.6 Assumptions for Modeling and Simplified Model

Consider a fiber reinforced cement composite tensile prism with randomly oriented and distributed fibers (Fig. 3.10(a)). The following sections describe a simplified model to predict the constitutive response of such composite.

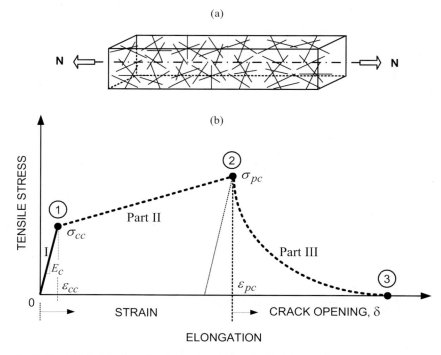

Fig. 3.10 (a) Model of tensile element considered. (b) Idealized stress-elongation response in tension of a strain-hardening FRC composite for modeling.

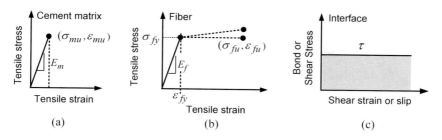

Figure 3.11 (a) Assumed tensile stress-strain response of the cement matrix. (b) Assumed tensile stress-strain response of the fiber. (c) Assumed bond stress versus slip response at very small slips.

The analytical modeling of fiber reinforced cementitious composites has to take into consideration at least two different states, the pre-cracking state (part I of Figs. 3.3(a) and 3.3(b)) and the post-cracking state of the matrix. The post-cracking state comprises two phases, a multiple-cracking phase (which may not exist – part II of Fig. 3.3(b)), and a pull-out or failure phase (part III of Fig. 3.3). Rational models have been developed to predict the key coordinates of the tensile stress-strain response, namely the stress and strain at first cracking $(\sigma_{cc}, \varepsilon_{cc})$, the elastic modulus of the composite (E_c), the stress and strain at the maximum post-cracking point $(\sigma_{pc}, \varepsilon_{pc})$, and the general pull-out response (part III) following crack localization. An idealized stress-elongation response curve with these key coordinates (points 1, 2, and 3) is shown in Fig. 3.10(b). A summary of the findings is provided in the following sections.

The following assumptions are generally made when modeling the tensile stress-strain response of cement based composites:

a. The matrix is brittle and best characterized by a linear stress-strain curve in tension; thus two of three parameters are sufficient to characterize the matrix: E_m, σ_{mu}, and ε_{mu} (Figs. 3.10(b) and 3.11(a)).

b. The fiber is either brittle (such as glass) or ductile with an initial elastic response (such as steel or polypropylene). For brittle fibers at least three parameters are needed: E_f, σ_{fu}, and ε_{fu}. For ductile fibers with well defined yielding behavior, the yield stress, σ_{fy} and the corresponding strain ε_{fy} are also needed (Figs. 3.9(b), and 3.11(b)).

c. The tensile strain at failure of the matrix is smaller than that of the fiber, that is, for brittle fibers: $\varepsilon_{mu} < \varepsilon_{fu}$ (or for ductile fibers $\varepsilon_{mu} < \varepsilon_{fy}$); this implies that cracking will occur in the matrix prior to failure of the fibers and the composite.

d. Although a weak, non-flowing, non-plastic bond generally exists at the fiber matrix interface, the properties of the interface are assumed characterized by an equivalent elastic perfectly plastic bond stress versus slip response, at small slips. (Fig. 3.11(c)). The value of bond strength selected is assumed to represent an average value over a reasonable range of slip between the fiber and the matrix.

e. While it is conceivable that the bond at the interface between fiber and matrix could fail prior to failure of the matrix in tension (thus nullifying composite action), this case is most unlikely and is not considered here. In fact, there is evidence that the bond is mobilized only partly prior to matrix cracking, and develops fully only after cracking.

3.7 Number of Fibers Per Unit Volume and Per Unit Area

The number of fibers per unit volume of composite and the number of fibers intersecting a unit area of composite are important parameters needed for deriving the tensile strength of the composite at first cracking and its maximum post-cracking strength. They have been derived elsewhere [13] and are given next for information. In the following equations, it is assumed that fibers are circular and of diameter d. Equivalent diameter could be used, as a first approximation, when slightly non-circular fibers are considered. Let us define:

N_v = average number of fibers per unit volume of composite assuming fibers of circular cross section

N_s = average number of fiber crossing a unit area or unit surface of composite. If a plane is cut through the composite, N_s is the average number of fiber intersections per unit area.

For a given fiber of length L and diameter d, added to a matrix in an amount corresponding to a volume fraction of composite, V_f, it can be shown that [13,29]:

$$N_v = \frac{4V_f}{\pi d^2 L} \tag{3.1}$$

Any plane cutting through the composite will intersect a number of fibers. The number of fibers, N_s, intersecting a unit area of the plane depends on the fiber orientation. It can be shown that:

$$N_s = \frac{4V_f}{\pi d^2} \alpha_2 \tag{3.2}$$

in which the orientation factor, α_2, is given by:

$\alpha_2 = 1$ for unidirectional fibers (1 Dimensional)

$\alpha_2 = \dfrac{2}{\pi}$ for fibers randomly oriented in planes (2 Dimensional)   (3.3)

$\alpha_2 = \dfrac{1}{2}$ for fibers randomly oriented in space (3 Dimensional)

The above suggested values of α_2 are derived from statistical considerations. Depending on the assumptions made, other numerical values of α_2 can be obtained [10,13]. For instance, other values of α_2 for the 3D case, as reported in the technical literature, are $1/\pi$ and $[2/\pi]^2$.

Note that, while the total number of fibers put in a given composite can be considered a deterministic value (V_f is an input parameter), the actual number of fibers per unit volume or number crossing a unit area are random variables with a probability distribution function of the Poisson's type [13]. For this chapter, the average values, N_v and N_s, are considered deterministic variables.

3.8 Stress and Strain in Composite at First Cracking of Matrix in Tension, $(\sigma_{cc}, \varepsilon_{cc})$

3.8.1 *Stress at first cracking*

Consider a fiber reinforced cement composite tensile prism with randomly oriented and distributed fibers (Fig. 3.10(a)). Assume that if a crack develops in the matrix, it will be normal to the direction of loading;

also assume that the fibers are circular in cross section (for non-circular fibers, see Section 3.8.4). It can be shown that the tensile stress in the composite at onset of first percolation cracking (that is, just before first cracking) of the matrix can be expressed as follows [13,29,30,15]:

$$\sigma_{cc} = \sigma_{mu}(1 - V_f) + \alpha\tau V_f \frac{L}{d} \qquad (3.4)$$

in which α is the product of several coefficients:

$$\alpha = \alpha_1\alpha_2\alpha_3 \qquad (3.5)$$

and:

V_f = volume of fraction of fibers

L = fiber length

d = fiber diameter

L/d = fiber aspect ratio

σ_{mu} = tensile strength of the matrix

τ = assumed average or equivalent bond strength at the fiber matrix interface

α_1 = coefficient describing the average contribution of bond at onset of matrix cracking. Its value is discussed in some details in [31].

α_2 = efficiency factor of fiber orientation in the uncracked state of the composite; it is equal 1 for unidirectional fibers; $2/\pi = 0.636$ for fibers randomly oriented in planes; and 0.5 for fibers randomly oriented in space (see also discussion in Section 3.7). This factor directly influences the number of fibers intersecting a unit area of composite, be it cracked or uncracked (Eq. 3.2).

α_3 = coefficient describing the reduction of bond strength at the fiber matrix interface due to an applied external stress radial or normal to the interface (Fig. 3.12); it is equal 1 for aligned fibers.

The advantage of using a single coefficient α in Eq. (3.4) is that the equation adapts itself to additional coefficients in case new parameters are discovered and added to the theoretical derivations. In the past [13,15], the author had only considered α_1 and α_2 but later added α_3 as a result of additional research.

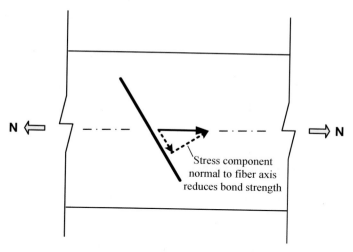

Fig. 3.12 Illustration to explain the bond reduction coefficient α_3 in the uncracked state of the composite.

Equation (3.4) can be put in the following normalized form:

$$\frac{\sigma_{cc}}{\sigma_{mu}} = (1 - V_f) + \alpha \frac{\tau}{\sigma_{mu}} V_f \frac{L}{d} \qquad \leq \text{ Eq. (3.9)} \qquad (3.6)$$

3.8.2 Strain at first cracking

Assuming linear elastic response up to first cracking, the strain at first cracking of the composite can be estimated from the following equation:

$$\varepsilon_{cc} = \frac{\sigma_{cc}}{E_c} \qquad (3.7)$$

in which E_c is the modulus of elasticity of the composite in the uncracked state. It is described in Section 3.9.

3.8.3 Upper bound stress in composite at cracking of matrix

The stress in the composite is limited to the value obtained assuming the composite is made of continuous fibers with equal strain distributions in the fiber and the matrix. Thus:

$$\sigma_{cc} \leq (\sigma_{cc})_{\text{upper-bound}} = \sigma_{mu}[1 + (n-1)V_f] \qquad (3.8)$$

which can also be put in the following normalized form:

$$\frac{\sigma_{cc}}{\sigma_{mu}} \leq 1 + (n-1)V_f \qquad (3.9)$$

in which

$$n = \frac{E_f}{E_m}. \qquad (3.10)$$

3.8.4 Case of non-circular fiber

For slightly non-circular fibers, Eqs. (3.4) and (3.6) which are derived for circular fibers can be used as a first approximation provided an equivalent diameter of fiber based on its cross sectional area is used, that is, $d = \sqrt{4A_f / \pi}$, where A_f is the cross-sectional area of the fiber and $\pi = 3.14$.

Generally, for a non-circular fiber, Eq. (3.4) can be rewritten as follows:

$$\sigma_{cc} = \sigma_{mu}(1 - V_f) + \alpha \times \tau \times V_f \times \frac{\psi \times L}{4A_f} \qquad (3.11)$$

in which: σ_{mu} is the tensile strength of the unreinforced matrix, V_f is the volume fraction of fibers, τ is the average bond strength at the fiber-matrix interface, ψ is the perimeter of one fiber, and A_f is the cross-sectional area of one fiber [32].

3.9 Elastic Modulus of the Composite in the Uncracked State

Similarly to the case of other elastic composites, the modulus of elasticity, E_c, of fiber reinforced cement composites assuming randomly oriented and distributed fibers can be estimated from a linear combination of its upper and lower bound values as follows:

$$E_c = mE_{c//} + (1 - m)E_{c\perp} \qquad (3.12)$$

where m is a coefficient between zero and 1, $E_{c//}$ is the upper bound modulus and $E_{c\perp}$ is the lower bound modulus.

Halpin and Tsai suggested a value of $m = 3/8$ for fibers randomly oriented in planes [33]. In a study on steel fiber reinforced concrete, a value of $m = 3/8$ was confirmed by Naaman and Najm from experimental tests [34], while a value of $m = 0.5$ was derived by Alwan and Naaman from analytical modeling and simulation [35]. Either value seems acceptable for practical applications of fiber reinforced concrete, since the fiber content is generally small.

The derivation of upper bound modulus (based on Voigt's model or equal strain model) assumes that all fibers are aligned in the direction of loading and equal strain exists in the fiber and the matrix. Its solution leads to the following equation [35]:

$$E_{c//} = E_m V_m + E_{f//} V_f$$
$$= E_m V_m + E_{f//}(1 - V_m) \qquad (3.13)$$
$$= E_{f//} V_f + E_m (1 - V_f)$$

where $E_{f//}$ is the modulus of elasticity of the fiber along its axis. While for steel fibers, the fiber modulus of elasticity is same in the longitudinal and transverse directions, this is not the case for polymeric fibers.

The derivation of the lower bound modulus (based on Reuss's model or equal stress model) assumes that all fibers have their axis normal to the direction of loading and that the stress in the matrix and the stress normal to the fiber axis are equal. Its solution leads to the following equation [35]:

$$\frac{1}{E_{c\perp}} = \frac{V_f}{E_{f\perp}} + \frac{V_m}{E_m} \qquad (3.14)$$

from which the following explicit solution is obtained:

$$E_{c\perp} = \frac{E_{f\perp} E_m}{E_{f\perp} V_m + E_m V_f}$$
$$= \frac{E_{f\perp} E_m}{E_{f\perp}(1 - V_f) + E_m (1 - V_m)} \qquad (3.15)$$
$$= \frac{E_{f\perp} E_m}{E_{f\perp}(1 - V_f) + E_m V_f}$$

where $E_{f\perp}$ is the modulus of elasticity of the fiber normal to its axis.

Equations (3.13)–(3.15) are purely theoretical. Another method can be used to predict the elastic modulus of the composite and offers the advantage to be easily verifiable by experimental tests. It assumes the following equation first introduced by Cox [36]:

$$E_c = \eta E_f V_f + E_m(1 - V_f) \qquad (3.16)$$

in which the coefficient η is the product of two theoretical coefficients, one dealing with orientation and the other dealing with the ratio of fiber length to critical fiber length. They are not discussed here for brevity. However, the approach proposed here for FRC composites is to determine η from tests. That is the matrix is tested first without fibers to determine E_m; then the composite (with same matrix composition) containing a reasonable volume fraction of fibers (about 1%) is tested to determine E_c. Tests should follow acceptable standard procedures. The average value of E_c is then used in Eq. (3.16) to determine the numerical value of η which can then be considered typical for a given fiber and valid for other volume fractions and aspect ratios. Tests by Patton and Whittaker [37] on steel fiber reinforced concrete led to a value of $\eta = 0.43$ in Eq. (3.16).

Additional details related to Eqs. (3.12 to 3.16) and examples can be found in [38].

3.10 Maximum Post-Cracking Stress: Composite Strength in Tension, σ_{pc}

Information on the basic mechanics of fiber reinforced concrete and its mechanical properties as influenced by addition of fibers, can be found in several books and symposia proceedings some of which are listed in the references [3,4,8,14,19]. In particular, high performance fiber reinforced cement composites are addressed in several symposia proceedings, Co-edited by Naaman and Reinhardt (1992, 1996, 1999, 2003, 2007, Refs. [14,16,17,18]). Next only a simplified approach is presented to provide a background to the key points of the discussion.

The typical stress-elongation response of a fiber reinforced cement composite indicate two properties of interest, the stress at first cracking, σ_{cc}, and the maximum post-cracking stress, σ_{pc} (Figs. 3.3, 3.4 and 3.10(b)). Extensive discussion on the meaning of these two properties and their relationship is given elsewhere as in Naaman and Reinhardt, 1996, and Naaman, 2002 [14,39]. While the cracking strength of the composite, σ_{cc}, is primarily influenced by the strength of the matrix, the post-cracking strength, σ_{pc}, is mainly dependent on the fiber reinforcing parameters and the bond at the fiber-matrix interface. Thus, improving the post-cracking strength is key to the success of the composite.

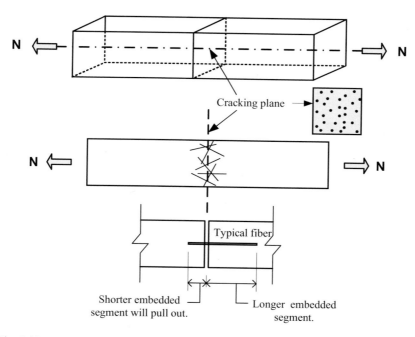

Fig. 3.13 Assumed model for general fiber pull-out after cracking of the matrix.

Consider first a tensile composite prismatic member with unidirectional fibers. Assume that cracking in the matrix has occurred and that the crack is normal to the axis of the fibers (Fig. 3.13). Three possible cases can occur:

    a.   Upon increased elongation or straining, all the fibers fail;

    b.   Upon increased elongation, all the fibers pull-out such as most commonly observed in fiber reinforced cement composites with high strength steel fibers;

c.   Upon increased elongation, a combined case occurs where some fibers fail and some fibers pull-out.

Cases *a* and *b* will be treated next. Clearly, case *c* should give results in between the other two. Because it involves a lengthy treatment based on probability and statistics, it is not treated here. The reader is referred to the following study for further information on case *c* [40]. Note that case *b* is most desirable since the pull-out of the fibers leads to increased energy absorption capacity.

Assume next that the fiber is circular.

3.10.1   *Composite strength assuming all fibers fail simultaneously*

In this case:

$$\sigma_{cu} \times (A_c = 1) = \sigma_{fu} N_s A_f \tag{3.17}$$

or

$$\sigma_{cu} = \sigma_{fu} N_s A_f \tag{3.18}$$

where:

σ_{cu} = ultimate strength of composite

σ_{fu} = ultimate tensile strength of fiber

N_s = number of fibers crossing a unit area of composite

(or of crack surface)

N_s can be obtained from Eq. (3.2); thus, for a circular fiber:

$$\sigma_{cu} = \sigma_{fu} \alpha_2 \frac{4V_f}{\pi d^2} \frac{\pi d^2}{4} \tag{3.19}$$

that is:

$$\sigma_{cu} = \alpha_2 V_f \sigma_{fu} \tag{3.20}$$

For unidirectional fibers $\alpha_2 = 1$; thus Eq. (3.20) is similar to the case of a cracked composite with continuous fibers assuming they all fail at their maximum strength.

3.10.2 *Composite strength assuming all fibers pull-out simultaneously*

For the derivations below, the following assumptions are made: (1) a critical planar crack exists across the entire section of the tensile member (Fig. 3.13), (2) the crack is normal to the tensile stress field, (3) the contribution of the matrix along the crack is negligible, and (4) the fibers crossing the crack are in a general state of pull-out. This is typically the case when steel fibers are used in concrete.

The detailed derivation of the equation to predict the post-cracking strength of the composite is time consuming. Some background is given in [13,15,38]. Here, it suffices to say that the equation can be most generally put in the following form:

For circular fibers:

$$\sigma_{pc} = \lambda \tau \frac{L}{d} V_f \qquad (3.21)$$

where λ is the product of several coefficients:

$$\begin{aligned} \lambda &= \lambda_1 \lambda_2 \lambda_3 \lambda_5 \\ \lambda_2 &= 4\alpha_2 \lambda_4 \end{aligned} \qquad (3.22)$$

Similarly to the approach followed in Eq. (3.4), the use of a single coefficient, λ, in Eq. (3.21) allows for future expansion of the equation, should new research justify the case. For instance, in the past [13, 15], the author used only $\lambda_1 \lambda_2 \lambda_3$ and later added λ_5 on the basis of new research.

The coefficients in Eqs. (3.21) and (3.22) are defined as follows:

λ_1 = average or expected value of the ratio of fiber shorter embedded distance from a forming crack to the length of the fiber. Its value is ¼ as derived from probability theory considerations. It is illustrated in Fig. 3.14. The shorter embedded length is assumed to be the one that will pull-out under load.

λ_2 = $4\lambda_4\alpha_2$; coefficient that takes into consideration orientation effect on pull-out resistance. It can be considered the efficiency factor of orientation in the cracked state.

α_2 = efficiency factor of fiber orientation in the uncracked state of the composite; it is equal 1 for unidirectional fibers; $2/\pi = 0.636$ for

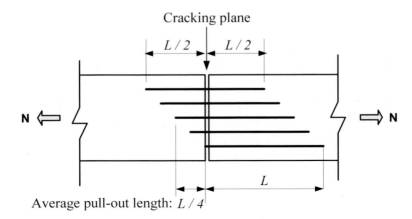

Fig. 3.14 Range of length of shorter pull-out segment, and average fiber pull-out length.

fibers randomly oriented in planes; and 0.5 for fibers randomly oriented in space. This factor directly influences the number of fibers intersecting a unit area of composite, be it cracked or uncracked (Eq. (3.3)).

λ_3 = group reduction coefficient for bond, to simulate the fact that the bond strength resistance per fiber decreases when the number of fibers pulling out from the same area increases [41,42].

λ_4 = expected value of ratio of maximum pull-out load for a fiber oriented at angle θ to maximum pull-out load of same fiber aligned with the direction of pull-out. The angle θ is the angle between the longitudinal axis of the fiber and the pull-out direction and varies between 0 and 90 degrees; thus $(\pi/2 - \theta)$ represents the angle between the longitudinal axis of the fiber and the cracking plane. $\lambda_4 = 1$ for fibers with longitudinal axis oriented in the direction of pull-out loading.

λ_5 = reduction coefficient to account for the fact that fibers inclined at an angle of more that about 60 degrees with the pull-out load direction contribute very little, due to spalling of the wedge of matrix at the cracking plane (Fig. 3.15). This is particularly true for stiff fibers such as steel. For aligned fibers, $\lambda_5 = 1$.

Note that the two coefficients α_2 and λ_4 are placed together since λ_4 and α_2 are correlated. For instance, if α_2 equals one, then λ_4 also

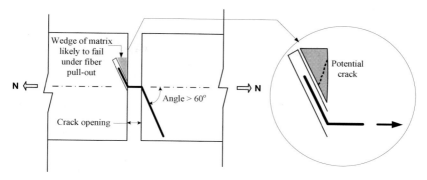

Fig. 3.15 Wedge of matrix likely to fail under fiber pull-out at large angles of fiber inclination to the loading direction.

equals one; that is the case of unidirectional fibers. Thus, the two coefficients are related and are both associated with orientation effects; α_2 affects the number of fibers at an angle to a plane, and λ_4 affects the pull-out force associated with an inclined fiber. Integration over the angle of orientation from zero to 90 degrees leads to the expected values of these coefficients.

In previous publications [13,15,29,], the author had not considered the coefficient λ_5, and the coefficient, λ_2, was simply termed the efficiency factor of orientation in the cracked state.

The coefficient, λ_4, was termed "snubbing coefficient" by Li and Wu [11]; using analogy with a pulley, they showed that λ_4 varies theoretically between 1 and 2.32. In analyzing the experimental results of Visalvanich and Naaman [43], Li and Wu back-calculated the values of λ_4 and obtained a value close to 2 for steel fibers.

The above form of Eq. (3.21) is convenient in allowing the experimental determination of a single coefficient λ from a direct tensile test [28]. Indeed, because of the many uncertainties associated with most of the above coefficients ($\lambda = 4\lambda_1\lambda_3(\lambda_4\alpha_2)\lambda_5$), one can undertake a tensile test and, assuming the bond strength τ known with relative accuracy, find out the coefficient λ by comparing the test value of σ_{pc} with the value predicted from Eq. (3.21). It is best, in such test, to use a notched tensile prism of sufficient cross-section to represent the real tensile member. The average bond strength, τ, used in the equation could be determined from separate pull-out tests.

Note that for fibers aligned in the loading direction, $\lambda_1 = 1/4$, $\alpha_2 = 1$, $\lambda_2 = 4$, $\lambda_4 = 1$, $\lambda_5 = 1$, $\lambda_3 = \lambda_3$. Thus for aligned circular fibers Eq. (3.21) becomes:

$$\sigma_{pc} = \lambda_3 \tau \frac{L}{d} V_f \tag{3.23}$$

3.10.2.1 Non-dimensional form

Equation (3.21) can be put in a non-dimensional form as follows:

$$\frac{\sigma_{pc}}{\sigma_{mu}} = \lambda \frac{\tau}{\sigma_{mu}} \frac{L}{d} V_f \qquad \leq \quad \text{Eq. (3.25)} \tag{3.24}$$

3.10.2.2 Upper bound limit

The postcracking strength of the composite derived in Eqs. (3.21 and 3.24) assumes that all fibers pull-out. It is thus limited by the case where all fibers fail. Hence, from Eq. (3.20), the following condition should be satisfied:

$$\sigma_{pc} \leq \alpha_2 V_f \sigma_{fu} \tag{3.25}$$

where α_2 is taken from Eq. (3.3). Replacing σ_{pc} from its value from Eq. (3.24) leads to the following condition:

$$\frac{\tau}{\sigma_{fu}} \frac{L}{d} \leq \frac{\alpha_2}{\lambda} \tag{3.26}$$

3.10.2.3 Example of 3D orientation

Consider a tensile FRC composite with randomly oriented and distributed steel fibers. Assume the following coefficients:

$$\lambda_1 = 1/4; \ \lambda_3 = 0.75; \ \lambda_4 = 1; \ \lambda_5 = 0.8, \text{ and } \alpha_2 = 0.5.$$

Thus:

$$\lambda_2 = 4\alpha_2\lambda_4 = 4 \times 0.5 \times 1 = 2$$

$$\lambda = \lambda_1 \lambda_2 \lambda_3 \lambda_5 = 0.25 \times 2 \times 0.75 \times 0.8 = 0.30$$

$$\sigma_{pc} = 0.30 \times \tau \frac{L}{d} V_f$$

Hence for $V_f = 2\%, L/d = 100$ and $\tau = 4$ MPa, we get:

$$\sigma_{pc} = 0.30 \times 4 \times 100 \times 0.02 = 2.4 \text{ MPa}.$$

This value is not very high and illustrates how difficult it is to achieve in practice a strain-hardening FRC composite.

3.10.3 Case of non-circular fibers

For slightly non-circular fibers, Eq. (3.21) which applies to circular fibers can be used as a first approximation provided an equivalent diameter of fiber based on its cross sectional area is used, that is, $d = \sqrt{4A_f / \pi}$, where A_f is the cross-sectional area of the fiber and $\pi = 3.14$.

If the fiber is not circular in cross-section, it can be shown that Eq. (3.21) takes the following form [32a,32b,28]:

$$\sigma_{pc} = \lambda \tau V_f \frac{\psi L}{4A_f} \tag{3.27}$$

in which: ψ is the perimeter of one fiber, A_f is the cross-sectional area of one fiber, V_f is the volume fraction of fibers, τ is the average bond strength at the fiber-matrix interface, and λ is as defined in Eq. (3.21). A non-dimentional form similar to Eq. (3.24) and an upper-bound solution similar to Eqs. (3.25 and 3.26) can also be developed for the case of non-circular fibers.

3.11 Strain at Maximum Post-Cracking Stress, ε_{pc}

At the time of this writing, no simple model exists in the technical literature to predict the strain at maximum post-cracking stress in strain-hardening FRC composites. The reader is referred to a study by Tjiptobroto and Hansen [21] where a prediction equation is given in terms of the fiber debonding energy; a related discussion can also be found in Ref. [15].

Numerical values of ε_{pc} depend on numerous parameters including not only the fiber reinforcing parameters, the matrix properties and the interface bond between fiber and matrix, but also the specimen size, the testing method, and the technique of measurement. Experimental values observed range from about 0.1% to 5%, with a large proportion being below 2%. For practical modeling purposes, a value of $\varepsilon_{pc} = 0.5\%$ is recommended for strain-hardening FRC composites with steel fibers [23], and 1% for strain-hardening FRC composites with Spectra fibers. These values are supported by various experimental tests [23,28,40,42, 44-50].

3.12   General Pull-Out Response – Part III of Fig. 3.10(b)

Several equations have been suggested to model the pull-out response of FRC composites after crack localization, that is, part III of Fig. 3.10(b) [12,43,51-53]. Assuming a single crack opening during the pull-out process, they predict the stress in the composite at a given crack opening or displacement δ. The maximum crack opening or displacement is taken equal to half the fiber length, as theoretically expected. The equation given next is recommended because of its very simple form as well as its theoretical derivation which was based on the assumption of a constant bond strength at the fiber-matrix interface; it was initially derived by Kosa and Naaman [52] and is given by:

$$\sigma_c(\delta) = \sigma_{pc}\left[1 - \frac{\delta}{k \times L/2}\right]^2 \quad \text{for} \quad \delta \le k \times L/2 \ \text{and} \ k \le 1 \quad (3.28)$$

where $\sigma_c(\delta)$ is the stress in the composite at crack opening (or face displacement) δ, L is the fiber length, and k is a damage coefficient. Equation (3.28) is a quadratic equation. When the damage factor k is taken equal to 1, the maximum value of displacement δ is equal to half the fiber length. However, should there be damage such as corrosion or bond deterioration or other, the damage can be accounted for by changing the value of k. Thus for $k = 0.5$, the maximum displacement reduces to $L/4$. The value of σ_{pc} in Eq. (3.28) can be predicted from Eq. (3.21).

Note that Eq. (3.28) applies to the general pull-out curve after crack localization for both a strain-hardening or a strain-softening composite.

3.13 Summary: Idealized Tensile Response for Modeling

Referring to Fig. 3.10(b) and assuming that the coordinates of the key points of the curve have been determined, an approximate stress-strain response can be determined for the composite as follows:

For a tensile strain in the composite $\varepsilon_c \leq \varepsilon_{cc}$, that is for part I of the curve in Fig. 3.10(b):

$$\sigma_c = E_c \times \varepsilon_c \qquad (3.29)$$

in which E_c can be calculated from Eqs. (3.12) or (3.16) or taken as a first approximation equal to the elastic modulus of the plain matrix. The value of ε_{cc} can be calculated from $\varepsilon_{cc} = \sigma_{cc}/E_c$ in which σ_{cc} is obtained from Eq. (3.4).

For $\varepsilon_{cc} \leq \varepsilon_c \leq \varepsilon_{pc}$, that is, for part II of the curve in Fig. 3.10(b):

$$\sigma_c = \sigma_{cc} + \left(\frac{\sigma_{pc} - \sigma_{cc}}{\varepsilon_{pc} - \varepsilon_{cc}} \right) \times \varepsilon_c \qquad (3.30)$$

in which σ_{pc} can be estimated from Eq. (3.21) and ε_{pc} can be obtained from experimental tests or estimated as suggested in Section 3.11.

Following crack localization, the general pull-out response can be modeled as a first approximation according to the discussion in Section 3.12 and Eq. (3.28). The stress is then a function of the crack opening and represents the condition around the critical crack only. For each stress value smaller than σ_{pc}, the strain in the other parts of the composite (on either side of the critical crack) can be calculated according to Eqs. (3.29) or (3.30).

The idealized response shown in Fig. 3.10(b) is simplest for modeling. A more accurate representation allowing some variability in shape for part II and part III of the curve is discussed in [23]. In all cases it is assumed that part I is linear; however, micro-cracking during loading may lead to a non-linear part I. In such a case one may need the three variables $\sigma_{cc}, \varepsilon_{cc}$, and E_c to better model part I.

3.14 Critical Volume Fraction of Fiber to Achieve Strain-Hardening Behavior in Tension

Prediction equations for the first cracking stress and maximum post-cracking strength of a fiber reinforced cement composite exists since the early 1970's all but with slightly different coefficients than in Eqs. (3.4) and (3.21) [13,15,38]. A first attempt to define the condition leading to strain-hardening behavior and multiple cracking in tension was suggested by Naaman in 1987 [30,54], by setting that the maximum post-cracking stress must be larger than or equal to the stress at first cracking. It appeared later in several other studies [15, 39, 55]. The derivation is illustrated below for circular fibers only, and can be easily extended to non-circular ones.

Using Eqs. (3.4 and 3.21), the following condition can be written:

$$\sigma_{pc} \geq \sigma_{cc} \tag{3.31}$$

$$\lambda \tau \frac{L}{d} V_f \geq \sigma_{mu}(1 - V_f) + \alpha \tau V_f \frac{L}{d} \tag{3.32}$$

Solving for V_f leads to:

$$V_f \geq (V_f)_{cri-tension} = \frac{1}{1 + \dfrac{\tau}{\sigma_{mu}} \dfrac{L}{d}(\lambda - \alpha)} \tag{3.33}$$

Assuming the value of V_f is relatively small such as is the case for fiber reinforced cement composites, Eq. (3.33) can also be written in the following form which illustrates not only the influence of the fiber volume fraction but also that of the other main causal variables:

$$V_f \frac{\tau}{\sigma_{mu}} \frac{L}{d} \geq \frac{1 - V_f}{(\lambda - \alpha)} \approx \frac{1}{\lambda - \alpha} \tag{3.34}$$

For relatively small values of V_f, Eq. (3.34) can be further reduced as a first approximation to:

$$V_f \frac{\tau}{\sigma_{mu}} \frac{L}{d} \geq \frac{1}{\Omega} \tag{3.35}$$

where:

$$\Omega = \lambda - \alpha = \lambda_1 \lambda_2 \lambda_3 \lambda_5 - \alpha_1 \alpha_2 \alpha_3 \tag{3.36}$$

Equation (3.35) has perhaps the simplest form to illustrate the direct influence of the independent variables leading to the development of multiple cracking. Assuming the right-hand side to be a constant for given conditions, Eq.(3.35) shows that the aspect ratio of the fiber and the ratio of bond strength to matrix tensile strength are as influential as the volume fraction of fibers.

Note that equations leading to the critical volume fraction of fibers can be derived from other models such as where fracture energy is considered; related derivations are discussed in [15].

3.14.1 *Graphical illustration*

Figure 3.16 initially developed in Refs. [15,30] illustrates the variation of the critical volume fraction of fiber, $(V_f)_{cri-tension}$ (Eq. (3.33)), versus the aspect ratio of the fiber, L/d, at different values of the ratio of bond

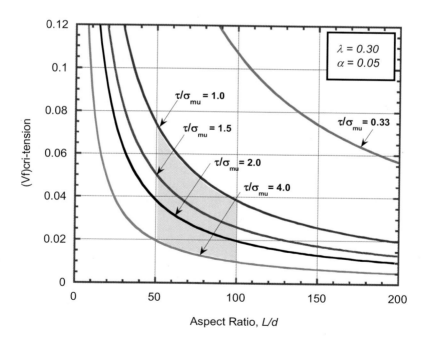

Fig. 3.16 Critical volume fraction of fibers to achieve strain-hardening behavior in tension.

strength to tensile strength of the matrix, τ / σ_{mu}. An estimated value of $\lambda = 0.30$ and $\alpha = 0.05$ for steel fibers is assumed (Eq. (3.33)). Similar figures can be developed for different values of λ and α. Each curve defines a boundary of feasibility above which strain-hardening behavior (such as in Figs. 3.3(b) and 3.4(b)) can be achieved. If, for instance, we assume steel fibers and a value $\tau / \sigma_{mu} = 2$, Fig. 3.16 indicates that in order to achieve strain-hardening behavior in tension, the volume fraction of fibers needs to exceed about 2% at an aspect ratio of 100, and 4% at an aspect ratio of 50. These values are almost doubled at $\tau / \sigma_{mu} = 1$. Since it is very hard in the field to mix more that 2% to 3% steel fibers by volume into a concrete mix, the importance of the ratio τ / σ_{mu} can be easily appreciated from the figure. It is clear that in the range of aspect ratios common for steel fibers for concrete (that is, $L/d = 50$ to 100) one needs a high value of τ / σ_{mu} in order to actually manufacture a strain-hardening fiber reinforced concrete composite. Note again that τ should be viewed as an average value of bond strength that can be maintained over reasonable slips.

3.15 Critical Volume Fraction of Fiber to Achieve Deflection-Hardening Behavior in Bending

In a derivation similar to the one leading to the critical volume fraction of fibers to insure strain-hardening behavior, the author has developed a similar expression leading to the critical volume fraction of fibers for which deflection-hardening is insured. Deflection-hardening implies that the maximum equivalent elastic bending stress (or modulus of rupture, MOR) after first cracking is larger than the stress at first cracking in bending, and that multiple cracking would generally occur after first cracking. Details are given in [39,55]. The following solution was obtained:

$$V_f \geq (V_f)_{cri-bending} = \frac{k}{k + \dfrac{\tau}{\sigma_{mu}} \dfrac{L}{d}(\lambda - k\alpha)} \qquad (3.37)$$

where the notation is same as above, and k is a coefficient less than 1. A value of $k = 0.4$ is recommended for practical applications of steel fiber reinforced concrete. For $k = 1$ Eq. (3.37) reverts to Eq. (3.33).

For the numerical results illustrated in Fig. 3.17, the following coefficients were used: $\lambda = 0.3$, $\alpha = 0.05$ and $\Omega = \lambda - k\alpha = 0.28$. Similar figures can be developed for different numerical values of these coefficients. Figure 3.17 is similar to Fig. 3.16 except that it illustrates the conditions for deflection-hardening fiber reinforced cement composites instead of strain-hardening ones. Note that for a typical steel fiber with an aspect ratio of 50 to 100 and reasonable bond strength ($\tau / \sigma \approx 1$ to 2), deflection hardening can be easily achieved with volume fractions of steel fibers in the range of 1% to 2%. Such fiber contents are feasible and practical in actual field applications.

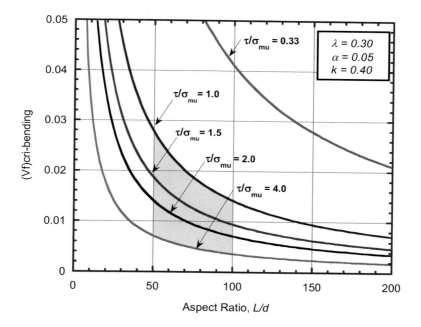

Fig. 3.17 Critical volume fraction of fibers to achieve deflection-hardening behavior in bending ($k = 0.4$).

3.16 Example: Critical Volume Fraction of Fibers

Assume the following coefficients for an FRC composite with randomly oriented and distributed fibers of circular section:

$$\alpha_1 = 0.1, \, \alpha_2 = 0.5, \, \alpha_3 = 1$$

$$\lambda_1 = 0.25, \, \lambda_3 = 0.75, \, \lambda_4 = 1, \, \lambda_5 = 0.8$$

$$\lambda_2 = 4\alpha_2\lambda_4 = 4 \times 0.5 \times 1 = 2$$

$$\frac{L}{d} = 100, \quad \frac{\tau}{\sigma_{mu}} = 1$$

Thus: $\quad \alpha = \alpha_1\alpha_2\alpha_3 = 0.1 \times 0.5 \times 1 = 0.05$

$$\lambda = \lambda_1\lambda_2\lambda_3\lambda_5 = 0.25 \times 2 \times 0.75 \times 0.8 = 0.30$$

For strain-hardening response in tension:

$$\Omega = \lambda - \alpha = 0.30 - 0.05 = 0.25$$

$$(V_f)_{\text{critical-tension}} = \frac{1}{1 + 1 \times 100(0.3 - 0.05)} = 0.0385 = 3.85\%$$

For deflection-hardening response in bending, assuming $k = 0.4$:

$$\Omega = \lambda - k\alpha = 0.30 - 0.4 \times 0.05 = 0.28$$

$$(V_f)_{\text{critical-bending}} = \frac{0.40}{0.40 + 1 \times 100(0.28)} = 0.0141 = 1.41\%$$

This illustrates that deflection-hardening occurs at a volume fraction of fibers significantly smaller than that needed for strain-hardening in tension. Examples of FRC materials where deflection hardening occurs at volume fraction of fibers between 1% and 2% abound in the technical literature, and support the range of results obtained from the above analysis.

Equations (3.33) and (3.37) infer that the larger the volume fraction of fibers is, the better are the chances to achieve strain-hardening or deflection-hardening response. However there is a practical limit beyond which proper mixing of the fibers is not possible, using standard mixing procedures, and a deterioration in mechanical properties may ensue, due to air entrapment and insufficient bonding at the fiber-matrix interface. Thus, optimization of composite performance should involve the manipulation of not only the fundamental composite parameters (matrix and fiber parameters), but also variables related to the production process, the rheology of the fresh mix, the properties of the hardening composite and the final application of the material.

3.17 Surface Energy in Tension

Surface energy is defined as the energy needed to create a unit surface of material. Typically a planar crack will create two new and equal surfaces. Surface energy is a material characteristic that describes how easy or difficult it is to break a material into two parts. In the present discussion we will assume that failure in fiber reinforced concrete occurs by general fiber pull-out, that is, no fiber failure.

To create a crack in a fiber reinforced concrete specimen subjected to tension, there is need to break the matrix and to pull-out the fibers entirely. As a crack has two opposing surfaces, the pull-out energy can be considered equal to twice the surface energy of pull-out, thus:

$$U_p = 2\gamma_p \tag{3.38}$$

The surface energy of the composite can be calculated from :

$$\gamma_{com} = \gamma_m + \gamma_p \tag{3.39}$$

where:

γ_m = surface energy of the matrix

γ_p = surface energy due to fiber pull-out

For normal concrete matrices, the surface energy of the matrix (γ_m) varies in a range of about 35 to 70 J/m^2 and an average value of 50 J/m^2 is reasonable for most common designs. Generally, the surface energy due to fiber pull-out is significantly larger than the surface energy of the matrix. Different formulae can be used to estimate such energy.

3.17.1 *Surface energy of pull-out*

In its simplest form the surface energy of pull-out of a fiber reinforced concrete material can be expressed as follows, assuming circular fibers [15,56]:

$$\gamma_p = \xi \bar{\tau} V_f \frac{L^2}{d} \tag{3.40}$$

where $\xi = \lambda/4$ if the post-cracking stress versus slip response varies linearly from σ_{pc} to zero, and $\xi = \lambda/6$ if the post-cracking stress versus slip response varies parabolically from σ_{pc} to zero, and λ is as defined in Eq. (3.22). In the above equation, $\bar{\tau}$ is assumed to be the average bond strength developed over large slips of the same order as half the length of the fiber.

It is important to observe that, beside the coefficient ξ, the surface energy increases with the bond strength, the volume fraction, the aspect ratio, and the length of the fibers. Moreover, the fiber length is to the square thus has an influence significantly higher than that of the other parameters.

The following equivalence is used in the following equations: 1 lb = 4.45 N; 1 in. = 0.0254 m; surface energy units: 1lb-in/in2 = 175.12 J/m2.

Direct tensile tests using notched steel fiber reinforced mortar prisms led to the following values of surface energy due to pull-out [56, 39]:

$$\begin{cases} \gamma_p = 0.173 \bar{\tau} V_f \dfrac{L^2}{d} = 0.173 \bar{\tau} V_f \dfrac{L}{d} L & \text{lb-in/in}^2 \\[3mm] \gamma_p = 0.175 \bar{\tau} V_f \dfrac{L^2}{d} = 0.175 \bar{\tau} V_f \dfrac{L}{d} L & \text{Joule/m}^2 \end{cases} \tag{3.41}$$

in which the value of average bond strength $\bar{\tau}$ was taken equal to 2.28 MPa (330 psi).

Consider an example using Eq. (3.41). For $\bar{\tau} = 330$ psi $= 2.28$ MPa, $V_f = 2\%$, $L/d = 100$ and $L = 1$ in. $= 25.4$ mm, Eq. (3.41) yields a surface energy equal 114 lb-in/in2 = 19950 J/m² = 19.95 kJ/m². This is about 400 times the surface energy of the plain matrix.

Tests using notched tensile prisms ($75 \times 75 \times 450$ mm) of SIFCON (slurry infiltrated fiber concrete) with fiber content of 12.8% by volume, led to surface energy values ranging from 57.44 to 100.52 kJ/m²; the maximum post-cracking stress observed in these tests was 27 MPa [46]. The strain at maximum stress ranged from 1% to 2%. In comparison to results predicted from Eq. (3.41), these results reflect some scale and group effects but are definitely within range of analytical predictions.

3.18   Experimental Observations of Tension, Compression, and Bending Response

Numerous investigations provide detailed information on experimental tests and experimental results obtained on the mechanical properties of fiber reinforced cement composites. The few selected examples below are given for illustration, but they, by no means, cover the wide range of responses available. It is noted that experimental tests depend not only on the method of testing, but also the size of the specimen, the testing equipment, the method of collecting the data, interpretation of the results and other important variables. Thus experimental results, as analyzed and presented in the technical literature, can be very subjective.

Figures 3.18–3.22 are taken from different investigations carried out at different times. They illustrate the behavior of fiber reinforced cement composites in tension, bending, and compression and are self-explanatory. Figure 3.18(a) illustrates the stress-elongation curve of typical strain-softening FRC composites [57] while Fig. 3.18(b) illustrates the stress-strain (strain is valid to the peak load only) of typical strain-hardening FRC composites [49].

Figure 3.19(a) shows a comparison between three fibers and fiber materials tested under similar conditions [44]; the maximum post-cracking point is indicated by a round marker which represents the limit between strain and crack opening. Figure 3.19b suggests that size effects can be significant especially for the peak strain values. Figures 3.20(a) and 3.20(b) describe how the load-deflection (or stress-deflection) response can be changed from deflection-softening to deflection-hardening by increasing the volume fraction of fiber. Figure 3.21 describes the influence fibers have on the compressive stress-strain response of mortar and concrete matrices; note that the tests shown are for different mixtures but have one common parameter, that is about the same compressive strength.

Figure 3.22(a) describes the tensile stress-strain response of a high fiber content SIFCON composite in comparison (inset) to the plain matrix. It is noted that while a high tensile strength and strain can be achieved, the volume fraction of fibers needed is very large. It is clear that the toughness of SIFCON in tension, as measured from the area under the stress-strain curve, can be one to two orders of magnitude that of plain concrete. Figure 3.22b describes the compressive stress-strain response of SIFCON in comparison to plain concrete.

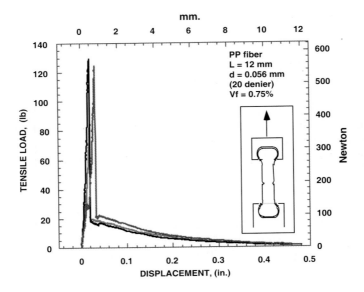

(a) Strain-softening FRC composite

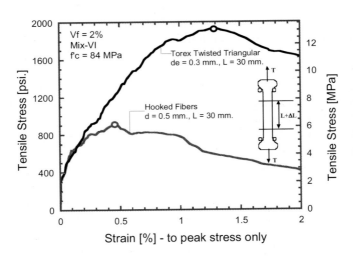

(b) Strain-hardening FRC composite

Fig. 3.18 Typical tensile stress-elongation curves.

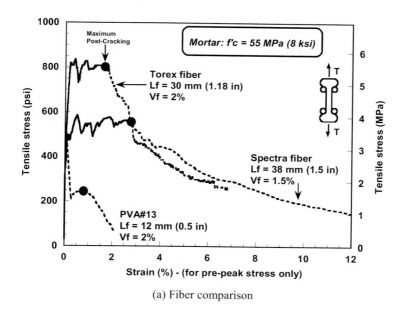

(a) Fiber comparison

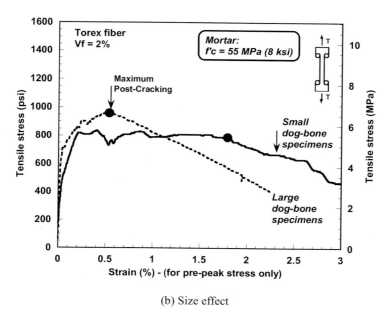

(b) Size effect

Fig. 3.19 Typical tensile stress-elongation curve of FRC composites.

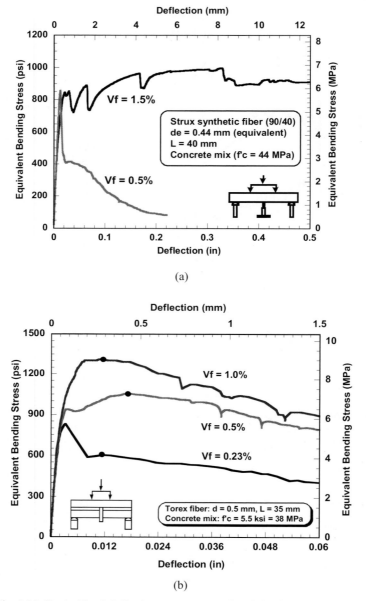

(a)

(b)

Fig. 3.20 Typical load-deflection curves comparing deflection-softening and deflection-hardening FRC composites.

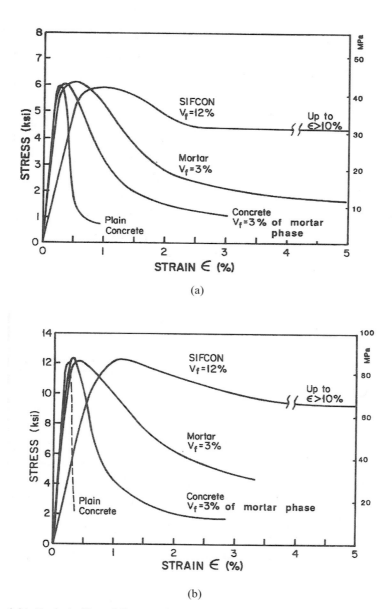

(a)

(b)

Fig. 3.21 Typical effect of fibers on the stress-strain curve in compression considering about equal compressive strength at two levels [46].

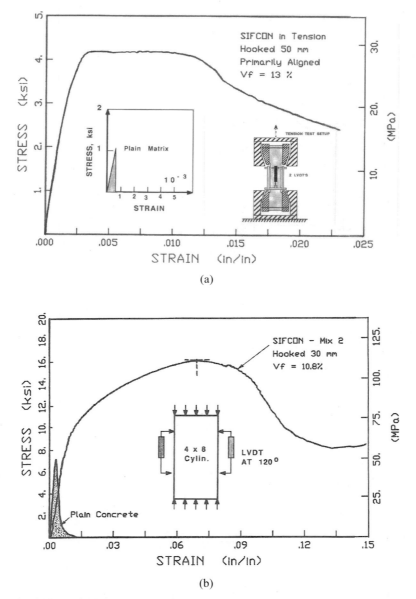

Fig. 3.22 Typical stress-strain curves of SIFCON in tension (a) and compression
(b) compared to the curves of plain concrete without fibers [46].

3.19 Fiber-Matrix Reinforcing Effectiveness

By its very definition a reinforcement (i.e., the fiber) is supposed to induce an increase in strength in the material to be reinforced (i.e., the matrix). Both analysis and experimental test results suggest that, in order to be effective in concrete matrices, fibers must preferably have the following qualitative properties (Fig. 3.23) [27,28,58]: (1) a tensile strength significantly higher that that of concrete (two to three orders of magnitude); (2) a bond strength with the concrete matrix preferably of the same order as or higher than the tensile strength of matrix; and (3) unless self-stressing is used through fiber reinforcement, an elastic modulus in tension significantly higher than that of the concrete matrix. Moreover, everything else being equal, a ductile fiber under tension is preferable to a brittle fiber, and a ductile or slip-hardening bond-stress versus slip response is preferable to a brittle or slip-softening one. The Poisson's ratio and the coefficient of thermal expansion should preferably be of the same order for both the fiber and the matrix. Indeed if the Poisson's ratio of the fiber is significantly larger than that of the matrix, detrimental debonding will occur under tensile load. However, these drawbacks can be overcome by various methods such as inducing surface deformation to create mechanical anchorage.

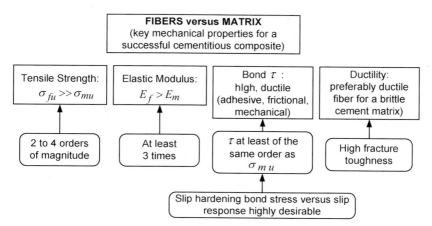

Fig. 3.23 Desirable fiber versus matrix properties for a successful cementitious composite [8].

3.20 Applications

Fiber reinforced cement and concrete composites have been used in numerous applications, either as stand-alone or in combination with reinforcing bars and prestressing tendons; they have also been used as support material in repair and rehabilitation work (Fig. 3.24). Stand-alone applications include mostly thin products such as cladding, cement boards, pipes, electrical poles, slabs on grades and pavements. Fibers are also used in hybrid applications to support other structural materials such as reinforced and prestressed concrete, and structural steel. Examples include impact and seismic resistant structures, jacketing for repair and strengthening of beams and columns, and, in the case of steel, encased beams and trusses to improve ductility and fire resistance. Particular applications of high performance fiber reinforced cement composites include bridge decks and special structures such as offshore platforms, space-craft launching platforms, super high rise structures, blast resistant structures, bank vaults, and other high-end structures.

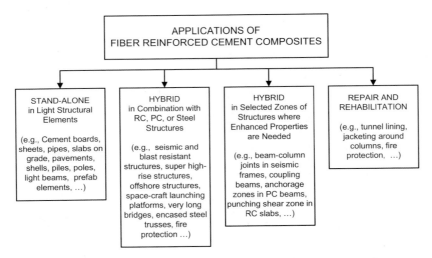

Fig. 3.24 Classes of applications of fiber reinforced cement composites.

Figure 3.25 illustrates typical applications of fiber reinforced cement composites either in stand-alone, or in combination with RC and PC structures, or in repair-strengthening situations [16]. Figure 3.26 illustrates the particular design property or properties that would call for their use in a particular application [27].

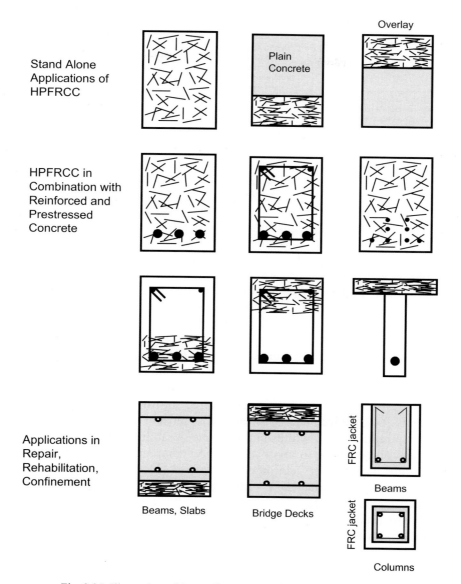

Fig. 3.25 Illustration of the applications of FRC composites in various concrete structures.

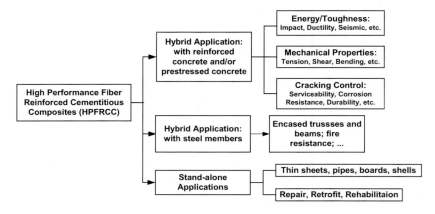

Fig. 3.26 Advantages of using HPFRC composites in structural applications.

The use of FRC composites, when considered as an alternative in design, is generally not necessary throughout the structure. Commonly, only a small part (selected zone) of the structure may be in need of strengthening or toughening. In such a case their use is often competitive and economically justifiable. Applications in selected zones of structures include: punching shear zone around columns in two-ways slab systems [47]; end blocks and anchorage zones in prestressed concrete beams; beam-to-column connections in seismic resistant frames (Fig. 3.27); beam to shear wall connections (Fig. 3.27); coupling beams for seismic-cyclic resistance [59]; out-rigger beams; in-fill damping structural elements [60]; lower end of shear walls; tension zone of RC and PC beams to reduce crack widths and improve durability; compression zone of beams and columns to improve ductility; compression zone of RC and PC beams using fiber reinforced polymeric (FRP) reinforcements to improve ductility and take advantage of the strength of FRP reinforcements [61, 62].

Fibers used in concrete structures are thought first to enhance several material properties, among which cracking and microcracking, resistance in tension, shear and bending, ductility, and energy absorption capacity. Even if only one property is sought, others are enhanced as well. Often not mentioned, but as important, is their contribution to structural performance in general, such as enhancing bond and the bond versus slip response between reinforcing bars and concrete under monotonic and cyclic loading, preserving the cover of concrete under large

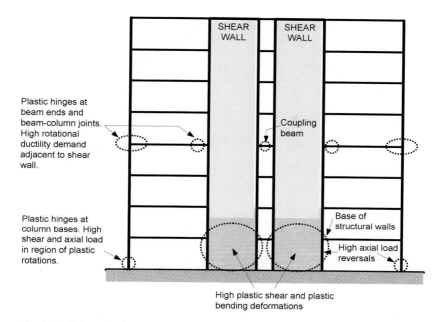

Fig. 3.27 Selected zones of RC building structure where HPFRC composites can be beneficially used [47, 14].

deformations, restraining spalling, and helping maintain the integrity of the structure by keeping reinforcing bars from buckling in columns. In short, as adequately put by Parra-Montesinos [47], fibers increase the damage tolerance of a structure.

At time of this writing, the use of fibers has just been approved in the ACI code to replace part or all the shear reinforcement in concrete members [59].

3.20.1 *Typical fiber contend and fiber volume fraction*

Due to the formulation of the mechanics of the composite (see Section 4 above), the fiber content in cement matrices is specified by volume fraction, that is, the volume of fibers divided by the total volume of the composite. Common ranges of fiber volume fraction in typical fiber reinforced concrete applications are shown in Table 3.4. Because of fiber materials of different densities, the same volume fraction of fibers of different materials leads to different weight fractions of fibers. Fibers

Table 3.4   Range of volume fraction of fibers for typical fiber
reinforced cement composites.

Material	Range of V_f	Remark
FRC – Fiber Reinforced Concrete (Strain-softening in tension, or "Conventional FRC")	$V_f \leq 2\%$	Fibers are premixed with the concrete matrix. Smaller coarse aggregates may be needed.
HPFRCC – High Performance Fiber Reinforced Cement Composites (Strain-hardening FRC in tension.)	$V_f \geq (V_f)_{cri-tension}$	Strain hardening and multiple cracking characteristics in tension. With proper design, critical V_f can be as little as 1%.
DHFRCC - Deflection-Hardening Fiber Reinforced Cement Composites (for bending).	$V_f \geq (V_f)_{cri-bending}$	Deflection-hardening and multiple cracking characteristics in bending. With proper design, critical V_f can be smaller than 0.8%.
Shotcrete (steel fibers)	$V_f \leq 3\%$	Applications in tunnel lining and repair.
Spray Technique (glass fibers)	$4\% \leq V_f \leq 7\%$	Applications in cladding and panels.
SIMCON (Slurry Infiltrated Mat Concrete – steel mat)	$4\% \leq V_f \leq 6\%$	A prefabricated steel fiber mat is infiltrated by a cement slurry.
SIMCON (Slurry Infiltrated Mat Concrete – PVA mat)	$V_f \approx 1\%$	A prefabricated PVA fiber mat infiltrated by a slurry matrix
SIFCON (Slurry Infiltrated Fiber Concrete)	$4\% \leq V_f \leq 15\%$	Fibers (mostly steel to date) are preplaced in a mold and infiltrated by a fine cementitious slurry matrix.

are purchased by weight, but mechanical properties of composites are based on volume fraction, not weight fraction of fibers. Typically a 1% volume fraction of steel fibers in normal-weight concrete amounts to about 80 kg/m³ of concrete; however, a 1% volume fraction of polypropylene fibers amounts to about only 9.2 kg/ m³.

Table 3.4 is purposely kept as general as possible with little reference to trade-type names. However, high performance fiber reinforced cement composites with strain-hardening tensile behavior have been given different names by their developers. Current names include SIFCON (Slurry Infiltrated Fiber Concrete), fiber reinforced DSP (Densified Small Particles systems), CRC (Compact Reinforced Composite), SIFCA (a form of SIFCON particularly suitable for refractory applications), SIMCON (slurry infiltrated mat concrete), RPCC (Reactive Powder Concrete Composites), ECC (engineered cementitious composites), Ductal [63], MMFRC (multi-modal FRC [48]), and other proprietary names. They have been shown to develop outstanding combinations of strength (up to 800 MPa in compression for RPC manufactured under temperature and pressure [64]) and ductility or energy absorption capacity (up to 1000 times that of plain concrete for SIFCON), while achieving substantial quasi-strain hardening and multiple cracking behavior.

3.20.2   Evolution in performance

The performance of fiber reinforced cement composites has consistently improved over time. While their compressive strength can be directly related to that of the plain matrix, their tensile strength, tensile strain capacity and ductility are more specific examples of their advance over plain concrete. Similarly to steel, an increase in tensile strength generally implies a trade-off for a decrease in strain capacity. Figure 3.28 illustrates examples of stress-strain curves with a wide range of tensile strength and strain capacities. They represent the available range technically possible at time of this writing.

3.21 Concluding Remarks

Fiber reinforced concrete has seen its first patent in 1874. Yet, for all practical purposes, progress in FRC composites has been almost at a standstill for more than 100 years, and picked up at an exceptional pace only during the 1960's. This may be partly due to fundamental research, better understanding of the reinforcing mechanisms of FRC composites, the need for materials with particular properties, developments in advanced materials, economic competitiveness, and global circumstances. A solid foundation has thus been built. It is likely that every area mentioned in the above discussion will see progress in the future. However, economic considerations will keep playing a major role.

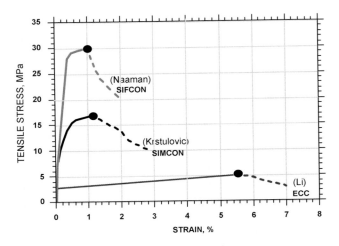

Fig. 3.28 Typical range of tensile stress-strain curves for HPFRC composites showing
the trade-off between strength and strain capacity.

Because of the particular strain-hardening behavior of high performance fiber reinforced cement composites, the dream that started in 1874, to mix fibers in concrete, like sand or gravel, to achieve a stand-alone structural material is closer today than ever before. It is satisfying to know that the next generation of civil engineers will have a structural concrete with a whole spectrum of new possibilities to explore in order to provide better structures for a continually challenging world.

3.22 Acknowledgments

The material described in this chapter is taken from yet unpublished teaching notes for a course titled "Fiber Reinforced Cement Composites" which the author has introduced and taught in the Department of Civil and Environmental Engineering at the University of Michigan, Ann Arbor, since 1985.

3.23 References

The following list of references is very incomplete due to space limitation and does not do justice to the thousands of studies available at the time of this writing.

1. Alwan, J., and Naaman, A.E., "New Formulation for the Elastic Modulus of Fiber Reinforced Quasi Brittle Matrices," ASCE Journal of Engineering Mechanics, Vol. 120, No. 11, November 1994, pp. 2443-2460.

2. Aveston, J., Cooper, G.A., and Kelly, A. "Single and multiple fracture — the properties of fiber composites," *Conference Proceedings of National Physical Laboratory,* IPC, Science and Technology Press, Ltd., 1971, pp. 14-24.

3. Balaguru, P., and Shah, S.P., *"Fiber Reinforced Cement Composites,"* McGraw Hill, New York, 1992.

4. Bentur, A., and Mindess, S., *"Fiber Reinforced Cementitious Composites,"* Elsevier Applied Science, London, UK, 1990.

5. Bolander, J., and Saito, S., "Discrete Modeling of Short Fiber Reinforcement in Cementitious Composites," Journal of Cement Based Materials, No. 6, 1997, pp. 76-86.

6. Bolander, J., "Spring Network Model of Fiber Reinforced Cement Composites," in *"High Performance Fiber Reinforced Cement Composites – HPFRCC 3,"* H.W. Reinhardt and A.E. Naaman, Editors, RILEM Pro 6, RILEM Publisations S.A.R.L., Cachan, France, May 1999, pp. 341-350.

7. Brandt, A., Li, V.C., and Marshall, I.H., Editors, "Brittle Matrix Composites 6, BMC-6," Woodhead Publishing Limited, Cambridge and Warsaw, October 2000.

8. Hannant, D.J., "Fiber Cements and Fiber Concretes," J. Wiley, 1978, 215 pp.

9. Karihaloo, B.L., and Wang, J., Micromechanical Modeling and Strain Hardening and Tensile Softening in Cementitious Composites," Journal of Computational Mechanics, Vol. 19, 1997, pp. 453-462.

10. Krenchel, K., *Fiber Reinforcement,* Akademisk Forlag, Copenhagen, 1964.

11. Li, V.C., & H.C. Wu, "Conditions for pseudo strain-hardening in fiber reinforced brittle matrix composites," *J. Applied Mechanics Review,* V.45, No. 8, August, pp. 390-398, 1992.

12. Li, V.C. & C.K.Y. Leung, "Theory of steady state and multiple cracking of random discontinuous fiber reinforced brittle matrix composites," ASCE *J. of Engineering Mechanics,* V. 118, No. 11, 1992, pp. 2246-2264, 1992.

13. Naaman, A.E., *"A Statistical Theory of Strength for Fiber*

Reinforced Concrete," Thesis presented to the Massachusetts Institute of Technology, Civil Engineering Department in partial fulfillment for the degree of Doctor of Philosophy, Sept. 1972, 196pp.

14. Naaman, A.E., and Reinhardt, H.W., Co-Editors, *"High Performance Fiber Reinforced Cement Composites: HPFRCC 2"*, RILEM, No. 31, E. & FN Spon, London, 1996, 505 pages.

15. Naaman, A.E., and Reinhardt, H.W., "Characterization of High Performance Fiber Reinfoced Cement Composites," in *"High Performance Fiber Reinforced Cement Composites – HPFRCC 2,"* A.E. Naaman and F.W. Reinhardt, Editors, RILEM Pb. 31, E. and FN Spon, England, 1996; pp. 1-24.

16. Naaman, A.E., and Reinhardt, H.W., Co-Editors, *"High Performance Fiber Reinforced Cement Composites – HPFRCC 4,"* RILEM Proc., PRO 30, RILEM Pbs., S.A.R.L., Cachan, France, June 2003; 546 pages.

17. Reinhardt, H.W., and Naaman, A.E., Editors, *"High Performance Fiber Reinforced Cement Composites,"* RILEM, Vol. 15, E. & FN Spon, London, 1992, 565 pages.

18. Reinhardt, H.W., and Naaman, A.E., Co-Editors, *"High Performance Fiber Reinforced Cement Composites – HPFRCC 3,"* RILEM Proceedings, PRO 6, RILEM Pbs., S.A.R.L., Cachan, France, May 1999; 666 pages.

19. Rossi, P., *Les Betons de Fibres Metalliques*, (Concretes with Steel Fibers), in French, Presses de l'Ecole Nationale des Ponts et Chaussees, Paris, France, 1998, 309 pages.

20. Shah, S.P., "Do Fibers Increase the Tensile Strength of Cement-Based Matrixe?" ACI Materials Journal, Vol. 88, No. 6, Nov.-Dec. 1991, pp. 595-602.

21. Tjiptobroto, P., and Hansen, W., "Model for prediction of the elastic strain of fiber reinforced composites containing high volume fractions of discontinuous fibers," *ACI Materials Journal,* V. 90, No. 2, March-April, 1993.

22. Naaman, A.E., and Reinhardt, H.W., "High Performance Fiber Reinforced Cement Composites (HPFRCC-4): International RILEM Report," Materials and Structures, Vol. 36, Dec. 2003, pp. 710-712. Also same in Cement and Concrete Composites, Vol. 26, 2004, pp. 757-759.

23. Naaman, A.E., and Reinhardt, H.W., "Proposed Classification of

FRC Composites Based on their Tensile Response " Proceeding of symposium honoring S. Mindess, N. Banthia, Editor, University of British Columbia, Canada, August 2005. Electronic proceedings, 13 pages. In print in Materials and Structures, Rilem, 2006.

24. Stang, H., and Li, V.C., "Scale Effects in FRC and HPFRCC Structural Elements," in 4th International Workshop on HPFRCC, Ann Arbor, Michigan, 2003, pp. 245-259.

25. Mehta, P. Kumar, and Monteiro, P.J.M., *Concrete: Structure, Properties and Materials*, 3rd Edition, McGraw Hill Professional, 2005, 669 pages.

26. Mindess, S., Young, J.F., and Darwin, D., *Concrete,* 2nd Edition, Prentice-Hall, Upper Saddle River, NJ, USA, 2003.

27. Naaman, A.E., "Fiber Reinforcement for Concrete: Looking Back, Looking Ahead," in Proceedings of Fifth RILEM Symposium on Fiber Reinforced Concretes (FRC), BEFIB' 2000, Edited by P. Rossi and G. Chanvillard, September 2000, Rilem Publications, S.A.R.L., Cachan, France, pp. 65-86.

28. Naaman, A.E., "Engineered Steel Fibers with Optimal Properties for Reinforcement of Cement Composites," Journal of Advanced Concrete Technology, *Japan Concrete Institute*, Vol. 1, No. 3, November 2003, pp. 241-252.

29. Naaman, A.E., F. Moavenzadeh and F.J. McGarry, "Probabilistic Analysis of Fiber Reinforced Concrete," *Journal of the Engineering Mechanic's Division*, ASCE, Vol. 100, No. EM2, April 1974, pp. 397-413.

30. Naaman, A.E., "High performance fiber reinforced cement composites," Proceedings of the *IABSE Symposium on Concrete Structures for the Future,* Paris, France, September 1987, pp. 371-376.

31. Bentur, A., Wu, S.T., Banthia, N., Baggot, R., Hansen, W., Katz, A., Leung, C., Li, V.C., Mobasher, B., Naaman, A.E., Robertson, R., Soroushian, P., Stang, H., Taerwe, L., "Fiber Matrix Interfaces," Chapter 5, in High Performance Fiber Reinforced Cement Composites: HPFRCC 2, A.E. Naaman and H.W. Reinhardt, Editors, RILEM, No. 31, E. & FN Spon, London, 1996, pp. 149-191.

32 (a) Naaman, A.E., "Fibers with Slip-Hardening Bond," in *"High Performance Fiber Reinforced Cement Composites – HPFRCC 3,"* H.W. Reinhardt and A.E. Naaman, Editors, RILEM Pro 6, RILEM

Publisations S.A.R.L., Cachan, France, May 1999, pp. 371-385.

32   (b) Naaman, A.E., "New Fiber Technology: Cement, Ceramic and Polymeric Composites," Concrete International, Vol. 20, No. 11, July 1998.

33.   Halpin, J.C., and Tsai, S.W., "Effect of Environmental Factors on Composite Materials," AFML-TR 67-423, June 1969.

34.   Naaman, A. E., and Najm, H., "Bond-Slip Mechanisms of Steel Fibers in Concrete," ACI Materials Journal, Vol. 88, No. 2, April 1991, pp. 135-145.

35.   Agarwal, D., and Broutman, L., Analysis and Performance of Composites, John Wiley and Sons, 1981.

36.   Cox, H.L., "The Elasticity and Strength of Paper and other Fibrous Materials," British Journal of Applied Physics, No. 3, 1952, pp. 72-79.

37.   Patton, M.E., and Whittaker, W.L., "Effect of Fiber Content and Damaging Load on Steel Fiber Concrete Stiffness," ACI Journal, Vol. 80, No. 1, Jan.-Feb. 1983, pp. 13-16.

38.   Naaman, A.E., unpublished notes for the course "Fiber Reinforced Cement Composites," Department of Civil and Environmental Engineering, University of Michigan, 1985-2006.

39.   Naaman, A.E., "Toughness, Ductility, Surface Energy and Deflection-Hardening FRC Composites," in Proceedings of the JCI Workshop on Ductile Fiber Reinforced Cementitious Composites (DFRCC) – Application and Evaluation, Japan Concrete Institute, Tokyo, Japan, October 2002, pp. 33-57.

40.   Maalej, M., "Fracture Resistance of Engineered Fiber Cementitious Composites and Implications to Structural Behavior," Ph.D. Thesis, University of Michigan, 1994, 223 pages.

41.   Naaman, A.E., and Shah, S.P., "Pull-Out Mechanisms of Steel Fiber Reinforced Concrete," ASCE, Journal of the Structural Division, Vol. 102, No. ST 8, August 1976, pp. 1537-1548.

42.   Markovic, I., Van Mier, J.G.M., and Walraven, J.C., "Tensile Response of High Performance Hybrid Fiber Concrete," in 5th International Conference on Fracture Mechanics of Concrete and Concrete Structures (FRAMCOS-5), Vail, Colorado, 2004, Vol. 2, pp. 1113-1121.

43.   Visalvanich, K. and A.E. Naaman, "A Fracture Model for Fiber Reinforced Concrete," Journal of the American Concrete Institute, Vol. 80, No. 2, March/April 1983, pp. 128-138.

44.   Chandrangsu, K., and Naaman, A.E., "Comparison of Tensile and Bending Response of Three High Performance Fiber Reinforced Cement Composites," in High Performance Fiber Reinforced Cement Composites (HPFRCC-4), A.E. Naaman and H.W. Reinhardt, Editors, RILEM Publications, Pro. 30, June 2003, pp. 259-274.

45.   Chanvillard, G., and Rigaud, S., "Complete Characterization of Tensile Properties of Ductal UHPFRC According to the French Recommendations," in *High Performance Fiber Reinforced Cement Composites (HPFRCC-4)*", A.E. Naaman and H.W. Reinhardt, Editors, RILEM Publications, Pro. 30, June 2003, pp. 95-113.

46.   Naaman, A.E., "SIFCON: Tailored properties for structural performance," in *High Performance Fiber Reinforced Cement Composites,* RILEM Proceedings 15, E. and FN SPON, London, 1992, pp.18-38.

47.   Parra-Montesinos, G., "High Performance Fiber Reinforced Cement Composites: an Alternative for Seismic Design of Structures," ACI Structural Journal, Vol. 102, No. 5, Sept.-Oct. 2005, pp. 668-675.

48.   Rossi, P., and Renwez, S., "High Performance Multi Modal Fiber Reinforced Cement Composites," 4th International Symposium on High Strength High Performance Concrete, Paris, France, 1996, pp. 687-691.

49.   Sujivorakul, C., and Naaman, A.E., "Tensile Response of HPFRC Composites Using Twisted Polygonal Steel Fibers", in Innovations in Fiber-Reinforced Concrete for Value," N. Banthia, M. Criswell, P. Tatnall, and K. Folliard, Editors, ACI Special Publication, SP216, American Concrete Institute, 2003, pp. 161-179.

50.   Sujivorakul, C., and Naaman, A.E., "Ultra High-Performance Fiber-Reinforced Cement Composites Using Hybridization of Twisted Steel Fibers and Micro Fibers," in Fiber Reinforced Concretes – BEFIB 2004, Edited by M. di Prisco, R. Felicetti, and G.A. Plizzari, RILEM Proceedings PRO 39, Vol. 2, 2004, pp. 1401-1410.

51.   Gopalaratnam, V., and Shah, S.P., "Tensile Failure of Steel Fiber Reinforced Mortar," Journal of Engineering Mechanics, ASCE, Vol. 113, No. 5, May 1987, pp. 635-652.

52.   Kosa, K., and Naaman, A.E., "Corrosion of Steel Fiber Reinforced

Concrete," ACI Materials Journal, Vol. 87, No. 1, January-February, 1990, pp. 27-37.

53.   Li, V.C., "Postcrack Scaling Relation for Fiber Reinforced Cementitious Composites," Journal of Materials in Civil Engineering, ASCE, Vol. 4, No. 1, 1992, pp. 41-57.

54.   Naaman, A.E., "Advances in High Performance Fiber Reinforced Cement Based Composites," Proceedings of the International Symposium on Fiber Reinforced Concrete, V.S. Parameswaran and T.S. Krishnamurti, Editors, Oxford IBH Publishing Ltd., New Delhi, India, December 1987, pp. 7.87-7.98.

55.   Naaman, A.E., "Strain Hardening and Deflection Hardening Fiber Reinforced Cement Composites," in *High Performance Fiber Reinforced Cement Composites (HPFRCC-4)*," A.E. Naaman and H.W. Reinhardt, Editors, RILEM Publications, Pro. 30, June 2003, pp. 95-113.

56.   Naaman, A.E., "Fiber Reinforced Concrete under Dynamic Loading," American Concrete Institute, International Symposium on Fiber Reinforced Concrete, Special Publication SP-81, Detroit, 1984, pp. 169-186.

57.   Naaman, A.E., and Guerrero, P., "A New Methodology to Determine Bond of Micro-Fibers Embedded in Cement Composites," in Brittle Matrix Composites 6, BMC-6, Edited by A. Brandt, V.C. Li and I.H. Marshall, Woodhead Publishing Limited, Cambridge and Warsaw, October 2000, pp. 42-51.

58.   Naaman, A.E., "Evaluation of steel fibers for applications in structural concrete," in Fiber Reinforced Concretes – BEFIB 2004, Edited by M. di Prisco, R. Felicetti, and G.A. Plizzari, RILEM Proceedings PRO 39, Vol. 1, 2004, pp. 389-400.

59.   Parra-Montesinos, G., "Proposed addition to ACI Code 318-05 on shear design provisions for fiber reinforced concrete memebers," personal communication, March 2006.

60.   Xia, Z., and A.E. Naaman, "Behavior and Modeling of Infill FRC Damper Element for Steel-Concrete Hybrid Shear Wall," *ACI Structural Journal*, Vol. 99, No. 6, Nov.-Dec. 2002, pp. 727-739.

61.   Naaman, A.E. and Jeong, S.M., "Structural Ductility of Beams Prestressed with FRP Tendons." Proceedings 2nd International Symposium on Non-Metallic (FRP) Reinforcement for Concrete Structures, L. Taerwe, Editor, Ghent, Belgium, August 1995; RILEM Proceedings 29, E & FN Spon, London, pp. 379-386.

62. Park, S.Y., and Naaman, A.E., "Shear Behavior of Concrete Beams Prestressed with FRP Tendons," PCI Journal, Vol. 44, No. 1, Jan.-Feb. 1999, pp 74-85.

63. Orange, G., Dugat, J., and Acker, P. "Ductal: New Ultra High Performance Concretes – Damage Resistance and Micro-mechanical Analysis," 5th RILEM Symposium on Fiber Reinforced Concrete, BEFIB-2000, pp. 781-790.

64. Richard, P., "Reactive powder concrete: a new ultra-high strength cementitious material," Proceedings of the 4th International Symposium on Ultilisation of High-Strength/High-Performance Concrete, F. de Larrard and R. Lacroix, Editors, Presses des Ponts et Chaussees, Paris, France, 1996, pp. 1501-1511.

Chapter 4

High Performance Steel Material and Structures for Earthquake Resistant Buildings

Keh-Chyuan Tsai,
Ying-Cheng Lin, Jui-Liang Lin,
Sheng-Lin Lin and Po-Chien Hsiao
National Center for Research on Earthquake Engineering,
Taipei, Taiwan

4.1 Introduction

4.1.1 *Background*

Iron and subsequently steel have played major roles in the history of mankind. Modern construction would not have been possible without steel as a primary material. Structural steel has been used in construction in the United States for more than 100 years, starting with the Eads Bridge in St. Louis (1867–1874). The performance of steel in recent earthquakes raised a number of questions related to design and fabrication of steel structures. Thus, historically acceptable criteria and especially connection geometries were questioned; it appeared that unrecognized factors lent themselves to unacceptable behavior. Further, the suitability of the properties as determined by the common uniaxial tension specimen was called into question, since several failure modes seemed to demand higher and better defined orthogonal strength characteristics [1].

On this background, significant research and design projects were undertaken to determine the necessary properties of steels that would satisfy the requirements for acceptable service under all conditions. All steels possess a combination of properties that determine how well a steel performs its intended function. Strength, weldability, toughness, ductility, corrosion resistance, and formability are all important to determine how well a steel performs. High-performance steel (HPS) can be defined as having an optimized balance of these properties to give maximum performance in bridge structures while remaining cost-effective.

Use of high-strength steels for bridge construction in Japan dates back to the 1960's [2]. Several hundred bridges have been constructed using 500 MPa and 600 MPa yield strength steel, and steel with a nominal yield strength of 800 MPa has also been used on several projects. The first Japanese project using HSS in structural elements of a building was the Landmark Tower in central Yokohama, using steel with a minimum tensile strength of 600N/mm2. Completed in 1989 it used HSS in the I-section columns fabricated from plates made with the thermomechanical process. Many researches in Japan have been conducted on the use of HSS in a building to change its natural frequency, and therefore potentially reducing the risk of earthquake damage, which is particularly important in Japan [3].

In the U.S., the Federal Highway Administration (FHWA) initiated a program to develop high-performance steels (HPS) for bridges in 1992. A 485 MPa grade of HPS was the first developed, and is specified in ASTM A709 as grade HPS-70W (HPS 485W) (the designation "W" stands for "weathering"). High Performance Steel, grade 70 (HPS-70W) became available for use in early 1996 for fabrication and testing in bridges [4]. Most HPS used in America cannot be classified as HSS as the yield strength is less than 460N/mm2, but cost savings and corrosion resistance show its merits.

The exact number of building projects using A992 is not known, although it is very large. Since the development of HPS in the mid-1990s, numerous bridges using HPS girders have been constructed, and many have been economically built. The benefits related to HPS include enhancements in: weldability, toughness, corrosion resistance, ductility, fatigue and fire resistance, formability, and strength. These factors combined lead to construction elements of higher economic efficiency, ease of maintenance, and longer service life. Because of the low carbon levels, minimum or no preheat would be required which allows increased productivity of fabrication and reduced cost. Recently Northwestern University [5] developed copper-precipitation-hardened, high-performance weathering steel (NUCu 70W Steel) that is produced by hot rolling after casting and then air-cooling. Due to simple processing the steel is fabricated at lower cost than competing high-performance steels that require quench and tempering.

4.1.2 *Structural design and specifications for high performance steel*

Since high performance steel have different performance from conventional steel, it needs to follow different design specification in order to fully take the advantages of high performance steel. Many countries and regions have developed their own guidelines and specifications. In US, the following four documents cover the design, fabrication and construction of steel bridges using high performance steels:

- AASHTO LRFD Bridge Design Specifications with Interims (AASHTO LRFD), 1998
- AASHTO Standard Specifications for Highway Bridges, 16th. Edition, 1996, 2000 Interim (AASHTO LFD)
- AASHTO Guide Specifications for Highway Bridge Fabrication with HPS 70W Steel (AASHTO HPS Guide), 2003
- ANSI/AASHTO/AWS D1.5-95 Bridge Welding Code with Addendums (AWS Code)

These documents reflect the findings and experiences on the applications of HPS by researchers, fabricators, manufacturers, owners and engineers working with high performance steels, and are the best references, as they are modified over time. The designers must make sure that all or parts of these documents are made a part of the contract document and add any supplemental requirements in the project special provisions.

HPS Designers' Guide [6] discusses the key elements of the above four documents as applied to high performance steels, identifies factors that should be considered, and provides sources and references where designers can obtain additional information to assure successful use of HPS in highway bridge construction.

In Europe, a variety of high-strength steels with yield strengths from 460 MPa to 690 MPa are available for applications. European structural steel standard EN 10025: 2004 is now published in six parts to bring together almost all the 'Structural Metallic Products' into one comprehensive standard. Grade S460ML, which has a nominal yield strength of 460 MPa, can be welded at room temperature for plate thicknesses up to 90 mm and has a specified minimum Charpy V-notch (CVN) energy of 27 J at –50°C.

Novel steel materials can be produced for particular requirements to provide steels with higher or lower yield strength (LYS) and with better mechanical properties. During an earthquake, steel members with very low capacity will yield first, reach the non-linear state and dissipate input seismic energy, while other members with higher yield strength stay at elastic state. Some examples of this seismic design concept are: Steel Plate Shear Walls (SPSWs) and Buckling Restrained Braces (BRBs).

The choice of using SPSWs as the primary lateral force resisting system in buildings has increased in recent years when structural engineers recognize its benefits. Traditional designs do not allow utilizing the post-buckling strength of the plate. They consider only elastic and shear yield behavior, and the designs often resulted in very thick plate sizes. This results stiffer structures that exhibits large accelerations during a seismic event. Also, surrounding frame members need additional strengthening to prevent mechanism formation due to the forces exerted by the panel at ultimate displacements. Therefore, the use of thinner plates with lower yield strength is advised by researchers in recent years.

The progress of researches about steel materials allows the energy dissipation device to be developed. In recent years, some researchers have confirmed that the *buckling restrained braced frame* (BRBF) is an effective system for severe seismic application. The BRB members can be made from several kinds of structural steel shape encased in steel tube and confined by infill concrete. An unbonding material placed between the core bracing and the infill concrete is required to reduce the friction while restraining the bracing from buckling. In addition, the all-metallic and the detachable BRBs are developed which could make the fabrication faster and the inspection more feasible.

4.2 High Performance Structural Steels

4.2.1 *TMCP: Thermo-mechanical controlled process steel*

The steel plates produced by on-line-accelerated cooling technique associated with specified controlled rolling process are known as *thermo-mechanical controlled processing* (TMCP) steels. The TMCP process is illustrated schematically in Fig. 4.1. The advantages in applying TMCP process lie in two folds: (1) increase of strength with the same composition of cold-rolling steel, and (2) reduction of alloy content to keep and/or enhance the mechanical properties of steel by changing the chemical structure from ferrite-pearlite to ferrite-bainite during accelerated cooling process, as shown in Fig. 4.2.

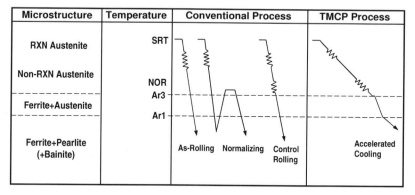

Fig. 4.1 Schematic illustration of TMCP process.

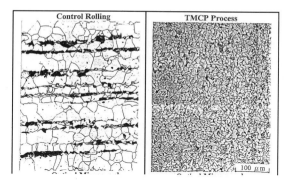

Fig. 4.2 Photo of CR and TMCP steels.

The comparison of chemical compositions used in conventional process (normalizing and controlled rolling) and TMCP process is illustrated in Fig. 4.3. Major TMCP steel plates may include A572/50, A572/60, SM570, AH/DH/EH 32~36, and API X52/X60 line pipe steels. Nowadays, high performance TMCP steel plates (such as high heat input welding steel, fire resistant steel, ULCB steel) have been developed according to their individual requirement as described in the following topics.

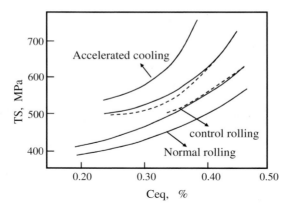

Fig. 4.3 TS-Ceq relationship of CR and TMCP steels.

4.2.2 *Characteristics of high performance steel plates*

■   High heat input welding steel

One of the most important parts of steel construction is the welding process. The idea of economy due to increase of domestic labor cost suggests obtaining a highly efficient welding ability for steel plates. In addition, the exploration of oil resources in underwater and colder regions like North Sea, Alaska, and Siberia, increases the demand for steel plates with higher strength, superior toughness, and which can be welded easily.

The principles of metallurgy and technology for improving the heat affected zone (HAZ) toughness under high heat input welding (HHIW) has been reviewed elsewhere. The application of HHIW technology and TMCP technology has led to the successful production of a heavy-gauged 80-mm thick SM570M structural steel plates capable of welding heat input up to 880 KJ/cm, and provide HAZ toughness better than 15J at −5°C. Such SM570M steels have been used for the construction of the tallest livable building in the world today, the Taipei 101 tower.

■   Fire resistant steel SN490CFR

For better resistance against fire, the steel members used for buildings are advised to be covered with a layer of refractory materials to withstand higher temperature up to 600 °C without neither yielding nor reducing and/nor eliminating the refractory layer. The key objective is to determine the strength of steel in the intermediate temperature, and to strengthen the weakness in that point by adding fine and stable precipitates. From literature, Mo-Nb-bearing complex precipitates are beneficial to the formation of fine and stable precipitates, and are found that the precipitate size is sensitive to the TMCP process.

■   Acicular ferrite-type Mn-Mo-Nb APIx70 steel

To explore efficiently the oil and/or gas resources in colder environment, the operation pressure and the diameter of steel pipe are increased. Such trends suggests that the modern line pipe steel must have higher strength (strength beyond X65), higher toughness (FATT<-20°C), easy-to-weld property, and good Hydrogen Induced Cracking resistance to oil or gas with very low-pH content.

Since the strength requirement is to reach the strength limit of conventional precipitation-hardened ferrite or pearlite steel, the aim is to find a suitable strengthening method without neglecting the toughness and weld ability of the materials. It is reported that steel with acicular ferrite structure exhibits excellent combination of strength, toughness, and weld ability due to its extremely fine structure (the size of ferrite-made lath is around 1-m). These acicular ferrite steels were introduced in 1970's using Mn-Mo-Nb alloying system subjected to conventional controlled rolling process. To obtain high strength and superior drop-weight tear test (DWTT) toughness, the Mn-Mo-Nb alloy associated with intensive TMCP process (intensive non-recrystallization region reduction) was adopted to ensure the formation of predominantly fine acicular ferrite structure. The high performance APIX70 line pipe steel with this structure has been produced successfully based from the advanced TMCP process.

■   ULCB steel HT590

Researchers tried to extend the strength capacity of the TMCP type ferrite or pearlite steels with tensile strength (TS) limit of SM570. To develop higher TS, other structural systems should be applied. The low temperature transformation products, such as bainite and martensite, are useful to increase strength. Since the resistance to fracture of bainite is better than martensite, bainitic steel is selected as a good alternative of the ferrite/pearlite steel. To balance the strength and weld ability, the carbon content is kept low. This develop the so-called, *"ultra low carbon bainitic*" (ULCB) steel with carbon content less than 0.06%. The hardening ability of ULCB is maintained by a certain amount of B-element together with ample Nb and Mn content. And, to minimize the addition of alloying elements, TMCP process is also applied to the production of ULCB steels. Such TMCP ULCB steel is used to produce the HT590 steel.

4.2.3 *Non-TMCP high performance structural steel plates*

The structural designers changed the safety design philosophy from "*seismic-resistant design method*" where the main objective is to avoid deformation in structures, to "*structure response control method*" which allows the steel members to deform and to absorb the large earthquake energy. The latter-type of structures, such as plate shear wall and bracing system, often contain a part made from ultra LYS steel, which will deform initially as the earthquake energy is being absorbed. To produce this material, the strengthening mechanics was reviewed to find out how to suppress the yield strength of the material. The method is

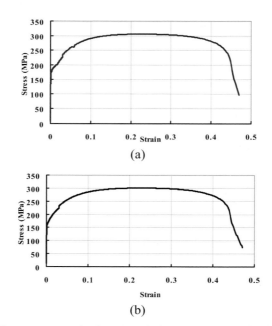

Fig. 4.4 Coupon test result of LYS steel plate : (a) in longitudinal direction; (b) in transverse direction.

achieved through the lowering of the strength of solid solution by adding Ti to decrease the content of interstitial Carbon and Nitrogen and to prevent the formation of fine precipitates and grain size, while applying heat treatment to coarsen those. Based from this, the coarse ferrite grain size and huge Ti-bearing precipitates required for the ultra LYS steel are obtained. Such steel, which possesses very low yield strength (< 170 MPa) and excellent elongation (> 45%), is termed as LYS 170. The plot of the stress versus the strain is shown in Fig. 4.4. It can be observed that performances of the ductility ratios in the transverse and the longitudinal directions are almost the same.

4.3 Steel Plate Shear Wall Building System

4.3.1 *General behavior of unstiffened SPSW*

Researches in Canada led to a new design philosophy of the steel plate shear wall (SPSW) building system (Fig. 4.5) that reduce plate thickness by allowing the occurrence of shear buckling [7]. Recent researchers on the behavior of SPSWs promote the use of relatively thinner plates with lower yield strength for the infill panels. After the thin LYS infill panel buckled and subsequently developed diagonal tension field, lateral load is carried by the infill panels. The smaller thickness of panel reduces forces on adjacent members resulting to a more efficient framing design.

Though infill panels provide their advantages in efficient design, there are some points needed to be improved. Recently, researchers have proven that the drawbacks of unstiffened SPSWs — the large bucking sounds, the pinched hysteretic responses (Fig. 4.6), and the unacceptable out-of-plane displacement (Fig. 4.7) — can be improved by attaching skillfully some restrainers without increasing the ultimate strength of the SPSW frame. The relevant design concept of buckling restrainers will be discussed in the following sections. Note that the single-pinched hysteretic response shown in Fig. 4.6(b) represents the following phenomena [8]:

Curve *ab*:   Unloading (elastic).
Curve *bc*:   The release of the tension field developed in previous excursion.
Curve *cd*:   The tension field redevelops in the opposite direction.
Curve *de*:   Yielding of various components of shear wall.
Curve *ea*:   Repeated unloading and reloading in the opposite direction.

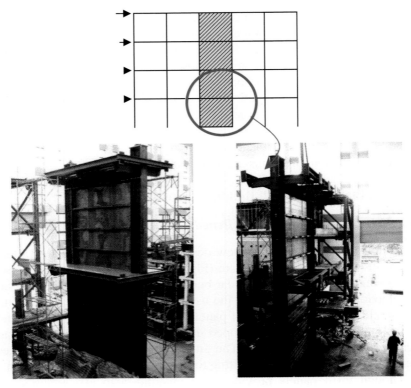

Fig. 4.5 Building of frame with SPSW.

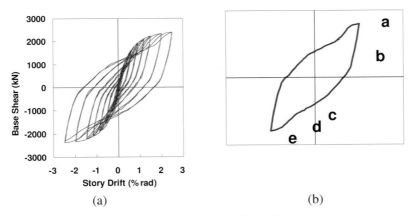

(a)   (b)

Fig. 4.6 Diagrams of pinched hysteresis: (a) Pinched hysteretic responses and
(b) Single pinched response.

Fig. 4.7 Out-of-plane displacement of SPSW.

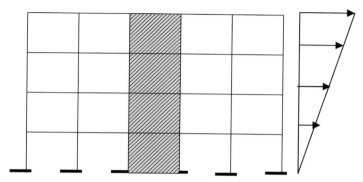

Fig. 4.8 Multi-story SPSW building frame.

4.3.2 *Seismic design of SPSW*

4.3.2.1 Design procedure

For a multistory SPSW shown in Fig. 4.8, neglecting the contribution of plastic hinges in beams and columns will give a conservative design in the part of rigid beam-to-column connections. The proposed procedure requires the designer to [9]:

1. Calculate the design base shear, and distribute it along the height of the building as described by the applicable building code;
2. Use the following equation to calculate the minimum plate thicknesses required for each story:

$$t_i = \frac{2 \cdot \Omega_s \cdot V_i}{F_y \cdot L \cdot \sin 2\alpha} \tag{4.1}$$

where

V_i = design story shear computed using the equivalent lateral force method (The yielding of boundary elements contributed approximately 25-30% of the total load strength of the system.),

Ω_s = the system overstrength factor, as defined by FEMA 369, and is taken as 1.2 for SSPWs [9],

F_y = design yield stress of the steel plate,

L = the distance between boundary column centerlines, and

α = assumed angle of inclination of the tension field measured from the vertical (Take $\alpha = 45°$ for initial calculation);

3. Develop the strip model described in Fig. 4.9. For elastic analysis using computer, use the equation below by Timler and Kulak to calculate the angle of inclination of the strips;

$$\tan^4 \alpha = \frac{1 + \dfrac{t \cdot L}{2 \cdot A_c}}{1 + t \cdot h_s \left(\dfrac{1}{A_b} + \dfrac{h_s^{\,3}}{360 \cdot I_c \cdot L}\right)} \tag{4.2}$$

where

h_s = story height,

L = bay width,

A_b = beam's cross-sectional area,

t = thickness of the plate,

A_c = the cross-sectional area of the bounding column, and

I_c = moment of inertia of the bounding column;

4. Design beams and columns according to capacity design principles or other rational methods using plate thicknesses specified; and

5. Check story drifts against allowable values from the applicable building code.

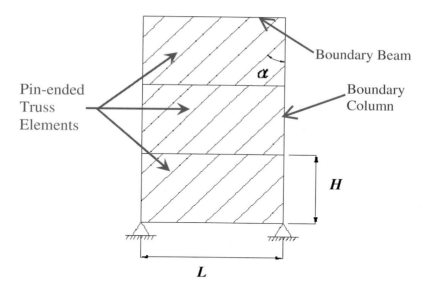

Fig. 4.9 Strip model of SPSWs.

4.3.2.2 Capacity design

■ Intermediate beams

In Fig. 4.8, the infill SPSW panels installed in the mid-bay of the SPSW-building frame develop tension field after buckling. If the difference in infill panel thickness is small or zero on adjacent floors, the forces from the yielding infill panels above and below the intermediate beam can be balanced or neglected. Otherwise, the design method should follow the approach of top and bottom beams.

■ Top and bottom beams

Considering the same SPSW frame pictured in Figure 4.9, if the beam-column connection in the mid-bay is pinned, the maximum moment M over the beam's length L, is:

$$M_{simple} = \frac{\omega_{max} \cdot L^2}{8}$$ (4.3)

where $\omega_{\max}$ is the maximum vertical load from the given infill panel's strength and thickness. Note that the simply-supported beams for the bay installed with SPSWs could decrease the demand in the beam size compared with the fix-ended beams, but the former could not provide any contribution towards seismic energy dissipation and lateral force resistance.

For a given infill panel, the maximum moment located at the end of a fully fix-ended beam is [10]:

$$M_{fixed-end} = \frac{\omega_{\max} \cdot L^2}{4} \qquad (4.4)$$

This means that a fully fix-ended beam will have twice the flexural strength of a simple one. In order to decrease the beam size and to provide better seismic energy dissipation, the *reduced beam section* (RBS) or "*dog-bone*" will be a good idea to enhance the performance of the SPSW anchor beam [11]. The design detail of RBS is well-described in FEMA 350.

Further, note that due to the tension field action, the maximum vertically distributed load of infill panel applied on top or bottom beam can be related to the panel properties by:

$$\omega_{\max} = F_u \cdot t_p \cdot \cos^2 \alpha \qquad (4.5)$$

where t_p is the panel thickness, α is the panel tension field angle of orientation with respect to the vertical, F_u is the panel's ultimate tension strength, say $1.8 F_{yp}$ for LYS170, and F_{yp} is the panel's yield strength.

■   Adjacent columns

Under monotonically increasing lateral forces and considering important details like the strong-column or weak-beam behavior, the yield drift limit of frame with SPSWs and the yield displacement of an infill panel [7], and neglecting the contribution from joint panel zone as well as frame member shear and axial deformations, the required boundary column section modulus for a new frame design situation is [10]:

$$S_{xc} \geq \left(\frac{t_p \cdot H^2 \cdot \sin^2 \alpha}{12} + \frac{12}{\sin 2\alpha \cdot \left(\frac{H}{I_c} + \frac{L}{I_b} \right)} \right) \frac{F_{yp}}{F_{yc}} \qquad (4.6)$$

where

H is frame's bay height between beam's centerlines; L, width,

I_c is the moment of inertia of the column; I_b, for the beam,

$S_{xc} \cdot F_{yc}$ is the column's section modulus and yield strength, respectively; $S_{xb} \cdot F_{yb}$, for the beam,

E is the elastic modulus of steel; μ_p is the target panel ductility, and

α is the assumed angle of inclination of the tension field measured from the vertical.

The above equation by Vian and Bruneau is derived from the concept that the panel yield drift is smaller than or equal to the yield drift of SPSW frame. The drift refers to the frame which will not yield while panels are yielding. If the target panel ductility, μ_p, is required at initial frame yield, then the (4.6) will become the following expression:

$$S_{xc} \geq \left(\frac{t_p \cdot H^2 \cdot \sin^2 \alpha}{12} + \frac{12 \cdot \mu_p}{\sin 2\alpha \cdot \left(\dfrac{H}{I_c} + \dfrac{L}{I_b} \right)} \right) \frac{F_{yp}}{F_{yc}} \qquad (4.7)$$

Moreover, the inequality of (4.6) can be solved to obtain the infill panel thickness for a retrofit situation if the other properties are known (Vian and Bruneau, 2005):

$$t_p \leq \left(S_{xc} \cdot F_{yc} - \frac{12 \cdot F_{yp} \cdot \mu_p}{\sin 2\alpha \cdot \left(\dfrac{H}{I_c} + \dfrac{L}{I_b} \right)} \right) \frac{12}{F_{yp} \cdot H^2 \cdot \sin^2 \alpha} \qquad (4.8)$$

4.3.2.3   Analytical method

Strip model

In the resistance of an SPSW structure against lateral forces, infill panels will develop the tension field action after they buckle. The simplified method to analyze SPSWs is the *"Strip Model"* presented by [7]. In the model, the infill panel can be represented as a series of inclined truss elements with zero compressive yield stresses as shown in Fig. 4.9. All strips have pin-ended connections and in the same orientation as the tension field. A minimum of 10 strips in each infill panel is recommended.

The cross-sectional area of a strip, A_s can be obtained by the following equation:

$$A_s = \frac{t_p \cdot (L \cdot \cos\alpha + H \sin\alpha)}{n} \tag{4.9}$$

where n (≥ 10) is the number of strips per panel.

According to the FEMA450, two of the analytical approaches are suggested to achieve capacity design and determine the same forces acting on the boundary columns, and these are represented as follows.

■   Non-linear push-over analysis

A model of SPSW (substituted by SSPW in FEMA450) can be constructed where bi-linear elasto-lastic panel elements of strength $RyFypAs$ are introduced in the direction α. Bi-linear plastic hinges can also be introduced at the ends of the boundary beams (substituted by horizontal boundary elements in FEMA450).In this model, standard push-over analysis conducted will provide axial forces, shears, and moments in the boundary frame when the infill panels are yielding. Separate checking is required to verify if plastic hinges do not develop in the boundary beams, except at their ends.

■   Combined linear elastic computer programs and capacity design concept

The following four-step procedure provides reasonable estimates of forces in the boundary elements of SPSW systems.

1. Lateral Forces: Use combined model of boundary elements and infill panels to come up with their demands based from the code's required base shear. The infill panels shall not be considered as vertical load-carrying elements.
2. Gravity Loads (Dead Load and Live Load): Apply gravity loads to the model with gravity frames only. Also, the infill panels shall not be considered as vertical load-carrying elements.

3. Without any overstrength factors, design the boundary elements using the demands based from the combination of forces in steps 1 and 2.
4. Boundary Element Capacity Design Check: Check the boundary element for the maximum capacity of the infill panels together with the maximum possible axial load due to over turning moment.
5. Use the axial force obtained from step 1 above and multiplies by overstrength factor Ω_o. Apply load from infill panels ($RyFypAs$) in the direction of α. For this capacity design check, use material strength reduction factor of 1.0.

Other details about lateral forces in the analytical method — Indirect Capacity Design Approach proposed by the CSA-S16-02 (CSA 2002) — is also described in FEMA450 Part2.

4.3.2.4 Seismic design of buckling restrainers

The buckling restrainers which are different from the welded stiffeners on the boundary frame may utilize steel or reinforced concrete members to be bolted on each side of the infill panels in any direction. The locations of bolts should be aligned with the direction of the tension field to exhibit the least loss of tension field strips. The required direction can be obtained by Eq. (4.2). Figures 4.10-4.12 depict the experimental applications of buckling restrainers on SPSW frames.

Fig. 4.10 SPSW specimen 2T with two horizontal steel-tube buckling-restrainers.

Fig. 4.11 SPSW specimen 3T with three horizontal steel-tube buckling-restrainers.

The maximum shear strength of a SPSW can be estimated by (Berman and Bruneau, 2003):

$$V_{base} = \frac{1}{2} F_u \cdot t_p \cdot L + \frac{4 \cdot M_p}{H} \qquad (4.10)$$

Fig. 4.12 SPSW specimen CP with reinforced-concrete buckling-restrainers.

where F_u, t_p, L, and H are the ultimate tension stress of LYS steel panel obtained by coupon test, panel thickness, width of the frame, and height of the frame, respectively. The width and the height are measured on center. M_p refers to the plastic moment of the reduced beam section (RBS).

It is approximated that the total out-of-plane force of panel is about 3% of the maximum shear strength of the SPSW. The tributary out-of-plane load on the restrainer, ω_p, is obtained by distributing first the total out-of-plane force to the whole panel area and then considering the tributary height of the horizontal restrainers. After calculating ω_p, the maximum flexural demand is computed as follows:

$$M_{\max} = \frac{\omega_p \cdot l_{res}^{\;2}}{8} \qquad (4.11)$$

where l_{rbs} is the length of the restrainer. After the restrainers are installed, the shear strength, V_b at which the buckling of a perfect plate begins is governed by the equation [12]:

$$V_b = \frac{\kappa \cdot \pi^2 \cdot E \cdot l_p \cdot t_p}{12(1 - \mu^2)(h/t_p)^2} \qquad (4.12)$$

where l_p is the panel's length, and h is the clear height measured between the restrainers. In addition, the shear strength of tension field action, V_{tf} is given by [9]:

$$V_{tf} = \frac{1}{2} F_{yp} \cdot t_p \cdot l_p \qquad (4.13)$$

where F_{yp} is the yield tension stress of LYS steel panel obtained from coupon test. The arrangement of the restrainers in Specimen 2T is meant to increase V_b so that its value falls between the V_{tf} and the shear yield strength, V_y, of the panel which is equal to:

$$V_y = 0.55 \cdot F_{yp} \cdot t_p \cdot l_p \qquad (4.14)$$

Whereas, the arrangement of the restrainers in Specimen 3T is meant to increase V_b so that its value exceeds V_y, while in Specimen CP is to restrain the entire infill steel panel.

4.3.3 *Experimental responses of buckling restrained SPSWs*

Several steel moment-resisting frames, consist of 345 MPa steel members, were 4000-mm wide and 2000-mm high measured on center (Fig. 4.13a). The LYS infill panels were 3-mm thick with initial yield strength of 131MPa. All specimens adopted RBS at each end of beams. A solid-type SPSW (Specimen S) is shown in Fig. 4.13.

Specimen 2T was constructed with two horizontal rectangular-tube (100 x 50 x 4.5 mm) restrainers on each side of the infill panel, as pictured in Fig. 4.10 while Specimen 3T was constructed with three horizontal restrainers on each side of the infill panel, as shown in Fig. 4.11. Specimen CP was restrained by three reinforced-concrete panels on both sides of the infill steel panel. In one side of the wall, these three reinforced concrete panels were separated by two welding tracks on the infill panel. The provision of restrainers is meant to increase the shear buckling strength obtained by Eq. (4.12) and to have a less pinched hysteresis during the cyclic loadings. The solid panel, Specimen S, as shown in Fig. 4.13(b), was compared with those constructed with restrainers.

All specimens were tested using a cyclic pseudo-static loading protocol similar to ATC-24. Loading history was displacement-controlled, and the loads were applied horizontally to the top beam using four actuators.

The shear force versus lateral drift deformation relationships of Specimen 2T are illustrated in Fig. 4.14(a) and Fig. 4.14(b) shows the buckled panel. Some plastic folds were formed and distributed on the panel. Small fractures were found in the corners of the panel after the test. The RBS connections localized all beams yielding to those regions as desired. The test ended up with cracks in RBS at 4% radian interstory drift. The restrainer performed well and the zones of panel restrained by steel tubes were still very flat after the test (Fig. 4.15).

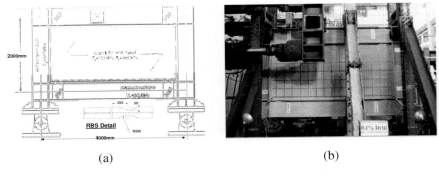

(a)   (b)

Fig. 4.13 Typical SPSW frame (Specimen S).

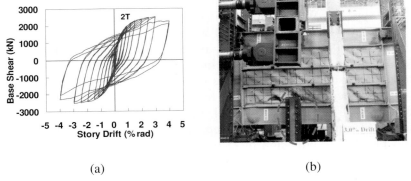

(a)   (b)

Fig. 4.14 (a) The hysteretic responses and (b) the buckled panel of Specimen 2T.

Fig. 4.15 Picture after removing restrainers
of Specimen 2T.

The relationship of force versus deformation of Specimen S is shown in Fig. 4.16(a). Figures 4.16(b) and 4.17 show the buckled panel and the yielded RBS connection following the test, respectively. Only four significant plastic folds were formed in the panel and near its corner where small fractures were found also. The RBS connections localized all beams yielding to those regions. The test ended up with cracks in RBS at 3% radian drift.

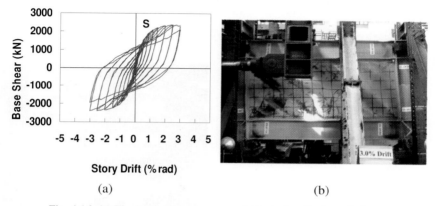

(a)   (b)

Fig. 4.16 (a) The hysteretic response and (b) the buckled panel after the test of Specimen S.

Fig. 4.17 RBS yielding of Specimen S.

Specimen 3T was restrained by three horizontal rectangular tubes on both sides of the panel. Its behavior of force versus deformation is shown in Fig. 4.18(a). Figure 4.18(b) shows the buckled panel of Specimen 3T. Plastic folds were formed and distributed in the panel's area. Small fractures were found also in its corners after the test. The

RBS connections localized all beams yielding to those regions, as designed. The test ended up with cracks in RBS at 4% radian drift.

Specimen CP was restrained by three reinforced-concrete panels on both sides of the infill wall. In Fig. 4.19 shows the relationship of force versus deformations. Figure 4.20 shows the buckled panel of Specimen CP. There were also plastic folds formed and distributed around the edge of restrained zone of the infill panel. Small fractures were found in the corners. The RBS connections localized all beam yielding to those regions as designed. The test ended up with cracks in RBS at 4% radian drift.

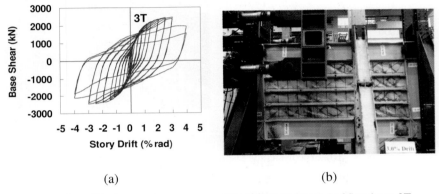

(a)   (b)

Fig. 4.18 (a) The hysteretic response and (b) the buckled panel of Specimen 3T.

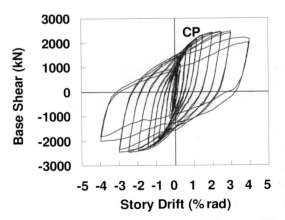

Fig. 4.19 Hysteretic response diagram of Specimen CP.

(a)

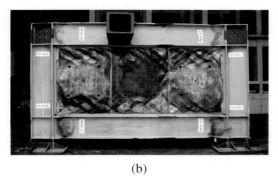

(b)

Fig. 4.20 Buckled panel of Specimen CP: (a) before disassembling the RC panels;
(b) after disassembling the RC panels.

Figure 4.21 shows the cumulative energy dissipation of the four specimens mentioned above. Comparing with Specimen S, Specimens CP, 2T, and 3T dissipated more energy. Figure 4.22 shows the differences of cumulative energy dissipation. In the first three cycles with 0.1% radian drift, the average energy dissipation of Specimens 2T and 3T were double than that of Specimen S. The decrease in the differences of cumulative energy dissipation for specimens 2T, 3T and CP (with the increasing story drift of 3% radian) in the last two cycles was still about 25% more than that for Specimen S. Interestingly, the cumulative energy dissipation of Specimens 2T and 3T were very similar noting that the additional short lateral brace was installed during the test of Specimen 2T (Fig. 4.14(b)) and was absent during the test of specimen 3T (Fig. 4.18(b)). After the Specimen CP's ninth cycle (0.3% radian story drift), the RC panels started to crack and developed consequently tension fields and started to dissipate more energy (twice that of energy dissipation in specimens 2T and 3T and more than 100% of energy dissipated in specimen S).

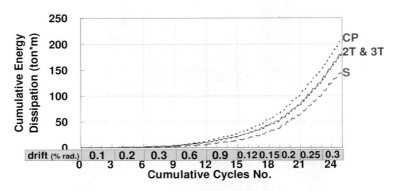

Fig. 4.21 The cumulative energy dissipation vs number of cycles.

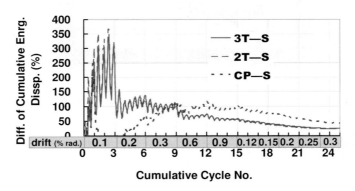

Fig. 4.22 Difference of cumulative energy dissipation.

From the results, Specimen CP possesses the best ability of energy dissipation because reinforced-concrete panels provide better effect on restraint of buckling in the infill panel. Though RC panels have better performance on restraining buckling, rectangular steel-tube restrainers are used on the specimens 3T and 2T and have shown better performance on the serviceability because of the less out-of-plane displacement than that of Specimens S or CP (Fig. 4.23).

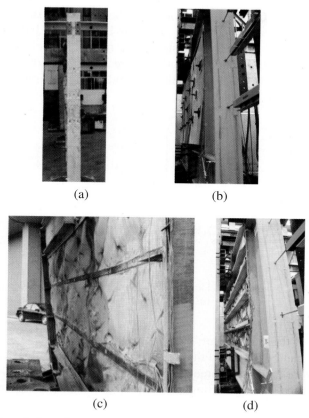

(a)   (b)

(c)   (d)

Fig. 4.23 Out-of-plane displacements after tests: (a) Specimen S, (b) Specimen CP, (c) Specimen 2T, (d) Specimen 3T.

4.4 Buckling-Restrained Braced Frame System

4.4.1 *General behavior of buckling-restrained braces (BRBs)*

Buckling restrained braced frame (BRBF) has been evolved as a very effective system for cases of severe seismic applications [13]. Buckling restrained braces (BRBs) or Unbonded Braces (UBs) are made from encasing a steel-core, cross-shaped or flat bar member, into a steel tube confined with infill concrete (Fig. 4.24(a)). The steel-core member is designed to resist the axial forces with a full tension or compression yield capacity without experiencing the local or global flexural buckling failure. When the brace is subjected to compression, an unbonding

material placed between the core member and the infill concrete is
required to reduce the friction. Thus, a BRB or an UB is composed
basically of three components: steel core member, buckling restraining
part, and the unbonding material. Figure 4.24b illustrates the typical
cross sections of the BRBs or UBs proposed by various researchers.

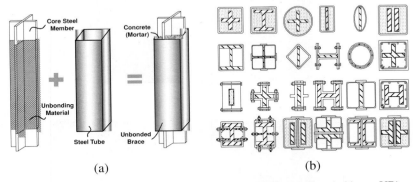

(a)   (b)

Fig. 4.24 (a) Scheme of buckling-restrained brace (BRB, or unbonded brace UB).
(b) Typical types of BRBs.

Most of these BRBs are proprietary but their concepts are essentially
similar. It can be observed that the cross section of the steel core member
is usually bi-axially symmetric which can be a cruciform, an *H* or a
flat-bar shape. The buckling restraining part can be constructed from
mortar-filled tube, reinforced concrete, reinforced concrete covered with
FRP, or all-metallic steel tubes. It has been found out that a single-core
brace-to-the-gusset connection is made typically of a butt joint using
several splice plates and two set of connecting bolts as illustrated in
Fig. 4.25.

In order to reduce the length and the number of bolts in the brace-
to-gusset connection, the *double-core buckling-restrained braces*
(DCBRBs) as illustrated in Fig. 4.26a have been developed [14] and
extensively tested [14-17] at the National Taiwan University (NTU) and
at the Taiwan National Center for Research on Earthquake Engineering
(NCREE) in the past few years. The proposed BRB members can be
connected conveniently to the gusset plate in the same manner as in the
traditional double-tee brace-to-gusset plate connections (Figure 4.26b).
Those researches include the investigation of the effective unbonding
material [14], and the cyclic and fatigue performance of the DCBRBs

employing mortar-filled double tubes for A36 [15] and A572 GR50 [17] steel cores. The design criteria of the brace-end connections and the inter-connecting ties between the two tubes have been established also through experimental tests [16]. Results have confirmed also that the self-compact concrete, much more cost-effective than the cement mortar, is a satisfactory alternative to restrain the core steel member in the BRBs [17].

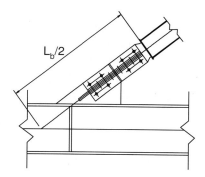

Fig. 4.25 Splicing detail for single-core BRB end connections.

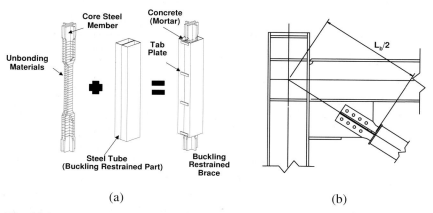

Fig. 4.26 (a) Scheme of DCBRB and (b) the detail of double-tee-to-gusset connections.

4.4.2 Seismic design of BRBF

4.4.2.1 Effects of various unbonding materials

In order to find out which kind of materials satisfactorily possesses unbonding effects, ten BRBs were tested under cyclically increasing displacements at NTU [14]. Table 4.1 summarizes the characteristics of unbonding materials and the corresponding cyclic loading protocol used for each specimen. The adopted standard loading protocol was referred

Table 4.1 Specimen schedule for the tests of unbonding materials.

Specimen (1)	Unbonding Material (2)	Unbonding Material Thickness (3)	Loading History (4)
AS-1	Asphalt Paint	N. A.	Standard
VF-1	Vinyl Sheet + Foaming Tape	2 mm	Standard
VK-1	Vinyl Sheet + Kraft Tape	2 mm	Standard
R2-1	Rubber Sheet	2 mm	Standard
R5-1	Rubber Sheet	5 mm	Standard
SR1-1	Silicone Rubber Sheet	1 mm	Standard
SR2-1	Silicone Rubber Sheet	2 mm	Standard
SR2-2	Silicone Rubber Sheet	2 mm	Low-Cycle Fatigue
SR2-3	Silicone Rubber Sheet	2 mm	Near-Fault
SR5-1	Silicone Rubber Sheet	5 mm	Standard

to that provided by SAC [1] (1997). The test setup is given in Fig. 4.27(a).The typical cyclic response of the specimen is given in Fig. 4.27(b). The results are summarized in Fig. 4.28.

For the purpose of discussion, the axial load difference Γ is defined as:

$$\Gamma = \frac{(C_{max} - T_{max})}{T_{max}} \qquad (4.15)$$

where C_{max} and T_{max} are the maximum compressive and tensile bracing forces at the same absolute axial deformation level, respectively.

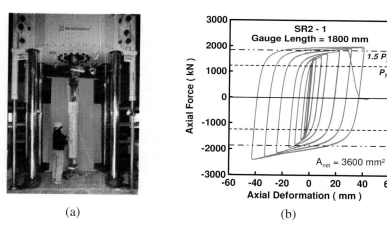

(a)   (b)

Fig. 4.27 (a) Experimental setup for BRB tests and (b) the cyclic response of the specimen SR2 using a 2-mm silicone rubber sheet unbonding material.

Theoretically, after the core member has yielded, the Poisson's ratio $\nu = 0.5$, can be applied in the following calculations and the volume of the yielding steel segment should remain constant, that is:

$$A_0 \cdot L_0 = A \cdot L \qquad (4.16)$$

where A_o and L_o correspond to the original core cross sectional area and length, respectively, while A and L correspond to those after the brace

---

[1] SAC is a joint venture of the Structural Engineers Association of California (SEAOC), the Applied Technology Council (ATC), and California Universities for Research in Earthquake Engineering (CUREe), formed specifically to address both immediate and long-term needs related to solving the problems of the Welded Steel Moment Frame (WMSF) connection. (*October 1997*).

deformed in either tension or compression. Therefore, it can be shown that the axial strain is:

$$\varepsilon = 1 - \frac{L_0}{L} = 1 - \frac{A}{A_0} \quad \text{and} \quad A = A_0(1-\varepsilon) \tag{4.17}$$

Thus, the absolute ratio of the compressive to the tensile brace forces for a given strain level could be:

$$\Gamma = \frac{(C_{max} - T_{max})}{T_{max}} = \frac{A_0(1+\varepsilon) - A_0(1-\varepsilon)}{A_0(1-\varepsilon)} = \frac{2\varepsilon}{1-\varepsilon} \approx 2\varepsilon \tag{4.18}$$

Eq. (4.18) suggests that Γ is about 4 % for a $\varepsilon = 2$ %. But the test results in Fig. 4.28 exhibit much higher Γ values (the maximum is about 30% for $\varepsilon = 2$ %). This may be due to the imperfect unbonding mechanism that creates a substantially-developed friction between the steel core member and the buckling restraining part. It can be observed in Fig. 4.28 that the 2-mm thick silicone rubber sheet has the least axial load difference (about 10% at an axial strain of 2%) under the cyclically increasing displacements. Therefore, for the subsequent tests of BRBs unless specified, the 2-mm thick silicone rubber sheets were adopted for the construction of most of the specimens. In Figure 4.27b, it should be noted that the strain hardening effects are evident comparing to the tensile yield capacity computed from the tensile coupon strength.

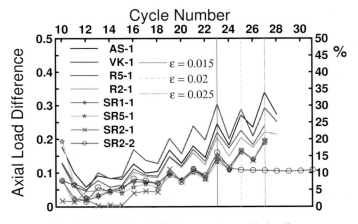

Fig. 4.28 Axial load difference under cyclic loading.

4.4.2.2 Key mechanical properties of the BRBs

In order to confine properly the BRB's inelastic deformations inside the restraining tube, the cross-sectional area (A_c) of the energy dissipation core segment (L_c) is smaller than that of the end-joint segments (L_j). A schematic illustration of a DCBRB in the frame is shown in Fig. 4.29(a), in which L_c and L_{wp} represent the core length and the work-point-to-work-point length, respectively. Between the end-joint and the core segment, a transition region (L_t) (as illustrated in Fig. 4.29(b)) can be devised.

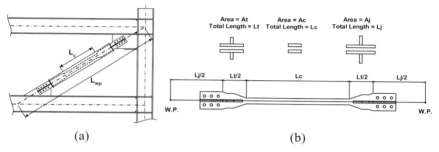

(a)   (b)

Fig. 4.29 (a) Dimensions of core length and work-point-to-work-point length and (b) profile of steel-core member of the double-core BRBs.

It is confirmed from the tests that the effective stiffness, K_e, of the BRB considering the variation of cross-sectional area along the length of the steel-core member can be predicted accurately by [15]:

$$K_e = \frac{EA_j A_c A_t}{A_j A_t L_c + A_c A_t L_j + A_c A_j L_t} \tag{4.19}$$

The relationships between the brace's overall strain (ε_{wp}) and the inter-story drift, θ can be approximated as:

$$\varepsilon_{wp} = \theta \sin 2\phi / 2 \tag{4.20}$$

where ϕ is the angle between the brace and the horizontal beam as illustrated in Fig. 4.30(a). The strain-to-drift ratio versus the beam angle ϕ relationships given in Eq. (4.20) are plotted in Fig. 4.30(b). The ratio of the core length and the work-point-to-work-point length is:

$$\alpha = L_c / L_{wp} \tag{4.21}$$

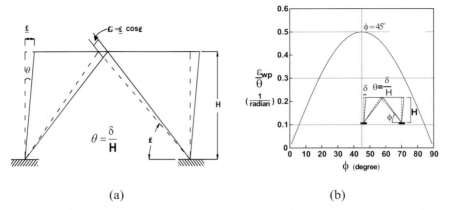

(a)   (b)

Fig. 4.30 (a) Schematic drawing of brace deformation versus inter-story drift angle
and (b) the relationship of brace strain to story drift ratio versus brace's angle.

Assuming the strain outside the core segment is negligible, the BRB core
inelastic upper-bound strain, ε_c can be expressed as:

$$\varepsilon_c \leq \varepsilon_{wp} / \alpha \qquad (4.22)$$

From Eqs. (4.20)–(4.22), it can be stated that if the inter-story drift
demand is 0.02 radian, then the peak core strain will be closed to 0.02 for
BRB with length aspect ratio $\alpha = 0.5$ oriented in a 45° angle.

The hysteretic responses of a typical DCBRB constructed with two
A36 flat-steel cores (Fig. 4.26(a)) and 2-mm silicone rubber unbonding
sheets are shown in Figure 4.31a. In the figure, the strain hardening
factor of about 1.5 is appropriate in estimating the peak tensile strength
for A36 core member. In addition, it appears that the following equation
should be applied in estimating the maximum compressive strength
developed possibly in a BRB:

$$P_{max} = \Omega \cdot \Omega_h \cdot \beta \cdot P_y \qquad (4.23)$$

where $P_y = A_c F_y$ is the nominal yield strength of the energy dissipation
core segment, Ω and Ω_h take into account the possible material's overstrength and the strain hardening effect of the core steel, respectively.
Also, the bonding factor β represents the imperfect unbonding behavior
of the material which causes the peak compressive strength to be greater
than the peak tensile strength during the cycles with large deformation.

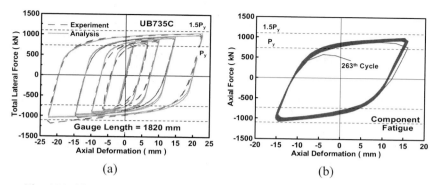

Fig. 4.31 (a) Experimental and analytical cyclic responses of a typical DCBRBs
and (b) its fatigue responses.

In order to prevent the BRB from a global flexural buckling, the required
stiffness of the steel casing is computed as follows [13]:

$$I_{tube} \geq FS \cdot \frac{P_{max}(kL)^2}{\pi^2 E} \qquad (4.24)$$

In Fig. 4.31(a), it is evident that for a properly fabricated BRB, a bonding
factor β of 1.1 is appropriate if the peak cyclic core strain demand is not
greater than 0.02. Further study, not described here, also indicated that
the experimental responses of the BRB can be represented accurately by
using the two surface plastic-hardening material model implemented for
the truss element in a general-purpose frame-response-analysis computer
program discussed by [18]. Figure 4.31(b) further confirmed that the
specimen can sustain a total of 262 cycles of large fatigue strain before
fracture after being subjected to SEAOC/AISC (2001) standard loading
protocols. Since the flexural buckling of a single-core BRB member
under large compressive strains could occur at a section near the end of
the steel tube as shown in Fig. 4.25 [19], it is recommended to meet the
following stability criterion for connection details shown in Figs. 4.25
and 4.26(b):

$$P_{e\_trans} = \frac{\pi^2 EI_{trans}}{(kL_b)^2} \geq P_{max} \qquad (4.25)$$

where P_{max} is given in Eq. (4.23) and EI_{trans} is the flexural stiffness of the
core member at a section near the end of the steel tube. As noted earlier,
the DCBRBs can be connected conveniently to the gusset plate as shown
in Fig. 4.26b so that the connection length is reduced to only one set of

connecting bolts required at each end of brace. The connection length can be reduced further if the three edges of each tee are welded to the gusset plate. An application example of the welded brace-end joint in a SRC building will be given later in this chapter.

4.4.2.3 Slip resistant bolted connection details in brace end double-tee- to-gusset joints

In 2002, six BRBs composed of double-plate core with double tee end details were tested at NTU [16] applying two different levels of roughness at the tee-to-gusset contacting surfaces. A set of cyclically increasing forces and displacements was applied to know the slip load of the brace-end bolted joints. Two additional sets, with similar cyclically increasing forces and displacements but reduced magnitude, were applied subsequently to confirm the cyclic degradation of the slip capacity. Test results suggest that the use of 1.5 times the strength of the bolt for design can prevent slip of the connection under cyclic loading reversals, expressed as:

$$R_{str} = N_s \cdot N_b \cdot (1.5 \cdot F_v \cdot A_b) \qquad (4.26)$$

where R_{str} is the slip-resistant strength of the connection, N_s is the number of slip surface, N_b is number of bolts, F_v is the nominal shear strength of the bolt, and A_b is the nominal tension area of the bolt. This research included investigations in both the longitudinal and transverse strain distributions in the tee's web connected to the brace-to-gusset joints, and concluded with several design recommendations for the end connection details of DCBRBs.

4.4.2.4 Tube-to-tube tie connection designs for DCBRBs

Since the proposed DCBRB is consisted of two independent units of bracing, the tie connection between the units can be made continuous or properly spaced. These ties may be composed of welded bars (tab plates) as illustrated in Fig. 4.32(a), or various possible bolting details as illustrated in Fig. 4.32(b). It is concluded that by using the Elastic Stability theory [20], the required strength P_{req} and stiffness β_{id} of the tie connection can be derived [16]:

$$P_{req} = \frac{3}{L_{tube}} (\frac{P_{max}}{2} - P_{cr}) \cdot (\frac{B \cdot \sigma_y}{E} + e) \qquad (4.27)$$

$$\beta_{id} = \frac{9}{2 \cdot L_{tube}} (\frac{P_{max}}{2} - P_{cr}) \cdot (1 + \frac{E \cdot e}{B \cdot \sigma_y}) \quad\quad 4.28)$$

where L_{tube} is the length of the buckling restraining tube, P_{max} is the maximum axial force suggested in Eq. (4.23) for the DCBRB, P_{cr} is the critical eccentric load of the single tube derived from Secant formula [20], B is the width of shorter side of the rectangular tube, E is Young's modulus of steel, σ_y is yield stress of steel tube, e is the eccentricity of load measured from the neutral axis of the single tube.

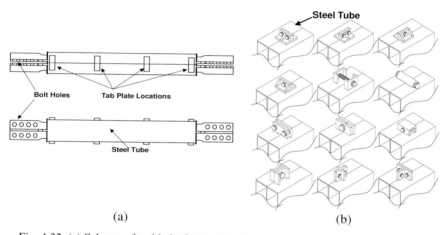

(a)   (b)

Fig. 4.32 (a) Scheme of welded tube-to-tube tie connections and (b) several possible tie-connection details.

For example, if the design strength P_y of the DCBRB is equal to 1960 kN, and the estimated maximum axial force $P_{max} = 4850$ kN, then according to the stiffness requirements noted in Eq. (4.28), the chosen size of each buckling restrained tube is 350×150×6 mm, with e = 32 mm, $L_{tube} = 4054$ mm, $B = 150$ mm, while P_{cr} can be calculated from the Secant formula as 810 kN. Based from Eqs. (4.27) and (4.28), the required strength, P_{req} and stiffness, β_{id} of the tie connections placed at every third of the tube's length are 38 kN and 306 kN/mm., respectively. Tests also suggested that the proposed strength and stiffness requirements developed for the tie connection elements between the twin tubes can be applied conveniently in the design and construction of the DCBRBs subjected to large inelastic strain reversals.

4.4.3 *Experimental responses of BRBFs*

4.4.3.1 Tests on single story V-shaped BRB frames

In order to assess the performance of the double-T-to-gusset connection details, three large scale single-bay V-shaped *buckling-restrained braced frames* (BRBFs) constructed with the proposed BRBs were tested at NCREE [15]. The primary objectives of the research include: (1) to investigate the experimental and analytical responses of the single-bay V-shaped BRBFs, each constructed with two BRBs in three different length aspect ratios, (2) to study the BRB steel-core strain versus inter-story drift relationships, and (3) to provide guidelines for the analysis and design of BRBF for severe seismic applications. Table 4.2 shows the schedule of the three frame specimen and one BRB component. Figure 4.33 shows the scheme of these specimens. Figure 4.34 shows the experimental setup for the frame test, and Fig. 4.35 shows the BRB-to-gusset connection details.

Table 4.2 Specimen schedule for BRB frame tests.

Frame (1)	BRB Specimen (2)	Yield Strength (kN) (3)	Total Bolts (F10T 24mm) (4)	α (5)	Peak Drift (% rad) (6)	Cycles for Fatigue Test (7)
NSYF	NSY880W	880	16	0.362	1.25	300
	NSY590E	590	12	0.362		
SYMF	SYM735W	735	16	0.371	1.5	300
	SYM735E	735	16	0.371		
SYMSCF	SYM735SCW	735	16	0.185	1.25	100
	SYM735SCE	735	16	0.185		
	UB735C	735	16	0.371	1.5	until Breaking

Figures 4.31(a) and 4.31(b) show the cyclic and the fatigue responses of the UB735, correspondingly. The experimental cyclic force versus deformation responses (Fig. 4.31(a)) of the BRB component exhibit

stable energy dissipation characteristics. Figure 4.34a also shows that experimental BRB responses can be predicted accurately using an inelastic truss element which incorporates two-surface plasticity model implemented in a general purpose finite-element computer program [18]. After a specimen undergone inelastic excursions as shown in Fig. 4.31(a), Fig. 4.31(b) indicates that this same specimen sustained a total of 262 cycles of fatigue strain in 0.0125 order before fracture. In Fig. 4.33(a), the arrangement of BRBs can be configured un-symmetrically in order to accommodate needs in architectural function, such as door or window openings. In this, the cross-sectional area and the length of the energy-dissipation segments of a pair of braces can be tailored specifically to avoid the potential unbalanced vertical load resultants while reaching the yielding point at the same time.

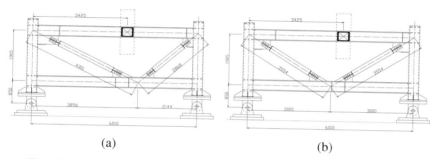

(a)   (b)

Fig. 4.33 (a) Elevation detail for Specimen NSYMF and (b) Specimen SYMF.

Fig. 4.34 Experimental setup for BRB frame test
(SYMF).

Fig. 4.35 Brace end double-tee-to-gusset connection details.

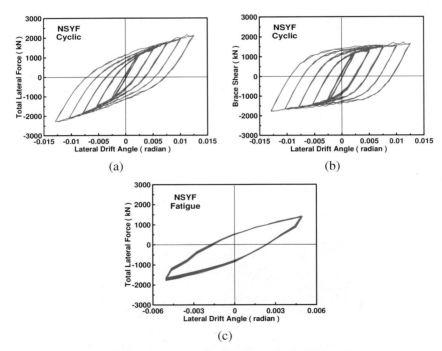

Fig. 4.36 (a) Total lateral force versus story drift, (b) total brace force versus story
drift and (c) fatigue performance (without failure) of NSYF BRB frame.

As shown from Fig. 4.36(a)–4.36(c), the cyclic or fatigue responses
of the frame systems and the BRBs indicate that the proposed BRB
components and framing system possess extremely stable characteristics

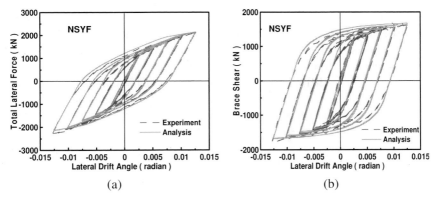

Fig. 4.37 Experimental to analytical relationships (a) between frame shear and lateral drift, and (b) between total brace shear and lateral drift in Specimen NSYF.

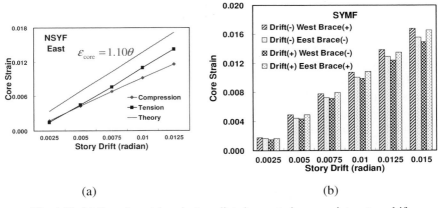

Fig. 4.38 (a) Experimental against predicted core strain versus inter-story drift relationships for the NSYF frames, and (b) Tensile strain versus compressive strain at peak inter-story drift for the SYMF frame.

(only Specimen NSYF is shown herein). In addition, as illustrated in Fig. 4.37, the inelastic finite element models predict accurately the frame and the BRBs' experimental lateral force versus story drift responses for all test frames (only Specimen NSYF is shown herein). In Fig. 4.38a, the three V-shaped BRBF tests confirm that the brace's steel-core strain demands can be predicted adequately by the story drift demands' geometry (Eq. (4.20)) and by including the ratio of the work-point-to-work-point length to the energy-dissipation core length (Eqs. (4.21) and (4.22)).

Test results in Fig. 4.38(a) and 4.38(b) reveal also that at a large story drift, the tensile strain in the tension brace is always greater than the compressive strain in the compression brace. This phenomenon is more likely when the inter-story drifts increase. This suggest that the peak compression and tension forces may have a tendency to self-equilibrate as story deformation increases and that these forces may reach a reduced unbalanced vertical force components resisted by the supporting beam member.

4.4.3.2   All-metallic and detachable BRBs

As cited above, single or double-core BRBs which are made from cement mortar or concrete filled in the steel casing exhibits satisfactory seismic resistant characteristics. However, as shown in Fig. 4.39(a), it does not allow visual inspection on the damage state of the inner core after an earthquake. To simplify the fabrication of BRBs made from infill mortar and unbonding coating, the all-metallic welded BRB (Fig. 4.39(b)) or the fully-detachable BRBs (Fig. 4.30) have been developed and tested extensively at NTU [17].

The detachable features as shown in Fig. 4.40 provide the possibility of disassembling the BRBs for inspection after an earthquake. To allow the extension and contraction of its two ends, a stopper (shown in Fig. 4.41) is locked in the restraining concrete or other buckling restraining part to prevent the restrainer from slipping off.

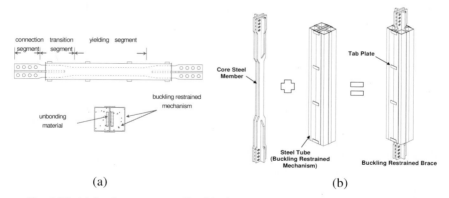

(a)   (b)

Fig. 4.39 (a) Steel cores are confined in the steel tubes without exposing the energy dissipating segments and (b) schematic diagram of welded all-metallic BRBs.

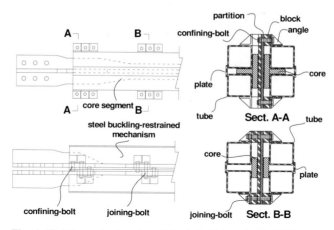

Fig. 4.40 Schematic drawing of the bolted detachable BRBs.

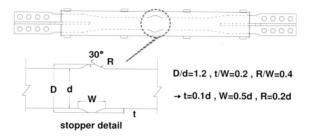

D/d=1.2 , t/W=0.2 , R/W=0.4

→ t=0.1d , W=0.5d , R=0.2d

Fig. 4.41 Details of the stopper (Peterson, 1974).

Instead of applying the unbonding coating, a 1-mm or 2-mm small gap has been placed experimentally between the surface of the inner core and the external restraining tubes. Test results indicate that the gap between the inner core and tubes is very effective in minimizing the difference between the peak BRB compressive and axial tensile force responses as depicted in Fig. 4.42(a). After applying a standard loading protocol, the specimen has never been fractured. The same specimen was subjected then to a constant fatigue strains until it failed. Test results suggest that the all-metallic and detachable BRBs can sustain stably the severely increasing cyclic and constant fatigue inelastic axial strain reversals. Since the specified constant strain varies, Fig. 4.43(a) shows the relationship of the number of fatigue cycles versus the fatigue strain

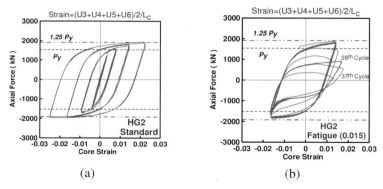

Fig. 4.42 Force versus strain for typical A572 GR50 BRBs in (a) the standard
cyclic loading test and (b) the fatigue test.

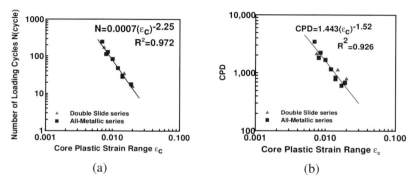

Fig. 4.43 (a) Number of fatigue-strain cycles versus core strain (A572 GR50 BRBs)
and (b) cumulative plastic deformation ($\times \Delta_y$) versus fatigue-strain relationships.

relationships. In addition, Fig. 4.43(b) shows the cumulative plastic
deformation (in terms of Δ_y). Tests confirmed that a typical fully-
detachable BRB can exhibit stable and excellent hysteretic behavior as
any other BRBs or UBs (restrained by mortar and steel casing) tested
before. It is noted also that the strain hardening factor for a BRB made
from A572 GR50 is about 1.25 for a peak core strain less than 0.025. In
this series of tests, it is found that the self-compact concrete (SCC) with a
compressive strength of 56 MPa is a cost-effective alternative to the
cement mortar used as filler in the steel casing.

4.4.3.3 Damage inspection and non-destructive testing of BRBs

As the energy dissipation segment of a conventional BRB is covered generally by the steel casing, direct damage detection may not be very straight forward. In order to inspect the BRBs after severe conditions, *non-destructive test* (NDT) techniques have been applied on more than 20 BRBs before, during, and after the load tests [17].

During various stages of each specimen's loading test, it is done by hitting simultaneously the BRB at one end and measuring the stress wave at the opposite end as illustrated in Fig. 4.44. The mentioned tests were conducted by hitting and observing the variation of stress wave form measurements collected from the BRB specimens. Receiver No. 2, shown in Fig. 4.44(b), is needed to normalize the impact of forces before comparing the impact and the response at the ends of a BRB. A typical set of stress wave form shown in Figure 4.45 was measured from Steps 1 to 5. The tests after hitting the specimen and measuring its wave form when the specimen was on the floor, on after connecting to the universal testing machine, and on after applying the standard loading protocol, were performed as Step 1, 2 and 3, respectively.

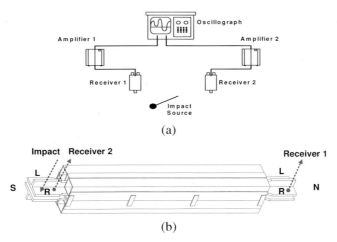

Fig. 4.44 (a) Schematic set-up of impact and stress wave
measurement and (b) locations of impact and measuring points.

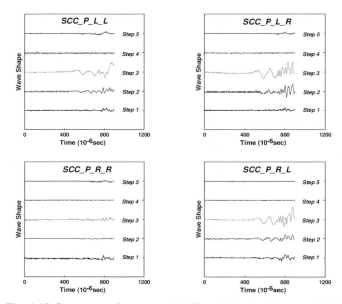

Fig. 4.45 Stress wave forms recorded from the impact tests on a BRB
using the self-compact concrete or SCC (Note: *P* stands for P wave,
L-R denotes the location of the Receivers 1 and 2 , see Fig. 4.67).

In addition, Fig. 4.45 suggests that the BRB has not been fractured after
the Steps 1 to 3. Step 4 was conducted after the same specimen has been
fractured under fatigue strain while Step 5 was done after the specimen
was removed from the load frame and examined on the floor. It is
evident in Fig. 4.45 that the stress wave forms are much calm after the
fatigue fracture test. These results suggest that the proposed NDT
technique can be a simple way to detect if inner core is fractured or not.
In addition, it is found in this study that the pressure wave (P-wave)
measurements are more satisfactory than those of the shear wave (S-
wave) for the stress wave NDT measurement for BRBs.

4.4.3.4   Pseudo-dynamic tests of a full-scale 3-bay 3-story CFT/BRB composite frame

The collaborative efforts of various researchers in Taiwan, Japan, and the
United States, a full-scale 3-story 3-bay RC column and steel beam RCS
composite moment frame was tested at NCREE in October 2002 using
networked pseudo-dynamic test procedures [21]. In October 2003, a full-

scale 3-story 3-bay composite braced frame (CFT/BRBF) consisted of CFT columns and BRB members have been tested similarly. The said model was designed for a highly seismic location either in Taiwan or in the United States using displacement-based seismic design procedures [22]. The typical bay width used was 7 meters and a height of 4 meters (Fig. 4.46(a)). The total height of the frame, including the footing, is about 13 meters. The 2150-mm wide concrete slab was adopted to develop the composite action of the beams. It measured 12 meters tall and 21 meters long and regarded as one of the largest frames ever tested (Fig. 4.46(b)). Three different types of moment connections, namely,

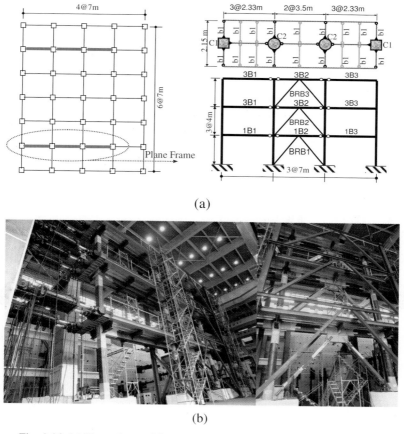

(a)

(b)

Fig. 4.46 (a) Floor plan and frame elevation and (b) experimental setup for pseudo-dynamic tests of a full- scale 3-story 3-bay CFT/BRB composite frame.

through-column beam, external diaphragm, and bolted-end plate, which were placed uniquely and as orderly from the first floor to the last, were fabricated for the exterior beam-to-column connections (Fig. 4.46(a)). Three types of BRBs, including the single-core, double-core and the all-metall BRBs, were adopted in the three different stories. In particular, two single-core unbonded braces, each consisting of a steel flat plate in the core, were donated by the Nippon Steel Company and were installed in the second story. The two BRBs installed in the third story are of double-core type constructed using cement mortar filled in the two rectangular tubes [22]. The frame has been tested using the pseudo-dynamic test procedures with input ground motions obtained from the 1999 Chi-Chi and 1989 Loma Prieta earthquakes which were scaled accordingly to represent 50%, 10% (DBE) and 2% (MCE) for 50-year seismic hazard levels.

Following the pseudo-dynamic tests and having none of the brace was fractured, quasi-static loads were applied to push cyclically the frame to have larger inter-story drifts until the braces failed. Being the largest and most realistic composite CFT/BRB frame ever tested in a laboratory, the test provides a unique data set to verify both computer simulation models and seismic performance of CFT/BRB frames. These two large scale frame tests have provided also great opportunities to explore international collaboration and data archiving envisioned for the

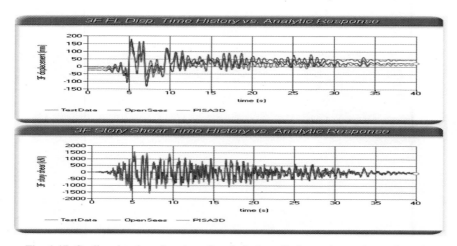

Fig. 4.47 On-line data broadcasting of analytical prediction and experimental results during and after the tests (http://cft-brbf.ncree.gov.tw).

Networked Earthquake Engineering Simulation (NEES) or the Internet-based Simulations for Earthquake Engineering (ISEE) research programs launched recently in USA and in Taiwan, respectively. The analytical predictions were broadcasted along with the real time experimental results during each test as can be seen in Fig. 4.47. Additional information can be found in the related web sites (*http://rcs.ncree.gov.tw* and *http://cft-brbf.ncree.gov.tw*).

4.5 Application of SPSWs and BRBFs for Seismic-Resistant Structure

4.5.1 *Application of SPSWs building*

Since 1970's, SPSWs have been used widely in Japan and in the U.S. for new buildings and for seismic retrofit of existing buildings. Some applications of stiffened or unstiffened SPSWs designed for structure as the primary lateral-load-resisting system are as follows:

- Nippon Steel Building shown in Fig. 4.48, a 20-story office building located in Tokyo completed in 1970, was the first building using steel plate shear walls. The SPSWs, 9' by 12'-2," are stiffened by horizontal and vertical steel channel stiffeners;
- The 53-story high-rise building in Tokyo with 10' × 16.5' SPSWs stiffened by T-shaped stiffeners welded on boundary frame;
- Also, a 31-story building using LYS (11.6-17.4ksi) SPSWs in Japan, and a 22-story office building (which SPSWs are primarily shop-welded and field-bolted) in Seattle, Washington;
- The new Sylmar Hospital — a 6-story hospital with reinforced concrete shear walls in the first two stories and SPSWs in the upper floors in Los Angeles, California; and
- A 30-story hotel (Fig. 4.49) in Dallas, Texas, with SPSWs in the transverse direction carrying about 60% of the tributary gravity load.[2]

---

[2] Astaneh-Asl, A. 2002. Seismic Behavior and Design of Composite Steel Plate Shear Walls. Steel TIPS Report, Structural Steel Educational Council, Moraga, CA.

Fig. 4.48 SPSWs are used in this Nippon Steel Building
(Photo from web page: office.mec.co.jp).

Fig. 4.49 A View of the 30-story SPSW-made building in Dallas, Texas
(Photo from web site: www.hyatt.com).

4.5.2 *Example applications of double-core BRBS in Taiwan*

In Taiwan, new projects that include retrofitting have selected the double-tube BRBs as the energy dissipating element to improve the seismic performance. The Shee-Hwa United World Tower, a 46-story office building in Tai-Chung designed before 1999, is being upgraded seismically to accommodate an increase in seismic hazard level as anticipated. About 80 pieces of double-tube BRBs are being installed in two opposite perimeter bays in the longitudinal direction of the frame. Similarly, a 10-story gymnasium at the Chinese Culture University gymnasium in Taipei has adopted a total of 96 pieces of large size double-core BRBs in its three-dimensional steel frame. The lateral-force-resisting structural system is consists of huge braced frames in the longitudinal direction and build-up truss moment-resisting frames in the transverse direction as shown in Fig. 4.50(a). In addition, the maximum length of the double-cored BRB is about 11 meters and for the A572 GR50 steel core plate size is 45 mm × 260 mm, placed in a concrete-filled 500 mm × 250 mm × 15 mm built-up steel box casing. The peak axial tensile and compressive strengths could reach almost 10,000 kN. In order to confirm the cyclic performance of the large size double-cored BRBs, four full-scale BRB components have been tested at NCREE (Fig. 4.50(b)). Test results illustrated in Fig. 4.51 confirm that the tested BRBs can sustain a large number of increasing inelastic cycles and constant fatigue strain reversals before failure. In addition, the

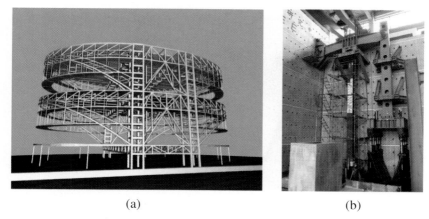

(a)   (b)

Fig. 4.50   (a) Three-dimensional model of the gymnasium, Taiwan.
(b) Experimental set up for a full-scale BRB tests.

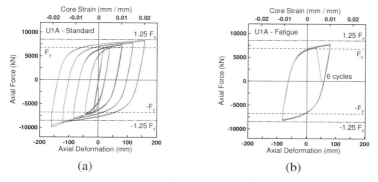

(a)   (b)

Fig. 4.51 (a) Cyclic force versus axial deformation
and (b) fatigue force versus core strain relationship.

Fig. 4.52 Construction of the Tzu-Chi Culture Building, Taiwan.

cumulative inelastic deformation exceeds $400D_{by}$ and meets well the prescribed requirements in SEAOC/AISC (2001).

Figure 4.52 shows the construction details of a 14-story Tzu-Chi Culture Building in Taipei. A total of 96 pieces of double-tube BRBs are being installed in the opposite perimeter bays along the transverse direction of this SRC structure. It is shown in Fig. 4.55 that the welded details have been adopted for the brace's end connections reducing effectively the size of the double tee-to-gusset plate joint.

References

1. Bjorhovde R., Development and use of high performance steel, Journal of Constructional Steel Research, Vol. 60, No.3, pp. 393-400, 2004.

2.   Miki, C., Homma, K., Tominaga, T. "High strength and high performance steels and their use in bridge structures." Journal of Constructional Steel Research, 58, pp. 3-20, 2002, Pocock, G., 2006, High strength steel use in Australia, Japan and the US, The Structural Engineer, Nov., pp. 27-30, 2002.

3.   Pocock, G., High strength steel use in Australia, Japan and the US, The Structural Engineer, Nov., pp. 27-30, 2006.

4.   Hunt, M. W., High-performance steel. (Editorial). Advanced Materials & Processes, Vol. 160, No. 11, pp. 4, 2002.

5.   Fine, Morris E. and Vaynman, Semyon, Development and Commercialization of High-Performance Steel, Northwestern University, Report No. A406, A426, A442, A454, 2003, 2003.

6.   Lwin, M. M., High performance steel designers' guide, 2nd ed., San Francisco (CA): Western Resource Center, Federal Highway Administration, US Department of Transportation, 2002.

7.   Thorburn, L. J., Kulak, G. L. and Montgomery, C. J., Analysis of Steel Plate Shear Walls, Structural Engineering Report No. 107, Department of Civil Engineering, The University of Alberta, Edmonton, Alberta, May 1983.

8.   Driver, R.G., Kulak, G.L., Elwi A.E., and Kennedy D.J.L., "Fe and Simplified Models of Steel Plate Shear Wall," Journal of Structural Engineering, ASCE, Vol. 124, No. 2, February 1998, pp. 121-130, 1998.

9.   Berman, J., and Bruneau, M., .Plastic Analysis and Design of Steel Plate Shear Walls. Journal of Structural Engineering, ASCE, Vol. 129, No.11, November 1, 2003, pp. 1448-1456, 2003.

10. Vian, D., Steel Plate Shear Walls for Seismic Design and Retrofit of Building Structures. Doctoral Dissertation, Department of Civil, Structural, and Environmental Engineering, State University of New York at Buffalo, September, 2005.

11. Vian, D., Lin, Y.C., Bruneau, M., and Tsai, K.C., Cyclic Performance of Low Yield Strength Steel Panel Shear Walls. Proceedings, The Sixteenth KKCNN Symposium on Civil Engineering, Kolon Hotel, Gyeongju, Korea, December 8-10, pp. 379-384, 2003.

12. Edwin H. G., Charles N. G., and James E. S., "Design of Steel Structure", 3rd Edition, McGraw-Hill, 1991.

13. Watanabe, A., Hitomoi, Y., Saeki, E., Wada, A., and Fujimoto, M., Properties of Brace Encased in Buckling-Restraining Concrete and

Steel Tube. Proceedings, Ninth World Conference on Earthquake Engineering, Vol. IV, Tokyo and Kyoto, Japan, August 2-9, pp. 719-724, 1988.

14. Lai, J.W. and Tsai, K.C., A Study of Buckling Restrained Brace Frames. Report No. CEER/R90-07, Center for Earthquake Engineering Research, National Taiwan University, Taipei, 2001. (in Chinese)

15. Huang, Y.C. and Tsai, K.C., Experimental Responses of Large Scale Buckling Restrained Brace Frames. Report No. CEER/R91-03, Center for Earthquake Engineering Research, National Taiwan University, Taipei, 2002. (in Chinese)

16. Weng, C.H. and Tsai, K.C., Experimental Responses of Double-Tube Unbonded Brace Elements and Connections. Report No. CEER/R91-02, Center for Earthquake Engineering Research, National Taiwan University, 2002. (in Chinese)

17. Lin, S.L. and Tsai, K.C., A Study of All Metallic and Detachable Buckling Braces," (in press) Center for Earthquake Engineering Research, National Taiwan University, 2003. (in Chinese)

18. Chang, L.C. and Tsai, K.C., The Platform and Visualization of Inelastic Structural Analysis of 2D Systems PISA2D and VISA2D. Report No. CEER/R90-08, Center for Earthquake Engineering Research, National Taiwan University, 2001. (in Chinese)

19. Tsai, K.C., Huang, Y.C., Weng, C.S., Shirai, T., and Nakamura, H., Experimental Tests of Large Scale Buckling Restrained Braces and Frames. Proceedings, Passive Control Symposium, Tokyo Institute of Technology, Tokyo, 2002.

20. Timoshenko, S.P., and Gere, J.M., 2nd Ed. Theory of Elastic Stability, McGraw-Hill Book Company Inc.:New York, 1961.

21. Chen, C.H., Lai, W.C., Cordova, P., Deierlein, G.C. and Tsai, K.C., Pseudo Dynamic Tests of a Full Scale RCS Frame: Part 1: Design, Construction, Testing. Structures Congress, ASCE, Nashville, 2004.

22. Tsai, K.C., Weng, Y.T., Lin, M.L., Chen, C.H. Chen, Lai, J.W., and Hsiao, P.C., Pseudo Dynamic Tests of a Full-Scale CFT-BRB Composite Frame: Displacement-Based Seismic Design and Performance Evaluations. Proceedings, International Workshop on Steel and Concrete Composite Constructions, National Center for Research on Earthquake Engineering, Taipei, Taiwan, 2003.

Chapter 5

Advanced Fiber Reinforced Polymer Composites

L. C. Hollaway

School of Engineering – Civil Engineering
University of Surrey, Guildford, Surrey, UK

5.1 Introduction

There has been great activity in the utilisation of advanced polymer composites (APCs) in the construction industry within the last fifteen to twenty years. The developments that have taken place during this period have been considerable and the requirements that have initiated this state have revolutionised some manufacturing techniques.

This chapter will demonstrate that the material has many advantages over the more conventional civil engineering materials. Furthermore, their mechanical and physical properties are such that when combined with reinforced concrete, steel, cast iron or timber structural members, they can produce a rehabilitated or retrofitted system superior to using most other conventional materials. 'All composite' unique structural systems can and have been produced which have decided advantages over other materials and forms of construction. Moreover, advanced polymer composites can be engineered to reflect their most advantageous properties. This chapter will discuss the properties of this material and will illustrate its advantages through practical examples

Before studying the subject of *advanced polymer composite materials*, it is essential for the reader to have a clear understanding of the meaning of that material. Therefore, the definition which was adopted, in 1989, by the Study Group [1], will be given here. It was developed, for the construction industry, from that produced by the

British Plastics Federation. The definition is as follows.

'Composite materials consist normally of two discrete phases, a continuous matrix which is often a resin, surrounding a fibrous reinforcing structure. The reinforcement has high strength and stiffness whilst the matrix binds the fibres together, allowing stress to be transferred from one fibre to others producing consolidated structures. In advanced or high performance composites, high strength and stiffness fibres are used in relatively high volume fractions whilst the orientation of the fibres is controlled to enable high mechanical stresses to be carried safely. In the anisotropic nature of these materials lies their major advantage. The reinforcement can be tailored and orientated to follow the stress patterns in the component leading to much greater design economy than can be achieved with traditional isotropic materials. The reinforcements are typically glass, carbon or aramid fibres in the form of continuous filament, tow or woven fabrics. The resins which confer distinctive properties such as heat, fibre or chemical resistance may be chosen from a wide spectrum of thermosetting or thermoplastic synthetic materials, and those commonly used are polyester, epoxy and phenolic resins. More advanced heat resisting types such as vinylester and bismaleimides are gaining usages in high performance applications and advanced carbon fibre/thermoplastic composites are well into a market development phase.'

The reinforcements are typically glass, carbon or aramid fibres but these names are generic names for fibres and it is vitally necessary to define which category of fibre is being used/described when discussing these materials. The fibres can be fabricated into various forms of continuous filament, tow or woven fabrics. The resins which confer distinctive properties such as heat, or chemical resistance may be chosen from a wide spectrum of thermosetting materials, and those commonly used are vinylester, epoxy, phenolic and sometimes polyester resins. More advanced heat resisting types such as bismaleimides are gaining usages in high performance applications and advanced carbon fibre/thermoplastic composites are well into a market development phase.

5.2 Reinforcement Mechanism of Fibre Reinforced Polymer Composites

5.2.1 *Introduction*

Advanced composite materials are manufactured by the managed distribution of two components, these are (i) a continuous matrix phase, (phase1), (ii) the fibre reinforcement phase, (phase 2). The boundary between the matrix and the reinforcement is controlled to obtain the desired properties from the given pair of materials.

5.2.2 *The polymer*

The matrix phase (the polymer) is an organic material composed of molecules made from many repeats of the same simpler unit called the monomer. The functions of the matrix phase are:

- To bind the fibres together and to protect their surfaces from abrasion, external influences and environmental degradation.
- To maintain the position of the fibres and to transfer stresses to the fibres by adhesion and/or friction. The adhesion to the fibres must be coupled with adequate matrix shear strength.

In addition, the following items must be fulfilled by the matrix:

- To maintain chemical and thermal compatibility with the fibre over the life span of the composite.
- To obtain complete wet out of the fibre by the matrix during the manufacturing process.
- To provide the colour and surface finish for architectural members.

Within the composite family there are two major types of matrices (polymers), both of which are used in construction. These polymers, the *thermoplastic* and *thermosetting*, require different procedures for their manufacture and the mechanical and in-service properties will be different; these are discussed in Section 5.2.3. Both types are composed of long chain molecule made by connecting many smaller molecules. Polymers will be either, (i) amorphous, which implies a random structure with a high concentration of molecular entanglement or (ii) crystalline, which has a high degree of molecular order or alignment. If heated, the random structure of the amorphous polymer will become disentangled and this will change the material from a solid to a viscous liquid, whereas heating the crystalline polymer will change it to an amorphous viscous liquid. A polymer which has a pure crystalline structure is difficult to attain because of the complex physical nature of the molecular chains, consequently, the polymers used in construction industry are a semi-crystalline material.

5.2.2.1   Thermoplastic polymer

The long chain molecules of this polymer are held together by relatively weak Van der Waals forces but the chemical valency bond along the chains is extremely strong, therefore, they derive their strength and stiffness from the inherent properties of the monomer units and the very high molecular weight. These polymers are generally of a semi-crystalline form.

The main thermoplastics polymers which are utilised in construction are the polyolefins and polyesters and these are used in geotechnical applications. The principal materials in this category are the geosynthetics in the form of geotextiles, geo-linear elements and geomembranes; reference can be found in [2,3]. The Polyacrylic fibre, which is a thermoplastic polymer, is used as the precursor for the manufacture of carbon fibres (Section 5.3.2).

5.2.2.2   Thermosetting polymers

This polymer is usually made from a liquid or a semi-solid precursor which harden irreversibly; this chemical reaction is known as polycondensation, polymerisation or curing and on completion, the liquid resin is converted to a hard solid by chemical cross-linking which produces a tightly bound three-dimensional network of polymer chains. The molecular units, the cross-links of the structure, forming the network and the length and density of these units will influence the mechanical properties of the material. The network and the length of the molecular units are a function of the chemicals used in the manufacture of the polymers and the cross-linking is a function of the degree of cure of the polymer. Furthermore, the degree of cure is a function of the temperature and length of the polymerisation period. The main polymers used in construction under this heading are the epoxies, the vinylesters, and occasionally, the unsaturated polyesters and the phenolics. These polymers will be used as the matrix material of the composite for structural components.

5.2.2.3 The elastomer

A third member of the polymer family is the elastomer used mainly in the form of a rubber, reinforced or un-reinforced. The material consists of long chain molecules, similar in form to the thermoplastic polymer, these long chains are coiled and twisted in a random manner and the molecular flexibility is such that the material is able to undergo very large deformations. Vulcanization is a curing process which enables it to recover from large deformations received under load. The molecules then become cross-linked in a similar way to that of the thermosetting polymer. The main use of this polymer is for elastomeric bridge bearings and resilient seatings in buildings.

5.2.2.4 Epoxies (Thermosetting polymers)

The most important epoxy resins are the low molecular weight polymers (oligomers), which are produced from the reaction of bisphenol A and epichlorohydrin. They range from the medium viscosity liquids through to high melting solids. In general, epoxies have:

- High specific strengths and dimensional stability.
- High temperature resistance and good resistance to solvents and alkalis but generally have weak resistance to acids.
- Superior toughness to that of the polyester resins and therefore they will operate at higher temperatures.
- Good adhesion to many substrates and low shrinkage during polymerisation. This allows mouldings to be manufactured of high quality and with good dimensional tolerance.
- Generally a high temperature resistance and can be used at temperatures up to 177°C. Some epoxies have a maximum temperature range up to 316°C.

Table 5.1(a), provides general and physical properties of epoxy polymers.

Table 5.1(a) General and physical properties of epoxy resins.

General information	General physical properties
Most common resin used in structural applications.	Broad range of formulations available.
Generally two part systems (Either hot or cold cure). (i) Base resin (Polymer) (ii) Hardener (Curing agent).	Can tailor to specific processing and cure conditions.
Mix ratios of manufacturers do vary and <u>must be followed carefully</u>	Higher performance requires an elevated temperature post-cure
	Superior mechanical properties and chemical resistance than other resins.
	Performance at elevated temperatures generally superior than other resins. (Depends upon post-cure history and glass transition temperature. (Tg).
	Lower degree of shrinkage on cure. (2-3%).
	Exotherm less problematic than other resins

5.2.2.5 Vinlyesters (Thermosetting polymers)

As vinylesters have unsaturated esters of epoxy resins, they have similar mechanical and in-service properties to those of the epoxies, but they have similar processing techniques to those of the polyesters. Generally, the vinylesters have:

- Good wetting characteristics and bond well to glass fibres.
- Good resistance to strong acids and strong alkalis.
- A processing method at both room and elevated temperatures.
- Reduced water absorption and shrinkage properties, as well as enhanced chemical resistance, compared to those of the polyester.

It is important to note that all polymers should be cured completely for maximum efficiency and this can be achieved by post curing the polymer (or composite if used as a structural member) at an elevated

Table 5.1(b) General and physical properties of vinylester resins.

General Information	General Physical Properties
Have an epoxy backbone, but polymerisation is similar to polyesters.	As with all polymer composites post cure should be undertaken.
Have improved mechanical properties over polyesters.	Has a high styrene content, which is given off during curing, must be managed. (Health & Safety Requirement).
Excellent chemical resistance.	More expensive than polyesters
High degree of shrinkage on cure.	Commonly used for the manufacture of FRP components to be used in harsh, chemical environments.

temperature for a few hours after its initial cure. Irrespective of the initial cure mechanism polymerisation generally reaches a level of cure no higher than about 90%, with the last part of cure cycle continuing very lowly. Incomplete cure can be the result of environmental conditions, incorrect stoichiometry of resin system components, or the failure to reach a sufficient temperature of cure. Incomplete cure will adversely affect mechanical properties, moisture absorption and susceptibility to moisture induced degradation of the resin and the fibre matrix interface.

Table 5.1(b), provides general and properties of vinylester polymers.

5.2.2.6 Unsaturated polyesters (Thermosetting polymers)

Currently, the polyesters are used in the manufacture of minor structural components for the construction industry; they have been largely replaced by the vinylester polymers for the manufacture of major structural components . However, they are readily processed and cured at ambient temperature. There are two commonly used polyesters; these are the orthophthalic and isophthalic resins. The former resin is the most commonly used but has low thermal stability and chemical resistance. The isophthalic resins, which contain isophthalic acids as an essential ingredient, are of superior quality with better thermal and chemical properties. There is a third polyester which is the bisphenol A, this is of a higher quality than the two former resins with a higher degree of hardness, chemical and thermal resistance and has a certain degree of flame resistance.

The polyester resin tends to have a high level of shrinkage at cure, particularly if styrene in monomeric form is used as a reactive diluent in the resin. This addition will have (i) an increase in hydrophobicity, thereby, effectively decreasing the level of moisture absorption and (b) an increase in shrinkage to levels of 5-19% by volume. This can result in significant micro-cracking in resin rich areas and high residual stresses in composites having high volume fractions. Table 5.1c, provides general and physical properties of polyester polymers.

Table 5.1(c) General and physical properties of polyester resins.

General Properties	General Physical Properties
Relatively inexpensive compared with vinylester or epoxy.	Cure at room temperatures, post cure important
Generally produce lower fibre volume fraction composites.	Rate of polymerisation (curing) reaction can be readily adjusted
Most commonly used with glass reinforcements.	Can be used to produce large, complex structures.
Catalyst and accelerator are added to initiate cure.	Exotherm issues limit the thickness that can be laminated in one shift
Monomers dissolved in styrene, and styrene emissions released during cure.	Can get high shrinkage on cure (7-8%)
Environmental legislation now more stringent on limits of emissions.	Have some chemical resistance but not as good as vinylesters.
Low emission resins have been developed but can be more difficult to work with.	

5.2.3 *The mechanical and physical properties of the polymer*

The mechanical properties of four polymers that are used in civil engineering are given in Table 5.2.

Before considering the mechanical characteristics of polymers (and polymer composites) a discussion will be given on their glass transition and curing procedures.

Table 5.2 Typical mechanical properties of four thermosetting polymers.

Material	Specific strength	Ultimate tensile strength (MPa)	Modulus of elasticity in tension (GPa)	Coefficient of linear expansion $(10^{-6}/^{0}C)$
Thermosetting				
Polyester	1.28	45-90	2.5-4.0	100-110
Vinylester	1.07	90	4.0	80
Epoxy	1.03	90-110	3.5	45-65
Phenolic	1.5-1.75	45-59	5.5-8.3	30-45

Glass Transition Temperature of Polymers The glass transition temperature (T_g) is the mid-point over the temperature range which an *amorphous* material changes from (or to) a brittle or vitreous state to (or from) a plastic state; *crystalline* materials do not have a glass transition temperature, they have a melting point. However, crystalline polymers will have some degree of amorphous structure. All polymers below the T_g, whether crystalline or not, are rigid and frequently brittle; they therefore, have both stiffness and strength. Above the T_g, amorphous polymers are soft elastomers or viscous liquids; consequently they have no stiffness or strength. If the polymer is crystalline it will range in properties from a soft to a rigid solid depending upon the degree of crystallinity. The epoxies used in construction would generally be in the amorphous state, with a small amount of crystalline structure. The following mechanical/physical properties of any polymer undergo a drastic change at the T_g.

(i)   hardness
(ii)   volume (volume ~ temperature is shown in Fig. 5.1).
(iii)   modulus of elasticity (Modulus of elasticity ~ temperature is shown in Fig. 5.1)
(iv)   percentage elongation-to-break.

Curing of Thermosetting polymers Section 5.4.2 discusses the manufacturing techniques of advanced polymer composites and gives details of the three main methods. The manual technique will invariably use a cold cure polymer with an ambient temperature cure; this category

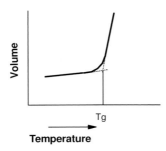 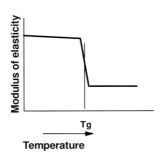

(i) Below the Tg (the glass transition temperature) the thermosetting polymers are glassy (rigid and frequently brittle) in natire but above the Tg the polymers are soft elastomers or viscous liquids.

(ii) Hot cured composites will have a higher Tg than the cold curing composites. The actual values will depend upon the temperature of cure. The Tg will be 10₀C - 15₀C above the curing temperature.

Fig. 5.1 Changes in properties at the glass transition temperature of any thermosetting polymer.

includes site fabricating. This initial cure will reach only 90% of full polymerization after some 10 days but this length of time will depend upon the ambient temperature, the lower this temperature the longer the initial cure. Ideally these polymers should be post cured at an elevated temperature for a certain length of time. The higher the temperature of post cure the less time is required for completion. Clearly, if the polymer has been undercured the long term properties of the polymer will be affected and the polymer will degrade. A further concern of the ambient cured polymer is that their T_g will be low. Typically this temperature will be about 15°C to 20°C above the cure temperature, which in some civil engineering environments is a low value.

The hot cure resin systems reinforced with fibres are fabricated in factory conditions and the temperature of operation of the composite forming machine will run at temperatures of 120°C to 130°C; this will be temperature at which the polymer will cure. Consequently, the T_g will have a value of about 150°C. If the ambient temperature of the final composite is placed in an environment greater than this value the polymer should also be post cured. It is noted that polymers have an upper limit for the T_g and if the post cure temperature were taken above this value the T_g could not be increased in value. This point is discussed in this Section under *Dimensional Stability*.

Dimensional stability is arguably the most important property of polymers. Thermosetting polymers will either soften, decompose or both when the temperature is raised above a critical value. As discussed earlier the thermosetting resins depend upon the cross-linking of their molecules for strength. If the temperature of the polymer increases, two phenomena occur; (i) as the polymer reaches it's Glass Transition Temperature (T_g) [see above] it begins to soften. The temperature at which this happens depends upon the detailed chemical structure of the polymer. The thermosetting polymers used in civil engineering are glassy (rigid and frequently brittle) in nature and will lose their dimensional stability above the glass transition temperature (T_g). (As stated above thermosetting polymers, used in construction, tend to be semi-crystalline). Consequently, polymer composite structural units should not be exposed to temperatures above the T_g value of the matrix material. Hot cured thermosetting polymers would normally have a higher T_g value than the room temperature cured thermosetting materials; their actual values would generally be about 10^0C-15^0C above the curing temperature. The T_g of *some* low temperature (ambient cured) moulded composites, can be increased in value by further post curing the polymer at a higher temperature than that of the original cure but there is a maximum value of the T_g, of all polymers irrespective of the post cure temperature value. Polymers with a high crystallinity will have a region of acceptable dimensional stability above T_g.

The stiffness of thermosetting polymers is influenced by the degree of cross-linking of that polymer.

The strength of thermosetting polymers will depend upon the type of loading applied to that polymer. The short term strength of a polymer is dependent upon the bonding between the molecular units and the length and density of these units, which in turn is dependent upon the degree of cross-linking of the structure of the polymer.

Toughness of a thermosetting polymer can be improved by blend filling or co-polymerising a brittle but higher stiffness polymer with a tough one. However, generally an increase in the toughness will decrease the stiffness of the polymer and a compromise between strength and stiffness of a material is always necessary.

Coefficient of thermal expansion of thermosetting polymers is an order higher than that of the more conventional civil engineering materials. This characteristic is extremely important in design when considering joining or bonding together two dissimilar materials. However, the bonding of polymers to civil engineering construction materials is not normally met in practice, but fibre/polymer composite materials are bonded to reinforced concrete or to steel structural members to enhance their stiffness and strength. With fibre/matrix composites the fibre, which has a much lower coefficient of thermal expansion compared to the matrix will stabilize the composite system and reduce the coefficient of thermal expansion to a value near to that of the conventional material. The coefficient of thermal expansion will vary with the environmental temperature range into which the polymer is placed. Furthermore, the degree of cross linking of the polymer will influence the rate of thermal expansion.

Thermal conductivity of all polymers is low and therefore they are good heat insulators. To improve the thermal conductivity still further the polymer can be used in the form of a foamed or aerated system in which cells are formed in the polymer by passing an aerating agent into the resin/hardener at the time of polymerisation. To reduce the thermal conductivity of a polymer it is possible to incorporate metallic fillers at the time of manufacturing the polymer.

Chemical resistance is the ability of a polymer to resist chemical attack and will depend upon the chemical composition and bonding in the monomer.

Solubility of a polymer is the property of the polymer which allows a solvent to diffuse into it. The polymer will not dissolve in a solvent unless the chemical structure of its monomer is similar to that of the solvent.

Permeability is the ability of polymers to allow gasses and other small molecules to permeate through it. It is likely that polymers with high crystallinity/density will have low permeability; a high degree of cross linking also reduces permeability. However, the ingress of moisture, if the polymer is permanently immersed in water or salt solution, will eventually permeate through the polymer [4-6].

Creep characteristics of polymers Thermosetting polymers have both the characteristics of elastic solids and viscous fluids and are classified as viscoelastic materials. Generally the ambient operating temperatures of many of these materials are close to their viscoelastic phase; consequently, the creep component of their long term carrying capacity becomes a significant consideration. In assessing the creep of a composite material it is vitally important to know its temperature and moisture environments and loading histories, the nature of the applied load, the dependence of the mechanical measurements on the rate at which these measurements are made. Furthermore, the direction of alignment, the type and the volume fraction of the fibres will affect the creep characteristic of a polymer composite. A basic requirement to minimise creep is to ensure that the service temperatures do not approach the glass transition temperature of the polymer. In addition, it is also dependent upon the time-dependent nature of the micro-damage in the composite material subjected to stress. Fibres such as glass, carbon and aramid have small if any creep component. Furthermore, the mechanics of creep in fibre/polymer composite materials are related to the progressive changes in the internal balance of forces within the materials resulting from the behaviour of the fibre, adhesion and load transfer at the fibre/matrix interface as well as from the deformation characteristics of the matrix. Figure 5.2 illustrates a family of creep curves consisting of the 100 s isochronous stress~strain creep curve (Fig. 5.2(a)), isostress creep curve (Fig. 5.2(b)), and Isostrain creep curve (Fig. 5.2(c)).

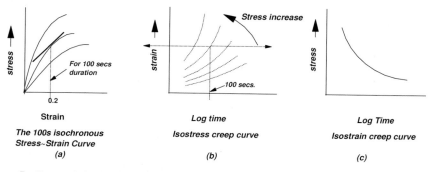

For fibre.matrix Composite Materials

As the fibre/matrix composite contains a visco-elastic polymer, this material will creep and similar forms of creep curves to those above will be formed but as the fibres do not creep, they tend to stabilise the polymer causing less creep in the composite.

Fig. 5.2 Creep curves — isochronous stress~strain curves (derived from BS 4618).

The creep strain would generally be measured as a function of time for a constant value of stress applied to the material. The recommended way to produce a set of creep curves in which the strains are measured at constant stress levels is to plot isostress creep curves. These curves are a function of time, in which a plot of strain against log time is produced (Fig. 5.2(b)), alternatively, a method which is recommended by BS 4618 is to produce isochronous stress/strain curves by cross-plotting, from (isostress) creep curves at constant times. This method yields a family of stress/strain curves, each relevant to a particular time of loading. As discussed in the book [2], this specification requires that the constant load tests are carried out under controlled conditions for the following durations 60 s, 100 s, 1 h, 2 h 100 h, 1 year, 10 years and 80 years. From Fig. 5.2(a), it will be seem that a family of creep curves, for any material, may be obtained by varying the stress. Isochronous stress/strain curves can then be drawn, and each will correspond to a specific loading direction. Thus a 100 s isochronous stress/strain curve implies that the total strain at the end of 100 s has been plotted against the corresponding stress level; the slope of this curve in not constant. The slope at any specific point on this curve will than define the creep modulus of the material.at that particular stress level.

To obtain creep data is time consuming and accelerated tests have been developed, one of these tests require the data from the creep experiments to be taken over a few days or weeks, and an extrapolation is then made beyond the test duration by mathematical formulation. Thus this procedure adopts the same law for the extrapolated period that existed at the commencement of the test; this assumption is not correct. Therefore, it is dangerous to extrapolate values greater than three times the length of the test duration. If a greater time value is required a more sophisticated approach should be used such as the time temperature superposition principle (TTSP) [7]. This approach has been discussed in the book [2]. By conducting a series of tests at different temperatures activation energy may be determined and through a kinetic approach of temperature-time superposition plots, a master curve can be deduced and predictions made. The time, applied stress superposition principle (TSSP) [8], can also be used for polymers and polymer composite materials.

5.3 The Fibre

The fibres that are mainly used in the construction industry for structural applications are the carbon, glass and aramid fibres. There are many other fibres that are on the market and are used under certain circumstances such as the synthetic fibres. These include the polyolefin family of polymers (viz. polypropylene, polyethylene, polyester and polyamide fibres). These fibres can be engineered chemically, physically and mechanically, at ambient temperatures, to produce applications for geotechnical engineering such as geosynthetic materials. These thermoplastic fibres and their applications will not be discussed in this chapter but further reading may be found in [2,9,10].

5.3.1 *The glass fibres*

Glass fibres are generally manufactured by the direct melt process, in which fine filaments of diameters 3-24 μm are formed during commercial production by continuous and rapid drawing from the melt; the temperature of the molten material is of the order of 1400^0C. During this process the molecules of the parent material are orientated into the direction of draw thus forming a series of filaments with exceptionally high specific stiffness and strength; it is the principal constituent of advanced composites. During the production stage strands, each consisting of 200 individual filaments, are produced and a surface treatment or sizing is applied before the fibres are gathered into strands and wound onto a drum. The production process of forming glass fibre is shown diagrammatically in Fig. 5.3. The strands are then formed into unidirectional fibres, chopped into short lengths of 50 mm to fabricate a randomly orientated mat of fibres or weaved into various woven fabrics.

Glass filaments are highly abrasive to each other and in order to minimize abrasion-related degradation of the glass fibre, particularly at the time of manufacture, a size is applied. The main functions of the sizing operation are thus, to reduce the abrasive effect of the filaments against one another, to minimise damage to the fibres during mechanical handling, and to act as a coupling agent to the matrix during impregnation.

Glass fibres are silica-based glass compounds that contain several metal oxides which can be tailored to create different types of glass. The main chemical composition of the most important grades of glass used in the construction industry is shown in Table 5.3. The main oxide is silica in the form of silica sand; the other oxides such as calcium, sodium and aluminium are incorporated to reduce the melting temperature and impede crystallization.

Glass fibres have excellent high strengths, and from a specific (strength) point of view are one of the strongest and currently provide the majority of structural composite components in the construction industry. The commercial grades of glass have strength values up to 4800 MPa.

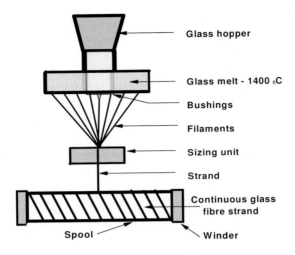

Fig. 5.3 Schematic representation for the manufacture of glass fibre strands.

Table 5.3 The chemical composition of E-glass and S-glass [11].

Glass material	CaO	Al_2O_3	MgO	B_2O_3	SiO_2
E-galss	17.5	14.0	4.5	10.0	54.0
S-glass	—	25.0	10.0	—	65.0

The fibre material is a relatively low cost material. The most important grades of glass are,

- *E-glass* which has a low alkali content of the order of 2%. It is used for general purpose structural applications and is the major one used in the construction industry; it also has good heat and electrical resistance.
- *S-glass* which is a stronger (typically 40% greater strength at room temperature) and stiffer fibre with a greater corrosion resistance than the E-glass fibre. It has good heat resistance. The S-2-glass has the same glass composition as S-glass but differs in its coating. The S-2-glass has good resistance to acids such as hydrochloric, nitric and sulphuric acids.
- *E-CR-glass* which has good resistance to acids and bases and has chemical stability in chemically corrosive environments.

- *R-glass* which has a higher tensile strength and tensile modulus and greater resistance to fatigue, aging and temperature corrosion to that of E-glass
- *AR-glass* which is an alkali-resistant glass and is used as the reinforcement for glass fibre reinforcement cement (GRFC). When glass fibres are used to reinforce cement, degradation in strength and toughness occurs when it is exposed to outdoor weathering, especially in humid conditions. This process can also take place with AR-glass, albeit at a much slower rate.
- *T-glass* which has improved performance compared to those of E-glass, with a 36% increase in tensile strength and 16% increase in tensile modulus. It has a 40% decrease in the coefficient of thermal expansion, and an increased heat resistance of 135^{0}C.
- *N-Varg* which is an alkali-resistant glass fibre. It is used mainly in composites with a cement based matrix.

The mechanical properties of the various types of glass fibre are given in Table 5.4

5.3.2 *The carbon fibre*

The carbon fibre is a high modulus and a high strength fibre which is used in the construction industry where the structural requirements of a system may need a material which has a higher modulus value and a higher fatigue strength value than those values which can be provided by the glass fibre. For such cases high performance fibres, such as carbon and aramid fibres, will be used. The two high performance carbon fibres used in construction are the high modulus (H-M) carbon and the ultra-high modulus (UH-M) fibres, these are the European definitions whereas other areas of the world may refer to these as intermediate modulus and high modulus fibres, respectively. The European definition will be used in this chapter. The basic manufacturing techniques for the H-M and the UH-M fibres are the same but the heat treatment temperature will be greater the higher the modulus of the fibres, thus, a more highly oriented fibre of crystallites will be formed for the UH-M fibre. Typical mechanical properties of carbon fibres are given in Table 5.4.

Carbon fibres are manufactured by controlled pyrolysis and crystallization of certain organic precursors and the manufacturing process consists of sequence of procedures these are (i) stabilisation,

Table 5.4 Typical mechanical properties of glass, carbon and aramid fibres. [2]

Material	Fibre	Elastic Modulus GPa	Tensile Strength MPa	Ultimate Strain %
Glass fibre	E	72.4	2400	3.5
	A	72.4	3030	3.5
	S-2	88.0	4600	5.7
Aramid fibre	49	125	2760	2.4
	29	83	2750	4.0
Carbon fibre *PAN- based fibre* Hysol Grafil Apollo	IM[1]	300	5200	1.73
	HM[2]	450	3500	0.8
	HS[3]	260	5020	1.93
PAN- based fibre BASF Celion	G-40-700	300	4960	1.62
	Gy 80	572	1860	0.33
Pan based fibre Torayca	T300	234	3530	1.51
Pitch based fibres Union carbide	P120	827	2200	0.27
	P100	724	2200	0.31
	P75-S	520	2100	0.40
	P55-S	380	1900	0.50
	P55-W	160	1400	0.90
Mild steel (for comparison)		210	370-700	2.50

[1] High modulus (Intermediate modulus)
[2] Ultra high modulus (High modulus)   [3] High strain

(ii) carbonisation and (iii) graphitisation and finally (iv) surface treatment. Figure 5.4 shows the various stages in the production process in diagrammatic form. During the manufacturing process most elements other than carbon are removed and carbon crystallites are preferably orientated along the fibre length. At manufacturing temperatures above 2000^0C the size of the carbon crystallites increase and their orientation improves thus causing the modulus to increase in value. This is shown

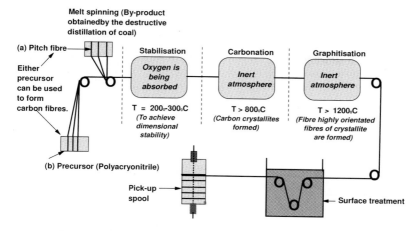

(a) Pitch fibre for production of ultra-high modulus carbon fibre (construction industry)
(b) Polyacryonitrile (PAN) fibre for production of high modulus-carbon fibre (construction industry)
    or for production of high modulus or ultra-high modulus fibre (aerospace industry).

Fig. 5.4 Schematic representation of the production of carbon fibre.

in Fig. 5.5 where the modulus value increases exponentially with rise in temperature; conversely, after 1,600°C of heat treatment the tensile strength falls. Consequently, to manufacture a high strength fibre a heat treatment of 1,600°C is required.

Carbon fibre filaments are available typically between 5 and 8 μm in diameter and are combined into tows containing 5000-12000 filaments. The 'tows' are equivalent to the strands of glass which are untwisted filaments of glass fibre contained in the strands. A common size untwisted carbon fibre 'tow' is called 12K which contains 12,000 filaments of carbon and are sold as (i) high modulus fibres, (intermediate modulus fibres) {stiffness 250-350 GPa} (ii) ultra high modulus fibres, (high modulus fibres) {stiffness 350 -1,000 GPa}; the value of stiffness of the UH-M carbon fibre used in construction is generally about 600 GPa. The tows can be twisted into yarns and woven into fabrics similar to that for glass fibre.

The precursor fibres that are used for the production of carbon fibres, currently, are (i) polyacrylonitrile (PAN) and (ii) pitch. The first fibre precursor is the basis for the vast majority of carbon fibres produced commercially and this is undertaken by spinning to produce a round cross-section fibre. About 50% of the original precursor fibre mass

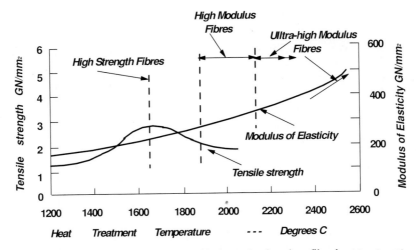

Fig. 5.5 Modulus of elasticity and tensile strength of carbon fibre heat treatment.

remains after carbonation. A new production system for the manufacture of PAN precursor carbon fibres is that of a melt assisted extrusion as part of the spinning operation. Cross-sections of rectangular, I-type and X-type are produced and with these geometric shapes a closer fibre packing in the composite can be obtained. These fibres are invariably used in the aerospace industry but are also used in the construction industry. The pitch precursor fibres are derived from petroleum, asphalt, coal tar and PVC., the carbon yield is high but the uniformity of the fibre cross-sections is not constant from batch to batch. Although this is a problem for the aerospace industry it is acceptable for the construction industry and consequently, the pitch fibre is used for the latter industry when the ultra-high carbon fibre is required; the former industry do not use the carbon fibres formed from the pitch precursor.

Not only does the construction industry require strong and high stiffness fibres but they require in-service resistance to withstand high temperature and aggressive environmental conditions, In addition, the fibres in general are not affected by moisture, solvents, bases and weak acids.

5.3.3 Aramid fibre (aromatic polyamide)

The aromatic fibre is produced by extruding and spinning a solid fibre from a liquid chemical blend. This relies on a co-solvent with an ionic component (calcium chloride) to occupy the hydrogen bonds of the amide groups. A solution of aromatic polymer in a suitable solvent (e.g., N-methyl pyrrolinidone) at a temperature of between -50°C and -80°C is extruded into a hot cylinder which is at a temperature of 200°C; this causes the solvent to evaporate and the resulting fibre is wound onto a bobbin. The fibre then undergoes a stretching and drawing process causing the polymer chains to orientate in the direction of the fibre thus increasing its strength and stiffness properties. Two grades of stiffness are generally available; one has a modulus of elasticity in the range of 60 GPa and the other has a modulus of elasticity of 130 GPa. The higher modulus fibre is the one that is used in polymer composites in construction.

The structure of the aramid fibres is anisotropic and gives higher strength and modulus values in its longitudinal direction compared with its transverse direction. The longitudinal strength is high and the fibre is heat and cut resistant. The aramid fibres retain their mechanical properties from cryogenic temperatures up to 400° C. They are resistant to fatigue, both static and dynamic. They are elastic in tension but it behaves non-linearly in compression and in addition has a ductile compressive characteristic. The fibre possesses good toughness and damage tolerance properties. It does not rust or corrode and its strength is not affected by immersion in water. Typical important mechanical properties of aramid fibres are given in Table 5.4.

5.4 Advanced Polymer Composites (Using Thermosetting Polymers)

It is convenient at this point to describe the various forms into which polymer composites fall and to understand the difference between ply, lamina, laminate and composite.

Ply – One of the layers of fibre that comprises a laminate. Also, the number of single yarns twisted together to form a plied yarn.

Laminate – The structure resulting from bonding multiple plies of reinforcing fibre or fabric.

Laminate ply – One fabric/resin or fibre/resin layer that is bonded to adjacent layers in the curing process.

Composite – A material that combines fibre and a binding matrix to maximise specific performance properties. Neither element merges completely with the other. Advanced polymer composites use only continuous, orientated fibres in a polymer matrix.

The advanced polymer composites (APC) are manufactured by the joining together of the fibre and matrix materials to form a composite system that can be/is designed simultaneously with the design of the complete structure. The fibre/matrix composite can have the properties of anisotropy or isotropy by virtue of the arrangement and direction of the fibres in the matrix, as illustrated in Section 5.4.3; this will be one factor which determines the mechanical properties of the composite. The choice of the relevant fibre material for construction (carbon, aramid and glass) have vastly different mechanical properties and these will also highly affect the mechanical property of the composite. Consequently, when considering new APC materials for civil/structural use their mechanical properties and fibre arrays are of paramount importance to the designers contemplating their use. Furthermore, the production methods for the manufacture of fibre/matrix composites will influence the mechanical properties. Likewise, of equal importance in choosing the materials and systems for use in the civil infrastructure applications are the environmental, durability and external loading conditions and a consideration of the duration of the applied loads. Exposure to a variety of adverse and sometimes harsh environmental conditions in construction could degrade the FRP composite material and thus this degradation would alter their mechanical performance. Exposure to high and low temperature variations, moisture and salt solution ingress, ultra-violet rays from the sun and fire will all lead to reduced mechanical performance. The curing of the APC systems is a further concern when considering their long term durability. If a composite is cured at ambient temperature a relatively low value of glass transition temperature will result, in addition, the polymer may not reach its full polymerization if it had not been post cured before the industrial use, thereby, making it more susceptible to degradation.

The properties of the fibre/polymer composite will have a wide spectrum of values depending upon

- The relative proportions of fibre and matrix materials, (the fibre/matrix volume or weight ratio).
- The method of manufacture.
- The mechanical properties of the component parts, (carbon, aramid glass fibres or a hybrid of the different fibres)
- The fibre orientation within the polymer matrix. (whether the composite is anisotropic or isotropic).

- The long term durability of the composite system. (this will involve any reduction in property values from those of the pristine condition of the composite)

5.4.1 *The relative properties of fibre/matrix materials*

The greater volume of fibre contained in the composite the greater will be the carrying capacity of the composite. However, there is a limit to the maximum fibre volume fraction that can be achieved and this depends largely upon the manufacturing techniques, generally, the fibre volume fraction (fvf) of the manual fabrication methods is lower than that of the automated methods due to the greater compaction that can be achieved with the latter technique. Nevertheless, there is a limit to the fvf for automated techniques as the resin must wet out the fibre completely with the minimum of air voids contained within the laminate at the completion of the polymerisation.

5.4.2 *Manufacturing methods of advanced thermosetting polymer composites and their properties.*

5.4.2.1 Introduction

Advanced polymer composites are manufactured by the wetting of the fibres by the thermosetting matrix, in its molten (or low viscosity) state to form a composite material. The coupling between these two components, (viz. the interface) allows stress transfer to take place between fibres. The manufacture of composites, which will be discussed in this section, are those which form the major ones utilised in construction.

It is stressed that which ever manufacturing technique is used to produce the composite, it should be post cured. Furthermore, each manufacturing method will apply varying degrees of compaction to the composite which is turn will have an overall effect upon the micro-structure and internal structure developed during the fabrication procedure all of which will affect their finished quality.

The various manufacturing processes can be divided into three major divisions, (i) the manual production process, (ii) the semi- automated process, and (iii) the automated process. A short discussion will be given here but for a further in-depth description of these systems the following references may be consulted [2,12].

5.4.2.2   Manual techniques

The manual processes used currently are a variation of the general Wet Lay-up method. These various procedures are:

- The X-X-sys Technologies method
- The REPLARK method
- The Dupont method
- The Tonen Forca method

(a) *The wet lay-up method*

This method requires one mould, either male or female, on which the matrix and fibre are fabricated. For a small number of specimens to be manufactured, the mould would probably be made from GFRP composites. This mould would be shaped and fabricated from a suitable wooden master pattern. If, however, the composite is to be manufactured on site for, say, the strengthening/stiffening of a bridge, the bridge soffit would then act as the mould. For a structural unit a gel coat would be employed and this is brushed onto the required finished surface of the mould, this gel coat would *not* be applied to the rehabilitation of structural members on site. The gel coat which can be reinforced with a surface tissue, will give added protection against severe external environments. On to the gel coat is brushed the first layer of laminating resin and hardener, followed by the first layer of fibre material which is rolled into the resin for compaction and to expel air voids. This operation is continued until the required thickness of the composite is achieved. To increase the extraction of air voids and thus obtain a good compaction of the matrix and fibre, a pressure or a vacuum bag may be used. The mould is encapsulated inside the vacuum bag and air is extracted to a vacuum of 1 bar. In the case of the pressure bag, the polymer membrane of the bag which is in direct contact with the composite places a pressure of up to 3 bar onto the surface of the composite; this allows a greater percentage of fibre to be used in the composite and a greater reduction of air voids in the manufactured laminates. Further description of this technique can be found in the book [2].

5.4.2.3   The semi-automated processes

The semi-automated processes used currently are (1) the resin infusion under flexible tooling (RIFT) process, and (2) the hot melt factory made pre-impregnated fibre with polymer matrix (prepreg).

(b)   *The resin infusion under flexible tooling technique*

This technique was developed to strengthen/stiffen metallic structures. Dry pre-formed fibres, manufactured in a factory, are attached to the structure in the area which requires strengthening/stiffening and a resin system is channelled to the preform. The whole is enclosed in a vacuum bag system. The resin flows through the dry fibre perform and acts as the matrix component of the composite and also as the adhesive which bonds the composite to the metallic substrate. The system provides a high volume fraction of composite but it is difficult to operate on site and is therefore not usually used.

(c)   *The hot melt factory-made pre-impregnated fibre (prepreg) (developed for the civil engineering industry)*

The pre-impregnated technique for the manufacture of FRP composites is a factory made prepreg utilised by the aerospace industry for patch repair of aerospace systems. This hot-melt factory-made prepreg has been considerably modified to enable the technique to be used in the civil engineering industry. The system can be manufactured as a rigid plate in the factory and then used as a structural member on site to rehabilitate RC or steel structural members using a two part adhesive (see Sections 5.7.2.2 and 5.7.2.3). It can also be produced as a pre-impregnated composite for upgrading/rehabilitating a concrete or steel structural member used in conjunction with a compatible film adhesive (see Sections 5.7.2.2 and 5.7.2.3). The pre-impregnated composite and the film adhesive immediately after manufacture are stored at -20^{0}C until used. After transportation to site both components are thawed for two hours and are then fabricated onto the field structure. To cure the on-site material a halar sheet and a breather blanket are places over the composite. A heater blanket covers the whole component in order to apply an elevated temperature cure of 65^{0}C applied for 16 hours or 80°C for 4 hours; a vacuum assisted pressure of 1bar is applied for compaction of the composite and the film adhesive [13,14]. This technique has been used in practice [15]. This procedure enables the composite and film adhesive to intermix and to be compacted and cured in one operation. Furthermore, a vacuum of 1 bar will reduce the volume of voids in the composite to a minimum value. An advantage of this system is that the polymer materials are cured at an elevated temperature which ensures the T_g of the system to be as high as is practically possible on site. This is a major advantage if a rapid cure is required for, say, the rehabilitation of a railway bridge structure which can be closed to train movement for only four hours during night time; any longer period of closure may incur heavy financial penalties to the contractor. Furthermore, the cure temperature of 65°C for 16 hours has been shown to be relevant for these site loading conditions [16]. As in this case the composite and adhesive film are under a vacuum assisted pressure the vibration set up by the train motion during curing of the composite/adhesive system could be an advantage in assisting the extraction of any further air voids.

5.4.2.4 The automated processes

The methods which are available to the construction industry are:

- The pultrusion technique
- The filament winding
- The resin transfer moulding (RTM)

(d) *The pultrusion technique*

The pultrusion technique is the manufacturing technique that is extensively used by civil engineers. It consists essentially of a resin bath, heated die and pullers. Continuous fibrous reinforcement roving and strand mat, or other designed reinforcement, are pulled through a reservoir of 'hot cure' resin and curing agent and then through a heated die, the temperature of which is generally between 120°C and 135°C. It is necessary to have sufficient longitudinal fibres in the fibre array during the pull as these fibres take the tensile load as the composite passes through the heated die. The fibres can, alternatively, be impregnated with resin by injection through port holes in the heated die as the fibres pass through it. The important design parameters which are established during the design/analysis of the product are the fibre placement, resin formulations, catalyst level, die temperature and pull speed, the last two items will be monitored at the time of manufacture. As the cured profile

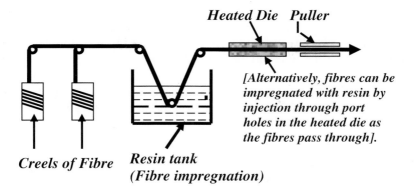

Heated Die   Puller

[Alternatively, fibres can be impregnated with resin by injection through port holes in the heated die as the fibres pass through].

**Creels of Fibre   Resin tank
(Fibre impregnation)**

Fig. 5.6 Schematic diagram of the pultrusion technique.

exits the die it is cut to a pre-set length by a saw synchronized to the puller speed. The composite should then be given a post-cure. Figure 5.6 illustrates the process.

Pultrusion products can be formed having various geometrical cross-section shapes and are generally straight in the longitudinal direction, although products can be manufactured which are curved in plan. Care must be taken to ensure that the fibres are well compacted into any bends in the cross-section (thus preventing voids forming), that there is complete wetting of the fibres in the pultrusion unit (again, preventing voids forming) and that the fibres are well distributed in all cross-sections.

Surface veil can be fabricated into structural components to provide a resin rich exterior surface. In addition, a peel-ply fabric can be attached to the exterior surface of the composite during the manufacturing process. Peel-plies are an important addition to the cured pultruded unit if the latter is to be bonded to the external surface of a structural member. From the time the pultruded unit leaves the factory to the time of bonding, the surface of the pultruded unit will remain free of any dirt, grease etc. The peel-ply is removed immediately before the composite unit is bonded to the structural member. Furthermore, complicated geometrical sections, including those which incorporate right angled bends can be manufactured. The limiting factor on the complexity of the cross-section and the size of the unit is the pull force required to draw the pultruded section through the die; the more complex the section the greater will be the force due to friction of the unit within the die.

Hydraulically driven systems capable of pulling up to 50 tonnes and 2.5 metres by 275 mm thick are currently operating in the USA.

The carbon, aramid and glass fibres and the epoxy, vinylester and polyester materials have all been used for the production of pultruded units. The epoxy polymers are probably the most difficult to pultrude, but they have low shrinkage during polymerisation (3-4%). The polyester polymer has a large shrinkage during polymerisation (12-19%) compared to the vinylesters, which have a shrinkage of 6-10%.

(e) *The pull-winding process*

The pull-winding process is an extension of the pultrusion technique for the manufacture of mainly closed sections. The pull-winder can be used in conjunction with standard pultrusion machines. It is designed to wind hoop or angled fibres at a constant wind pitch between layers of unidirectional fibres. The unidirectional fibres are drawn through a resin tank in conjunction with a standard type dipper system. The heads rotate about a hollow spindle through which passes the mandrel and the impregnated unidirectional reinforcements. Each winding head can accommodate a number of externally unwind spools of fibre. Further information on these two techniques can be obtained in other publications [2,12].

(f) *The filament winding technique*

The filament winding technique consists of continuous strands or roving of dry fibres which are passed through a bath of 'cold cure' resin and curing agent on to a 'pay-out eye' which is mounted on a moving carriage along the length of a constant speed rotating mandrel. The roving fibre delivery system reciprocates along the length of the mandrel and is controlled relative to the rotation of the mandrel to give the required fibre orientation. The speed of reciprocation and rotation are synchronized to hold a pre-set winding angle typically 7° to 90°. The machine has the ability to lay fibres in any direction and to employ as many permutations of movements as is required by the structural design. If resin pre-impregnated fibres, prepregs, are used, they are passed over a hot roller until tacky to the touch and are then wound on to the rotating mandrel. The composite unit is then removed from the mandrel, after completion of the initial polymerization, and is then

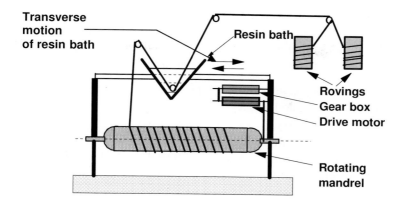

Speed of reciprocation and rotation are synchronised to hold a pre-set angle 7^0 up to 90^0.

Fig. 5.7 Diagrammatic representation of filament winding technique.

fully cured by placing it into an oven enclosure at 60^0C for eight hours. Figure 5.7 shows a schematic representation of the filament winding technique.

(g) *The resin transfer moulding (RTM)*

There are a number of industrial RTM processing techniques available to fabricate and manufacture composite units. In the RTM procedure, a fibre preform is placed on a tool or inside a mould cavity and is encapsulated in a vacuum bag. The thermosetting resin is then injected into the mould where it saturates the preform and fills the mould, this operation takes several minutes; the polymer is then cured. The impregnation and filling the mould is a critical procedure; void forming must be prevented and complete wetting of the fibre must be attained. If any voids are present defects in the composite will form and this will diminish the strength and quality of the cured part.

The resin transfer moulding process (SCRIMP), vacuum assisted resin transfer moulding (VA-RTM) and thermal expansion resin transfer moulding (TERTM)™ are all industrial processes which are a variation of the general technique discussed above. Over a number of years these liquid moulding methods have been investigated with the aim of developing fabrication techniques for thermosetting composites. The composites in construction which would be manufactured by this technique would be mainly street furniture. The benefits that they can offer over other composite processing techniques are the:

- ability to tailor reinforcement type, level and orientation
- ability to fabricate large, complex parts to net shape
- integration of parts
- low capital investment
- ability to encapsulate a core material and the possibility of incorporation inserts into the composite for attachment to other civil engineering units.

The processing systems, which have been considered in this section, are the most likely ones to be used in the civil infrastructure applications. The wet lay-up, the pultrusion, and the new two stage curing of factory made prepregs which are given their second stage final cure on site, would be used for the rehabilitation of structural members and the manufacture of new buildings and structural members. The filament winding system would be considered for gravity and high pressure pipes, tanks etc. and the RTM for street furnishings. Both the epoxy and vinylester polymers are likely to be widely used in all applications and the polyester in a few applications.

5.4.3 *Properties of advanced polymer composites*

The mechanical properties of the fibre components of the composite have been discussed in Section 5.3 and it was shown that the stiffness and strengths of the three fibres (carbon, aramid and glass) are different, although the latter property for the three fibres is of the same order. Consequently, the higher stiffness fibres will provide the greater stiffness to the composite. A hybrid of fibres can be formed to provide the required stiffness in any particular direction of the composite in order to satisfy the design criteria.

(i) *Composites under Tensile Loads.* The fibre orientations can take the form of unidirectional, bi-directional, various off-axis directions and random arrays. Table 5.5 gives the typical tensile properties of composites manufactured using long directionally aligned fibre reinforcement of glass, aramid and carbon with a fibre/matrix ratio

Table 5.5 Typical tensile mechanical properties of long directionally aligned fibre reinforced composites (fibre weight fraction 65%) manufactured by automated process (the matrix material is epoxy) [2].

Material	Specific Weight	Tensile Strength MPa	Tensile Modulus GPa	Flexural Strength MPa	Flexural Modulus GPa
E-glass	1.9	760-1030	41.0	1448	41.0
S-2 glass	1.8	1690.0	52.0	—	—
Aramid 58	1.45	1150-1380	70-107	—	—
Carbon (PAN)	1.6	2689-1930	130-172	1593	110.0
Carbon (Pitch)	1.8	1380-1480	331-440	—	—

Table 5.6 Typical tensile mechanical properties of glass fibre composites manufactured by different fabrication methods [2].

Method of manufacture	Tensile strength MPa	Tensile modulus GPa	Flexural strength MPa	Flexural modulus GPa
Hand lay-up	62-344	4-31	110-550	6-28
Spray-up	35-124	6-12	83-190	5-9
RTM	138-193	3-10	207-310	8-15
Filament Winding	550-1380	30-50	690-1725	34-48
Pultrusion	275-1240	21-41	517-14448	21-41

Table 5.7 Typical tensile mechanical properties of Glass fibre/vinylester polymer (compression moulding – randomly orientated fibres) [2]).

Fibre/matrix ratio %	Specific weight	Flexural strength MPa	Flexural modulus GPa	Tensile strength MPa	Tensile Modulus GPa
67	1.84-1.90	483	17.9	269	19.3
65	1.75	406	15.1	214	15.8
50	1.8	332	15.3	166	15.8

by weight of 65%. Table 5.6 shows typical tensile mechanical properties of glass fibre composites manufactured by different techniques; it clearly illustrates the effect that the methods of fabrication have on the properties. Table 5.7 shows the variation of the composite tensile properties when the fibre matrix ratio is changed, the method of manufacture and component parts of the composite remaining constant.

(ii) *Composites under Compressive Loads.* In compression loading, the fibres are the principal load bearing elements and the matrix protects the fibres and supports them from becoming locally unstable and undergoing a micro-buckling type failure. Under a compressive load situation the integrity of both component parts of the composite are far more important than under tensile loading. In addition, local resin and interface damage caused by compressive loading leads to fibre instability which is more severe than the fibre isolation mode which occurs in tensile loading. The mode of failure for FRP subjected to longitudinal compression may include transverse tensile failure, fibre micro-buckling or shear failure; the mode of failure will depend upon the type of fibre, the fibre volume fraction and the type of resin. In general, compressive strengths of composite materials increase as the tensile strengths increase, the exception is that of AFRP composites where the fibres exhibit non-linear behaviour in compression at a relatively low level of stress. The value of the compressive modulus of elasticity of FRP materials is generally lower than their tensile value. Test samples containing 55% to 60% weight fraction of continuous fibres in a vinyl-ester resin have compressive modulus of elasticity values of approximately 80%, 100% (but the compressive strength is very low) and 85% of the tensile value for GFRP, AFRP and CFRP respectively [17], have shown that by adding nanoclays to the polymer the compressive strengths of GFRP composites do increase.

5.4.4 *Fibre orientation*

The direction of orientation of the fibre in the composite will determine whether the composite is anisotropic, bidirectional, angle ply or isotropic. The direction and volume fraction of the fibres in the composite will determine the strength and stiffness of the composite. Figure 5.8 illustrates the relationship between stiffness (or strength) of the composite in the coordinate axes direction relative to the fibre axes.

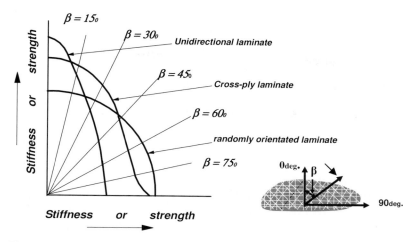

Fig. 5.8 Diagrammatic relationship between stiffness at angle β or ultimate strength at angle β between rotational angle β varying between 0^0 and 90^0 for different laminates.

5.4.5 *Durability*

All engineering materials are sensitive to environmental changes in different ways and any degradation of FRP materials, which will be discussed in this section, are not unique to these materials. Indeed, composite materials do offer some significant durability advantages over the more conventional construction materials, one such advantage is associated with their use as external reinforcement for RC or metallic structural members as they have good resistance to corrosion and have magnetic neutrality.

In the early period of utilising FRP composites in construction, structures tended to be architecturally orientated to building systems using only FRP composite materials specifically shaped to provide strength and stiffness to that structure. In addition, FRP composite infill panels to the conventional steel or reinforced concrete skeletal beam and column structural units were also manufactured [18]. These constructions now provide a reasonable time span in which to study any degradation of the FRP composite material. At the time of manufacturing these composites, the matrix component was generally the polyester but this material has now been superseded by the vinylester or epoxy polymers, as stated in Section 5.2. Nevertheless, these early polymers can give a good indication of what can be expected from the materials used currently; these latter materials should have a more robust durability property than that of the former. Dirt accumulation has been the main problem with the external surfaces of these early types of structures; there is always reluctance on the part of the owners to keep surfaces clean. After 30 years in service chalking on the surface of exposed panels tends to be the major degradation characteristic.

One of the problems in acquiring data relating to the durability properties of any material and in particular with that of FRP composites is the length of time involved in gathering the relevant information. This is particularly difficult with respect to construction polymer composites as there are many different polymers on the market and the majority will have additives which may have been incorporated to enhance curing or to improve some specific mechanical or physical property. In effect the addition of additives to composites is equivalent to adding impurities to the polymer. Consequently, many of the FRP composite structural components which are being, or have recently been, erected are being monitored for any changes in their stiffness characteristics or any signs of degradation. These results will eventually reveal important data on their long term behaviour but the disadvantage of this method of obtaining information is the length of time involved. Reference [19] has discussed many field surveys which have been and are being undertaken throughout the world and the results from these tests have revealed or are revealing interesting discoveries regarding the resistance of FRP composites to the natural or specific environments to which civil engineering composites were exposed; environments which at one time were thought to be a risk to that material. The following section was first reported by Hollaway [19] and due to the importance that the author attaches to the understanding and interpretation of the accelerated test results is repeated here. For instance, Hollaway [19] has discussed the previous work [21-23] that accelerated laboratory test results of GFRP in a simulated concrete pore water solution of high pH values and at elevated temperatures up to 80^0C have indicated that there is a decrease in the tensile, shear and bond strengths,; these results would suggest that there is a case for not using GFRP rebars in concrete [24]. However, Tomosawa and Nakatsuji [25] have shown that after 12 months exposure to alkaline solutions at temperature between 20-30°C, and [26], likewise, after 2 years exposure to a tropical climate on a test platform off the Japanese coast, have reported that there had been no material or physical deterioration to the GFRP composite. Furthermore, Sheard et al. [27]

reported that the overall conclusions of the work of the EUROCRETE project were that GFRP is suitable in a concrete environment.

Polymers have been modified by chemists over the years to improve their performance and as a consequence it would be hoped their durability performance; it will be realized that modern structural FRP composites have been on the market for some 15 years only. However, structures such as, for instance, the Lancashire classroom system UK and the Mondial House on the north bank of the river Thames UK have been in-service for some 30 years. These early structures were manufactured using polyester resin and randomly orientated E-glass fibre and their recent visual inspection have shown them to be in excellent condition; the durability of these structures have been discussed in the publication [19]. Nevertheless, information on durability is rather limited in terms of the duration and realism of testing conditions. In the past scientific data from field testing conditions have been based primarily on the observations made over several months or years, rather than more pertinent timeframes of the order of decades.

Clearly, it is necessary to obtain the durability characteristics of a composite as quickly as possible. Consequently, sometimes test specimens or structural units are exposed to an accelerated test environment which generally involves the specimen being subjected to an environment many times more severe than that which would be experienced in practice, (c.f. the example stated above of immersing GFRP rebar into a simulated concrete pore water solution of high pH values and at elevated temperatures up to 80°C). This extreme method, although allowing information to be obtained very rapidly from that one known environmental situation, will generally not be equivalent or relevant to the more gradual degradation effect had the environment been applied in a less rigorous manner. Materials used in construction would normally be exposed to many different environments acting simultaneously but in a less harsh way, and each environment possibly having an effect upon the other. In addition, in a typical substrate/FRP hybrid material system, there are multiple, simultaneous, interactive degradation mechanisms in operation, each having different rate kinetics with respect to a given acceleration factor. Furthermore, accelerated testing sometimes involves applying an elevated temperature to the polymer, as was used in the above example; this elevated temperature method should be used with caution. Temperatures in regions above 60^0C are not relevant in the context of making service life predictions for most civil engineering structures that operate in normal conditions, as the degradation mechanism in FRP materials are different under the lower practical temperatures. As the temperature rises towards the glass transition temperature the polymer will lose some of its stiffness and strength with the result that the investigation will not be analysing the original material.

With the increase in the number of FRP composite structural members being constructed currently, and a sparse knowledge of the long term property characteristics of composite materials, it is advantageous to monitor the performance of such structures with time, although this does increase the overall cost of the construction. This procedure might involve: (i) Test loading the structural system at certain periods of time after construction to obtain stress and deflection values against time under specified external loading; such tests would characterise the overall behaviour of the structural system. (ii) Visual inspections of the structure, at specified times, for any signs of distress or deterioration of the material from the point of view of environmental degradation or from exposure to long term application of loads, which might influence the creep rate of the polymer component of the composite, (iii) Mechanical and physical/chemical material tests on specific areas of the exposed FRP structure to investigate any possible degradation effects between the component parts of the FRP material.

5.4.5.1   Methods to improve the durability of FRP composite materials in the civil infrastructure

The major concern with the *Durability* of FRP composites used in the civil engineering environments are: (i) moisture and aqueous solutions, (ii) alkaline environments, (iii) fire, (iv) thermal effects, and (v) ultra-violet radiation.

The long term loading conditions in which the material may have to function are: (i) fatigue and (ii) creep

The composite systems used in construction can be engineered to provide resistance against moisture and aqueous solutions diffusing into them, however, moisture eventually will diffuse into all organic polymers leading to changes in mechanical and chemical characteristics. This process may cause reversible and irreversible changes to a polymer. If these changes do occur, there will be a lowering of its T_g value and in the composite deleterious effects may occur to the fibre/matrix interface bond resulting in a loss of integrity. The following solutions will help to reduce diffusion through the polymer, by:

(i)   Applying a protective coating/gel coat on to the system.
(ii) Curing the product at an elevated temperature.
(iii) Manufacturing the composite under factory fabrication, (controlled conditions and elevated curing temperatures), such as the pultrusion. This ensures low void content, full cure, high levels of overall integrity, a high glass transition temperature and greater resistance to moisture degradation (compared with the cold curing polymers).
(iv) The use of appropriate sizing/finishes on the fibres.
(v)   The addition of nanoparticles into the polymer. This system has the potential for lowering the permeability and thus providing a barrier property to polymers to reduce the ingress of moisture and salt solutions into them and to improve the mechanical strengths and the fire resistance of the polymer [6].

5.5 Adhesives

Over the last two decades the utilisation of adhesives and sealants to bond dissimilar materials to concrete and metallic surfaces has become a subject of great importance to the civil engineer. This is due to several reasons but the main one is that the transportation infrastructure in many regions of the world is decaying. As concrete ages and steels corrode more repairs become necessary. Consequently, applications of adhesives have expanded to include bonding aids between two similar and dissimilar adherents, between hardened concretes, between fresh and hardened concretes, and between fresh concrete and other construction materials such as steel and advanced polymer composites. Furthermore, currently a significant number of new bonding materials and processes are available for repair of materials of construction.

The nature of repairs to roadway surfaces, and to bridge and building structures has changed over the past two decades. Traffic volumes have increased, and there is a demand for repair and replacement options that can be undertaken with a minimum of lane-closure. To satisfy this demand adhesive and sealant systems for these applications have also continued to develop. Polymeric materials and sealants are used to

refurbish or bond old concrete and metallic structures to new civil
engineering materials.

Structural adhesives used in the civil environment can be both cold
cured and hot cured epoxy polymers and all are cross-linked, this renders
the polymer insoluble and infusible and these characteristics greatly
reduce creep of the adhesive. The cold cured epoxy polymer will cure at
room temperature and *should be post cured at a temperature of 50^0C.* To
avoid brittle behaviour, using additives toughens most modern epoxy
adhesives. All amorphous epoxy polymer adhesives have a glass-
transition temperature (T_g), below this value of temperature they are
relatively hard and inflexible and are described as glassy. Above the
value of T_g they are soft and flexible and are then described as rubbery. It
is unacceptable for adhesive polymers to pass from one state to another
during service. Cold cured epoxy polymer adhesives generally have T_g
between 50°C and 65°C; they will soften at this temperature when
exposed, for instance, to the sun's rays. (In some circumstances, such as
bonding to metals, or when bonding FRP materials to the top surface of a
bridge deck that is to receive hot bituminous surfacing, an adhesive with
a higher T_g would normally be required) [28,29].

5.5.1 *Adhesive bonding of concrete surfaces*

Typical adhesives for bonding polymer composites to concrete surfaces
consist of solvent-free polymers, such as epoxies and polyurethanes.
Table 5.8 provide mechanical properties of adhesives that are used on
civil engineering construction. Epoxy adhesives and their hybrids (e.g.,
epoxy-polysulfides, epoxy-urethanes) are the ones that are generally used
for bonding polymer composites to concrete structures. These adhesives
are generally formulated from epoxy resin, an amine or polyamide curing
agent, reactive diluents, and organic fillers and thixotropic agents.

The strength of an adhesive bond depends upon:

(i)   Adhesion of the adhesive to the substrate material (bond strength)
(ii) Cohesive strength of the adhesive material.
(iii) Cohesive strength of the substrate materials.

The cohesive strength of the concrete is weaker than that of the
adhesive polymer and therefore this area will generally be the failure
criterion.

Table 5.8 Typical physical properties of common adhesives used with concrete, based upon reference [30].

Property		Epoxy	Polyester*	Polyurethane
Tensile strength, (ASTM D638)	MPa psi	28-90 4000-13,000	4-90 600-13,000	1.2-70 175-10,000
Tensile elongation, % (ASTM D638)		3-6	2-6	100-1000
Compressive strength, (ASTM D695)	MPa psi	105-175 15,000-25,000	90-205 13,000-30,000	140 20,000
Comp. modulus, 10^3 MPa(ASTM D695)	MPa psi	—	2-3 300-400	70-700 10-100
Heat deflection temperature, °C (ASTM D648)		45-260	60-200	—
Coefficient of thermal expansion, 10^{-6} / °C (ASTM D696)		45-65	55-100	100-200

*Rarely used in adhesives for civil engineering structures

5.5.2 *Adhesive bonding of steel adherents*

The failure modes of a rigid CFRP composite plate/steel adherent joint using a two part adhesive system varies between a cohesive, an interlaminar and an interfacial one. For a pre-impregnated CFRP composite and compatible film adhesive/steel adherent joint, (see Section 5.4.2) it is likely that an interfacial mode or the failure strain of the CFRP composite would be exceeded; the film adhesive used with the pre-impregnated composite gives a higher test failure result compared with the two part cold setting adhesive when testing a double lap joint [31,32]. The reason for the variation of the failure modes between these two methods of bonding is that a better compacted join and an elevated temperature cure provides a better formed joint. Theoretically, the adhesive layer should not be the weak link in the joint and wherever possible the joint should be designed to ensure that the adherent fails before the bond layer. However, in a steel/CFRP composite joint the adhesive is much weaker than the FRP composite or steel adherents as the thickness of the adherents are thick compared to the adhesive and therefore the bond stresses become relatively large until the bond failure occurs at a lower load than that for which the adherents fail. Consequently, in a well-bonded steel/FRP composite joint, failure should occur within the adhesive (cohesive failure) or within the adherent (FRP inter-laminar failure). Failure at the adherent-adhesive interface (interfacial or bond-line failure) generally indicates that a stronger bond should be made.

The characteristic properties that are important for an adhesive to successfully bond concrete are discussed in Table 5.9. These are properties of the cured adhesive and refer to the two part adhesive system.

Further information on the technique and analysis of bonding like and dissimilar materials may be obtained from, [31,33-37].

5.6   Preparation of substrate surfaces for Bonding Like and Dissimilar Adherents

To ensure a successful bonding of the FRP composite plate to the adherent, a high standard of surface preparation of both adherents is essential. The surface of the substrate should be free of all contaminants including grease and oil which inhibit the formation of the chemical bonds [37]. The FRP composites are highly polar and hence very receptive to adhesive bonding, whereas the metals and aluminium adherents will range from a physical to a chemical method; metals will require solvent degreasing, abrasion and grit blasting, whereas the pre-treatment of the aluminium includes etching and anodising procedures thus causing chemical modification to the surfaces involved. The pre-treatment of the concrete substrates should ensure that surface laitance, unevenness, sharp ridges or formwork marks are removed. Sudden changes in level should be less than 0.5 mm. To achieve these requirements grit blasting is used. In the UK 'Turbobead' grade 7 angular chilled iron grit [38], of nominal 0.18 mm particle size is generally used. The grit would be applied at a blasting pressure of 80 psi. With this operation the surface cement layer (the laitance) must be removed, providing a uniform exposure of the underlying aggregate; this operation is followed by vacuum and solvent cleaned. If grit blasting is not possible, scabbling, needle gunning or grinding may be used, although these techniques are not recommended as they will almost certainly leave a much rougher surface than grit blasting.

Table 5.9 Important physical properties of cured adhesives bonding to civil engineering materials to concrete.

Property	Characteristics
Tensile strength and elongation	Tensile strength of thermosetting, reactive adhesives is generally greater than concrete, and this property is therefore not a controlling factor. Tensile elongation is an indication of the relative stiffness of the adhesives and its movement capability.
Compressive strength	It is advisable that the adhesive material should have a compressive strength of the same order as that of the concrete substrate.
Flexural strength	The flexural strength of most adhesives is greater than that of the concrete, therefore, it is not generally a critical factor but it provides an indication of the rigidity of the adhesive
Shear strength	The shear strength is the strength of the bond when the joint is under a shear force; it should not be refereed to as the 'bond strength'. Bond strength is often determined by "slant-shear tests" as per ASTM C882. The bulk shear and tensile strength at 20°C should be $\geq$ 12MPa. The minimum shear strength at 20°C, measured by the Thick Adherend Shear Test (TAST), should be 18 MPa
Modulus of elasticity	The modulus of elasticity is a measure of the stiffness of the adhesive and is highly affected by temperature (see Appendix 1), and most thermosetting adhesives have modulus values of 2-20 percent of that of the concrete. A repair material with a lower modulus of elasticity will exhibit lower internal stress thus reducing the potential for cracking and delamination of the repair. The flexural modulus of the material should fall within the range of 2 to 10 GPa at 20°C. Generally held views are that the lower bound to this range might be reduced to 1GPa.
Creep	*See Section 5.2.3* – Creep characteristics of polymers.
Coefficient of thermal expansion	*See Section 5.2.3* – Coefficient of thermal expansion of polymers. The coefficient of thermal expansion of most adhesives is two to ten times that of concrete. Differences in thermal expansion between the adhesive and concrete can induce stresses in the joint during temperature excursions.
Chemical, permeability and moisture content	*See Section 5.2.3* The chemical and permeability of polymers do vary from adhesive family to adhesive family and even within certain families. This is generally measured as a loss of physical property or a weight gain after immersion in the salt solution at a specified temperature for a specified period of time. See Section 5.4.5 – Durability regarding accelerated testing. If impermeable materials are used for large patches or overlays, moisture that diffuses through the base concrete can be trapped between the substrate and the impermeable repair material. This can lead to degradation of the concrete. The equilibrium moisture content (M_∞) of the adhesive should not exceed 3% by weight after immersion in distilled water at 20°C. Its coefficient of permeability should not exceed 5×10^{-14} m²/s. The adhesive should not be sensitive to the alkaline nature of concrete (if present) and its potential effect on the durability of joints.

The pre-treatment of steel substrates requires a clean rough surface on which to bond the FRP composite. To achieve this and a greater durability of the joint, solvent degreasing and grit blasting (e.g. to the Swedish Code SA $2^{1/2}$ Grade 3 Dirk grit) in conjunction with a silane is often used. Silanes have been shown to enhance the durability of bonded steel structures, but compatibility between the adherent and the silane must be achieved. The substrate would then be finally solvent degreased again immediately before the adhesive is applied.

5.7 The Current Developments of Fibre Reinforced Composites in the Civil Infrastructure

Since they were first seriously considered for use in the construction industry some thirty-five years ago, advanced polymer composites have made great advances compared with the more conventional materials, which have been evolving over the last century. The first major use of FRP composites in the building industry commenced in the early 1970's. The form of construction was the non-load bearing and the semi-load bearing infill panels manufactured by the hand lay-up process. The advanced polymer composites did not enter the civil engineering construction industry until the late 1980s early 1990s. The utilisation of FRP composites over these past thirty years can be conveniently divided into specific areas, which will be discussed briefly.

1. *'All FRP composite' structures (Building and civil structural systems).*
2. *External reinforcement for metallic and RC structures (Civil engineering structural systems) including Seismic retrofit of RC and masonry structures.*
3. *FRP composite rebar reinforcement (Civil engineering structural systems).*
4. *Replacement of degraded bridge deck systems (Civil engineering structural systems).*

5.7.1 *'All FRP composite' structures*

The 'all composite' structural systems are manufactured entirely from the advanced polymer composite material and can be divided into two specific categories, (a) the substitution of the traditional materials with the new FRP material, (b) the new material-adapted concepts. These systems produced to date have, in the main, tended to be single prestigious structures; one exception to this statement is the FRP composite bridge deck, where this construction is only a part of the complete bridge structure. The most advantageous way to construct large 'all composite' structures is using the 'building block' technique, in which the structure is built from a series of identical smaller units.

The advantages of the 'building block' construction may be stated as:

(a) Construction of identical units manufactured on a production line.
(b) Controlled mechanised/manual factory manufacture and fabrication of identical structural units. The transportation to site of the light-weight units should present no particular problem.

The Lancashire GFRP Class Room System, UK, 1974 [19], is an 'all composite' structure and comes under the new material adaptive concept category This is the first 'all composite' structures, and is an excellent example of a manual, factory manufactured FRP composite system, which uses identical lightweight *'building blocks'* manufactured from randomly orientated glass fibre in a polyester resin. Figure 5.9 shows a picture of the Class Room.

Bonds Mill Bridge, Gloucestershire, England 1996, falls into the substitution category. The bridge was manufactured from ten identical pultruded Advanced Composite Construction System (ACCS), 'Maunsell' planks, [*automated building blocks*]. These planks were manufactured mainly from unidirectional E-glass fibre/vinylester

Fig. 5.9 GFP Class Room, Mill Lift, Lancashire, UK.

Fig. 5.10 GFRP Bonds Bridge, Gloucestershire, UK. (From Ref. 2, by permission of Elsevier).

polymer composite pultruded units, with a fibre/matrix weight ration of 68% to form the box cell beam (the ACCS units forming the walls of the box beam). It is the first example of the automated building block (Fig. 5.10).

5.7.2 *External reinforcement to concrete and steel structures using FRP plate bonding*

The cost of repair and rehabilitation of structures is generally much less expensive and less time consuming than the cost of replacement. Consequently, the repair, upgrading and strengthening or stiffening, deteriorated, damaged and substandard infrastructure has become one of the fast growing and important challenges confronting the civil engineer worldwide. Changes in use of a structure include, (i) *Increased live load* – increased traffic on a bridge, changes in use of a building resulting in greater imposed live loads. (ii) *Increased dead loads* – additional load on the structure due to new construction. (iii) *Increased dead load and live loads* – widening a bridge to add an extra lane of traffic. (iv) *Modern design practice* – changes in techniques and updates. These changes will involve structural techniques including (a) flexural and shear strengthening/stiffening of RC, steel, timber and aluminium beams and RC slabs, (b) column confinement, using FRP jacket. (c) near surface mounted (NSM) FRP rods for strengthening/ stiffening RC, timber and masonry beams. To overcome some of the short comings that are associated with upgrading using steel plate-bonding, it was proposed in the mid-1980s that fibre reinforced polymer (FRP) plates could prove advantageous over steel plates in strengthening applications [39-41]. Unlike the steel plates, FRP composites are unaffected by electrochemical deterioration and can resist the corrosion effects, alkalis, salts and similar aggressive materials under a wide range of temperatures [2].

(a) *Upgrading Reinforced Concrete (RC) Structural Members*

The FRP material used for the composite plate bonding to upgrade structural beams made from RC concrete is generally the CFRP or GFRP composites and these composites will be fabricated by one of three methods, namely (a) the pultrusion technique, where the factory made rigid pre-cast FRP plate will be bonded on to the degraded member, (b) the factory made rigid fully cured FRP prepreg plate bonded to the degraded member, (c) the wet lay-up process [See Section 5.5.1]. Currently, the preferred system of upgrading is the pultruded method and this implies that, (i) the material cannot be reformed to cope with geometries of the structural member and (ii) a two part cold setting epoxy adhesive will be used for bonding the plate on to the substrate. There are three areas relating to this topic which require particular attention, these are. (a) The bond line thickness which is difficult to control and can vary considerably over the length of the plate because of the undulating nature of the soffit of the concrete beams, (b) The long term bonding characteristics of the adhesive. (c) The durability of the composite plate (See Section 5.4.5 for items (b) and (c)).

There are nine failure areas of a RC beam upgraded with an unstressed FRP plate; these are illustrated in the publication [42] and information on the technique and analysis of FRP composites to reinforced concrete may be obtained from other publications [28, 29, 42-44]. There are also several design guides for the design calculations for rehabilitation of FRP composites to reinforced concrete structures, used throughout the world [45].

(b) *Upgrading steel structural members*

Until recently only a limited amount of research work had been conducted on the application of these materials to metallic structures, but this situation is now changing [46-51]. The CFRP composites will normally be used when upgrading metallic structural units because of the relatively high stiffness values of early and modern steels and cast iron metals. The high modulus (HM) CFRP composites will have stiffnesses of the same order as that of the steel; this implies that, for plate bonding applications, substantial load transfer can only take place after the steel has yielded. With the ultra High-Modulus (UH-M) CFRP composites it has been shown in Section 5.3.2 that the unidirectional UH-M pitch carbon fibres have values in excess of 600 GPa. Therefore UH-M pitch CFRP prepregs at 60% F.V.F. could have modulus values approximately equal to 40 GPa. Consequently, the stiffness of this material could be twice as high as that of the steel and the load transfer will then take place before the steel has yielded; it must be remembered, however, that the strain to failure of the UH-M carbon fibres are very low (less than 0.4% strain, this value will depend upon the modulus of elasticity value). Consequently, the choice of the CFRP composite must be chosen carefully and fully understood. There are the three standard fabrication/adhesive bonding methods for upgrading metallic structural members (i) the pultruded rigid plate or the pre-impregnated rigid plate bonded with two part cold setting epoxy adhesive, (ii) the wet lay-up method, where the matrix material component of the composite also acts as the adhesive material of the up-grade, and (iii) the hot melt and adhesive film, (see Section 5.4.2) The current shortcomings of a well-designed FRP plate bonding system are a lack of long-term knowledge of the load carrying characteristics and the durability of the composite material. Hollaway [19] has reviewed the durability of some structures which have been in a civil engineering environment for some 30 years.

Further information on the technique, analysis and design of the rehabilitation of FRP composites to metallic structures may be obtained from other publications [52-58].

(c) *Seismic retrofit of RC columns*

FRP composite wraps are used for seismic retrofit purposes. The composite material can be applied to the columns in the form of a prepreg or by dry fibres which are then impregnated with a resin. This technique is also used to repair reinforced concrete columns where steel corrosion has taken place. For protection of the polymerised composite material, an appropriate post fabrication surface treatment is invariably applied by painting/rendering.

5.7.3 *Post-tension tendons*

A potential rehabilitation application procedure for prestressed concrete (PC) consists of prestressing bars and tendons made from FRP composites. A number of studies have been undertaken by researchers into this technique [59-66]. FRP reinforcement offers many advantages over steel, including, non-corrosive, non-magnetic, high strength and lightweight properties. Several design guidelines for structural concrete reinforced with FRP composites have been developed [67-71]. However, it seems unlikely that, in the immediate future, FRP tendons will gain acceptance in construction without an initial economic incentive to use them.

5.7.4 *Near surface mounted (NSM) FRP rods*

The introduction and utilisation of NSM, FRP composite rods to strengthen the flexural and shear components of RC beams has been described [72]; a section of an RC beam with a near NSM rod in its soffit is shown in Fig. 5.10. There are numerous applications where NSM rods have been employed, an example is the strengthening of the Myriad Convention Centre, Oklahoma City, OK (USA) in 1997-1998 [73].

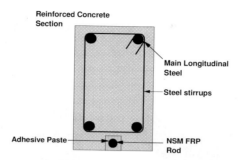

Fig. 5.10 Cross section of RC beam with NSM bar in position.

5.7.5 *FRP jacketing in confining RC members under axial compression*

As FRP composites have high specific tensile strength and stiffness and high corrosion resistance in most environments met in civil engineering FRP jacketing is an ideal technique to encase concrete columns. This method would be applied to earthquake-damaged structures and to structures for rehabilitation where containment will enhance all performance aspects of the upgraded members. Tinazzi et al. [74] discussed the value of FRP confinement as a means of repairing/ strengthening RC members with non-ductile material.

5.7.6 *Internal reinforcement to concrete members*

The pultrusion technique using the thermosetting polymer is the most common method used to manufacture FRP rebars for the reinforcement of concrete. Several techniques are used to improve the bond characteristics between the concrete and the rebar [45,75]. Advantages of utilising FRP rebars lie in the fact that they do not corrode and are not susceptible to chloride or carbonation initiated corrosion in a concrete environment. Although higher in cost, carbon and aramid fibre composites are considered to be inert to alkaline environment degradation and can be used in the most extreme cases.

5.7.7   *FRP deck replacement for bridge deck systems*

The GFRP composite material is a viable alternative material for replacing *short* span RC bridge decks, [9,76-79]. The demand for the development of more efficient and durable bridge decks is at the forefront of the priority of highway authorities worldwide. The bridge deck generally requires the maximum maintenance of all elements in a bridge superstructure.

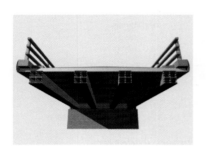

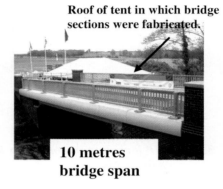

Roof of tent in which bridge sections were fabricated.

10 metres bridge span

Diagramatic section used at Bridge   Completed Bridge West Mill, Gloucester UK

Fig. 5.11 An all GFRP composite bridge deck (ASSET) was developed by an European consortium. Section of the deck unit and that of the completed bridge.
(Courtesy of Mouchel Parkman, West Byfleet, UK).

5.8   Miscellaneous

It will be clear from the previous sections that the advanced polymer composite material is a unique system and must be designed in such a way as to take full advantage of its extraordinary properties. It has many advantages over the conventional materials used in civil engineering and these advantages must be exploited by engineers' innovative skills.

Transmission towers are taken as an example where innovative skills have been utilised. These types of towers are usually manufactured from steel tubes but towers have been manufactured, using advanced polymer composites pultruded tubes, which are joined by specially designed structural elements. These composite skeletal structures provide immense advantages over metallic incumbent structures due to increased durability and ease of installation especially in areas which are difficult to access. Three types of fasteners will be discussed.

(i) The clip-on-fasteners have been developed by W. Brandt Goldworthy & Associates, Inc. The pultruded composite member (or other manufactured rigid member) can be joined on to another member by the Snap Joint technology method where one end is shaped as a 'fir-tree', and therefore has a large load bearing area, is snapped into another structural member. The 'fir tree' end of the joint is slotted along its longitudinal length to provide enough lateral flexibility to compress when entering the part to be joined. A hole is drilled into the section at the end of the slot to prevent crack propagation along the length of the pultruded member. The fibre architecture of the 'fir tree' end must be designed in such a way that the load bearing surfaces have high inter-laminar shear strength capacity. Further information may be obtained from [80].

(ii) A specially designed exterior end cap with fins to mate with slotted cover plates to mechanically join other skeletal members at a node point of either an axially loaded member or at skeletal joints. The end cap is manufactured by the injection mould technique (not described in this chapter) and consists of an inner core and outer sleeve, which are joined at the base of the cap. The outer surface of the inner core and the inner surface of the outer sleeve have a thread moulded into them. The precise amount of adhesive is poured between the outer sleeve and the inner core thus enabling the adhesive to work its way around the threads as the pultruded, or similar member, is inserted into the annulus. The cover plates then slots with the fins to provide the correct angle of the skeletal member. Further information may be obtained from [18].

(iii) The 'energy' loaded joint to deploy a skeletal structure, this system is described in [81].

References

1. Report of Study Group (1989) on 'Advanced Polymer Composites in Civil Engineering' produced by the Institution of Structural Engineers, UK.

2.   Hollaway, L C and Head P R (2001), *'Advanced Polymer Composites and Polymers in the Civil Infrastructure'*. Pub. Elsevier, Oxford.

3.   Fannin, J. (2004) 'AASHTO M288, Durability considerations in Atandard ASpecificatuion Doculments' *Proceedings 57th Canadian Geotechnical Conference, 5th Joint CGS/IAH-CNC Conference*. Session 4D pp. 21-26.

4.   Haque, A. and Shamsuzzoha, M. (2003), "S2-Glass/Epoxy Polymer Nanocomposites: Manufacturing, Structures, Thermal and Mechanical Properties", *Journal of Composite Materials*, 37(20), pp. 1821-1837.

5.   Liu, W., Hoa, S. and Pugh, H. (2005), "Epoxy-clay nanocomposites: dispersion, morphology and performance", *Composites Science and Technology*, 65, pp. 307-316.

6.   Hackman, I and Hollaway, L C (2006). 'Epoxy-layered silicate nanocomposites in civil engineering', *Composites Part A*, Vol. 37 No. 8, pp. 1161-1170.

7.   Aklonia, J J and MacKnight, W J (1983). Introduction to Polymer Viscoelasticy, 2nd edn., *Wiley, New York, NY*, pp. 36-56.

8.   Cessna, L. C. (1971) 'Stress-time superposition for creep data for polypropylene and coupled glass reinforced polypropylene', *Polymer Engineering Science,* Vol. 13, May 1971, pp. 211-219.

9.   Aboutaha, R S (undated). 'Investigation of Durability of Wearing Surfaces for FRP Bridge Decks' *Project No. 01-50 Syracuse University and Cornell University*

10. Liu, Chia-Nan (2003). 'Working-strain based design of Land-fill Final Cover Systems' *Journal of the Chinese Institute of Engineers*. Vol. 26, No. 2 pp. 249-253 (In English).

11. Phillips, N L. (1989) (Editor) Design with Advanced Composite Materials. The Design Council, London Publised Springer-Verlag, London, New York.

12. Starr, T F (2000) (Editor). *'Pultrusion for Engineers'* Pub. Woodhead Publishing Ltd. Cambridge.

13. Hulatt, J., Hollaway, L. & Thorne, A. (2003), 'The use of advanced polymer composites to form an economic structural unit' in *International Journal 'Construction and Building Materials'*.

14. Hulatt, J., Hollaway, L C. and Thorne, AM (2004), 'A novel advanced polymer composite/concrete structural element' *Proc. of the Institution of Civil Engineers' Special Issue: Advanced*

Polymer Composites for Structural Applications in Construction. Feb. 2004, pp. 9-17.

15. Garden, H N and Shahidi, E G. (2002). 'The Use of Advanced Composite Laminates as Structural Reinforcement in a Historical Building' In: *Proceedings of the International Conference 'Advanced Polymer Composites for Structural Applications in Construction'* Editors R A Shenoi, S J Moy and L C Hollaway, held at University of Southampton , UK, 15-17 April 2002. pp. 457-465. Vol. 17, No. 1, pp. 55-68.

16. Zhang, Lei., Photiou, N K., Hollaway, L C., Teng, J G., Zhang, S S. (2007). 'Advances in Adhesive Joining of Carbon Fibre/Polymer Composites to Steel Members for Rehabilitation of Bridge Structures' under review.

17. Subramaniyan, A. K., Bing, Q., Nakaima, D., and Sun, C. T. (2003), 'Effect of Nanoclay on Compressive Strength of Glass Fibre Composites", *Proceedings of the 18th Technical Conference, American Society for Composites*, Gainesville, FL, Oct. 20-22.

18. Hollaway, L C(1993). 'Polymer Composites for Civil and Structural Engineering' *Blackie Academic and Professional*, Glasgow.

19. Hollaway L C (2007), 'Survey of Field Applications' Chapter 12, *'Durability of Composites for Civil Structural Applications'*, Edited by V. M. Karbhari published by Woodhead Publishing, Oxford.

20. Hollaway, L C (2008) Section 7 of 'Manual of Construction Materials' Ed. M.J.Forde to be published by Thomas Telford, London.

21. Bank, L. C. and Gentry, R.T. (1995) 'Accelerated test methods to determine the long-term behaviour of FRP composite structures: Environmental effects', Journal of reinforced Plastics and Composites, Vol. 14, pp. 559-587.

22. Sen, R., Mullins, G. and Salem, T. (2002). 'Durability of E-glass/ Vinylester Reinforcement in alkaline solution', *ASI Structural journal*, Vol. 99, pp. 369-375.

23. Bank, L. C., Gentry, R.T., Barkatt, A., Prian, L., Wang, F and Mangla, S. R. (1998). 'Accelerated aging of pultruded glass/vinylester rods', *Proceeding 2nd International Conference on Fibre Composites in Infrastructure (ICCI), (1998)*, Vol. 2, pp. 423-437.

24. Uomoto, T. (2000). 'Durability of FRP as reinforcement for concrete structures' *Advanced Composite Matrerials in Building and Structures 3rd International Conference,Ottawa, Ontario, Canada, (2000)* pp. 3-14.

25. Tomosawa, F and Nakatsuji, T (1997) 'Evaluation of ACM reinforcement durability by exposure tests', Non-metallic (FRP) reinforcement for concrete structures, *Proceedings 3rd International Symposium, Sappoto, (1997)* Vol. 2, pp. 139-146.

26. Clarke, J. L. and Sheard P., (1998). 'Designing durable FRP reinforced concrete structures', *Proceedings 1st International Conference on Durability of Fibre reinforced polymer (FRP) Composite for Construction (CDCC 1998), Sherbrooke, Quebec, Canada, (1998),* pp. 13-24.

27. Sheard, P., Clarke, J.L., Dill, M., Hammersley, G. and Richardson, D. (1997). 'EUROCRETE – Taking account of durability for design of FRP reinforced concrete structures'- non-metallic (FRP) reinforcement, for concrete structures, *Proceedings of 3rd International Symposuim, Sapporo, (1997),* Vol. 2, pp. 75-82.

28. Concrete Society (2000) *'Design Guidance for Strengthening Concrete Structures Using Fibre Composite Materials'*, TR55, 2nd ed., Camberley UK.

29. Concrete Society (2003) *'Strengthening Concrete Structures using Fibre Composite Materials: Acceptance, Inspection and Monitoring'* TR57, Camberley UK.

30. Modern Plastics Encyclopedia, *McGraw-Hill, New York*, October 1988, Vol. 65, No. 11, pp. 576-619.

31. Photiou, N.K., Hollaway, L.C., and Chryssanthopoulos, M.K. (2006) 'Strengthening of an artificially degraded steel beam utilising a carbon/glass composite system' In *Construction and Building Materials,* Vol. 20 Nos.1-2 Feb./March 2006

32. Zhang, L, Hollaway, L C, Teng, J-G, Zhang, S S. (2006). 'Strengthening of Steel Bridges under Low Frequency Vibrations' Proceedings of the 3rd International Conferenceon FRP Composites in Civil Engineering (CICE 2006), December 13-15 2006 Miami, Florida, USA.

33. Mays G C. and Hutchinson A R. (1992). Adhesives in Civil Engineering, *Cambridge University Press.*

34. Hutchinson A R (1997). Joining of Fibre-reinforced Polymer Composites Materials' *Project Report 46, CIRIA, London.*

35. Hollaway, L C (2005) 'Advances in Adhesive Joining of Dissimilar Materials with Special Reference to Steels and FRP Composites' *Proceedings of the International Symposium on Bond Behaviour of FRP Structures' (BBFS 2005) Ed. J F Chen and J G Teng*, 7-9 December 2005, Hong Kong, China.

36. Clarke, J and Hutchinson, A R. (2003), Eds, Strengthening concrete structures using fibre composite materials: acceptance, inspection and monitoring, Technical Report No. 57. The Concrete Society, Crowthorne, UK

37. Kinlock, A J., (1987). *'Adhesion and adhesives: Science and technology' London, Chapman and Hall*

38. Guyson, (1989). 'Manual of Blast media', *Guyson Data Sheets.*

39. Meier, U. (1987), 'Bridge Repair with High Performance Composite Materials', *Material und Tcchnik*, Vol. 15, pp. 125-128, 1987 (in French and German).

40. Kaiser, H.P. (1989), 'Strengthening Reinforced Concrete with Epoxy-bonded Carbon-fibre Plastics', *Doctoral Thesis*, Diss. ETH, Nr. 8918, ETH Zurich, Ch-8092 Zurich, Switzerland, 1989 (in German).

41. Meier, U. and Kaiser, H.P. (1991), 'Strengthening of Structures with CFRP Laminates', *Proc. Advanced Composite Materials in Civil Engineering Structures'*, Mats. Div., ASCE, Las Vegas, Jan. 1991, pp. 224-232.

42. Hollaway, L. C. and Leeming, M B. (1999). 'Strengthening of Reinforced Concrete Structures, – using externally-bonded FRP composites in structural and civil engineering'. *Pub. Woodhead Publishing Ltd. Cambridge, England.*

43. Teng, J G., Chen J F., Smith, S T. and Lam, L. (2001) 'FRP Strengthened RC Structures' *Pub. John Wiley, England, USA, Germany, Australia, Canada, Singapore.*

44. Oehlers, D J. and Seracino, R. (2004). 'Design of FRP and Steel Plated RC Structures' – Retrofitting Baems and Slabs for Strength, Stiffness and Durability. *Pub. Elsevier.Amsterdam, London New York, Sydney.*

45. Bank, L C. (2006). 'Composites for Construction Structural design with FRP Materials'.Publisher John Wiley & Sons, New Jersey.

46. Mertz, D. and Gillespie, J. (1996), 'Rehabilitation of steel bridge girders through the application of advanced composite material'

NCHRP 93-ID11, Transportation Research Board, Washington, D.C, pp. 1-20.

47. Mosallam A.S. and Chakrabarti, P.R. (1997), 'Making connection', *Civil Engineering, ASCE*, pp. 56-59.

48. Luke, S. (2001), 'Strengthening structures with carbon fibre plates. Case histories for Hythe Bridge, Oxford and Qafco Prill Tower', *NGCC first annual conference and AGM – Composites in Construction through life performance, 30th-31st October 2001, Watford, UK.*

49. Tavakkolizadeh, M. and Saadatmanesh, H. (2003), 'Strengthening of steel-concrete composite girders using carbon fibre reinforced polymer sheets', *Journal of Structural Engineering, ASCE*, January 2003, pp. 30-40.

50. Luke, S and Canning, L. (2004).'Strengthening Highway and Railway Bridge Structures with FRP Composites – Case Studies.' *Proceedings 'Advanced Polymer Composites for Structural Applications in Construction'* Second International Conference, held at University of Surrey, Guildford , UK on 20th-22nd April 2004, Editors L C Hollaway, M K Chryssanthopoulos and S J Moy pp. 747-754.

51. Luke, S., and Canning, L., (2005). Strengthening and Repair of Railway Bridges Using FRP Composites, In: Parke, G.A.R., and Disney, P. (Eds), *Bridge Management 5*, Thomas Telford Ltd., 2005, pp. 684. ISBN 0727733540.

52. Hill, P S, Smith, S and Barnes, F J. (1999), 'Use of High Modulus Carbon Fibres for Reinforcement of Cast Iron Compression Struts within London Underground: Project details. *Conference on Composites and Plastics in Construction, Nov. 1999, BRE, Watford, UK. RAPRA Technology, Shawbury, Shrewsbury, UK.* Paper 16 1-6.

53. Liu X., Silva P.R. and Nanni, A. (2001), 'Rehabilitation of steel bridge members with FRP composite materials', *Proc. CCC 2001, Composites in Construction, Porto, Portugal, Oct. 2001*, Editors, J. Figueiras, L. Juvandes and R Furia. pp. 613-617.

54. Moy SSJ,(2001) editor. 'FRP composites – Life Extension and strengthening of Metallic Structures' (2001). London,: *Institution of Civil Engineers,* 33-35.

55. Leonard, A R. (2002). 'The Design of Carbon Fibre Composite Strengthening for Cast Iron Struts at Shadwell Station Vent Shaft'

In: *Proceedings of the International Conference 'Advanced Polymer Composites for Structural Applications in Construction'* Editors R A Shenoi, S J Moy and L C Hollaway, held at University of Southampton , UK, 15-17 April 2002. pp. 219-227.

56. Cadei, J M C., Stratford, T J. Hollaway, L C. and Duckett, W G. (2004). 'Strengthening Metallic Structures Using Externally Bonded Fibre-reinforced Polymers' *Pub. CIRIA, London.*

57. Photiou N.K., Hollaway L.C. and Chryssanthopoulos M.K. (2004)[A], 'Strengthening of an Artificially Degraded Steel Beam Utilising a Carbon/Glass Composite System' in, *Conf. Proc. 'Advanced Polymer Composites for Structural Applications in Construction'* – *ACIC 2004,* held at Univ. of Surrey 20th-22nd April, 2004

58. Photiou N.K., Hollaway L.C. and Chryssanthopoulos M.K. (2004)[B], 'Selection of CFRP Systems for Steelwork Upgrading' *Conf. Proc. IMTCR 04 held at Lecce 7th – 9th June 2004.*

59. Arockiasamy, M, Shahawy, MA, Sandepudi, K, and Zhuang, M. (1996). 'Application of High Strength Composite Tendons in Prestressed Concrete Structures'. *Proceedings of the First InternationalConference on Composites in Infrastructure: Fiber Composites in Infrastructure,* University of Arizona, Tucson, Arizona, USA, pp. 520-535

60. Balázs, G L, Borosnyói, A and Almakt, M M. (2000). 'Future Possibilities in Using CFRP Tendons for Pre-stressed Pre-tensioned Girder', *Proceedings of the Bridge Engineering Conference 2000: Past Achievements, Current Practice, Future Technologies, Egyptian Society of Engineers,* Sharm El-Sheikh, Egypt. 8pp (CD-Rom).

61. Burke, C R and Dolan C W. (2001). 'Flexural Design of Prestressed Concrete Beams using FRP Tendons, *PCI Journal,* 46(2), pp. 76-87.

62. Fam, A Z, Rizkalla, S H. and Tadros, G. (1997). 'Behaviour of CFRP for Prestressing and Shear Reinforcements of Concrete Highway Bridges', *ACI Structural Journal,* 94(1), pp. 77-86.

63. Grace, N F.(1999). 'Continuous CFRP Prestressed Concrete Bridges, Concrete Bridges', *Concrete International,* 21(10) pp. 42-47.

64. Grace, N F. (2000). 'Transfer Length of CFRP/CFCC Strands for Double-T Girders, *PCI Journal,* 45(5), pp. 110-126.

65. Lu, Z, Boothby, T E, Bakis, C E, and Nanni, A. (2000). ,Transfer and Development Lengths of FRP Prestressing Tendons', *PCI Journal*, 45(2), pp84-95.

66. Saadatmanesh, H and Tannous, F E. (1999). 'Relaxation, creep and fatigue behaviour of carbon fiber reinforced plastic tendons', *ACI Materials Journal*, 96(2), pp. 143-153.

67. Committee 440 (1996) 'State-of-the-Art Report on Fiber Reinforced Plastic (FRP) Reinforcement for Concrete Structures', *American Concrete Institute, Farmington Hills, Michigan, USA*.

68. ACI Committee 440 (2000) 'Guide for the Design and Construction of Concrete Reinforced with FRP Bars', *American Concrete Institute, Farmington Hills, Michigan USA*.

69. BRI (1995) 'Design Guidelines for FRP Prestressed Concrete Members' (in Japanese), *FRP Reinforced Concrete Research Group, Building Research Institute, Japanese Ministry of Construction, Tokyo, Japan.*

70. CSA (2000). Canadian Highway Bridge Design Code (CAN/ CSAS6), *CSA International, Toronto, Ontario, Canada.*

71. ISIS Canada (2001) Reinforced Concrete Structures with Fiber-Reinforced Polymers(ISIS=-MO4-00), The Canadian Network of Centres of Excellence on Intelligent Sensing and Innovative Structures (ISIS Canada), *University of Manitoba, Winnipeg, Manitoba, Canada.*

72. De Lorenzis, L. Tinazzi, D. and Nanni, A (2000), 'Near Surface Mounted FRP Rods for Masonry Strengthening: Bond and Flexural Testing' *Proceedings of the National Conference, 'Meccanica delle strutture in muratura rinforzate con FRP materials: modellazione, sperimentazione, progetto, controllo;'*, Venezia, Italy, December 7-8 (in Italian).

73. Hogue, T., Cornforth, R. C. and Nanni, A. (1999) 'Myriad Convention Centre Floor System Reinforcement', *Proceedings of the Fourth International Symposium on Fiber Reinforced polymer Reinforcement for reinforced concrete Structures*, Editors, C.W. Dolan, S. Rizkalla and A Nanni, American Concretre Institute SP-188, pp. 1145-1161.

74. Tinazzi, D. Modena, C. Nanni, A. (2000), 'Strengthening of Masonry Assemblages with Fiber Reinforced Polymer Rods and Laminates', *Proceedings: Advancing with Composites 2000, Milan, Italy.*

75. Gdoutos, E E, Pilakoutas, K and Rodopoulos. (2000). 'Failure Analysis of Industrial Composite Materials', Published by McGraw-Hill New York, London, Lisbon Madrid.

76. Luke, S., Canning, L., Collins, S., Brown, P., Knudsen, E. and Olofssoof (2002). 'The development of an Advanced Composite Bridge Decking System – Project ASSET', *Structural Engineering Interrnational* Vol. 12, No. 2 pp. 76-79.

77. GangaRao, H V S, Thippeswamy, H K, Shekar, V and Craigo, C. (1999). 'Development of Glass Fibre Reinforced Polymer Composite Bridge Deck', *SAMPE Journal* Vol. 34 No. 4. pp. 12-24.

78. Kim, H-Y, Choi, Y-M, Hwung Y-K and Cho, Y-M. (2005) 'Optimum Design of a Pultruded FRP Bridge Deck' *Conf. proc. '2003 ECI Conference on Advanced Materials for Construction of Bridges Buildings and other Structures III'*. Edited, V Mistry, A Azizinamini and J M Hooks (2005).

79. FHWA Bridge Program Group (2001), 'Count of Deficit Highway Bridges' (http://www.fhwa.dot.gov/bridge). (March 20, 2002). *The Office of Bridge of Bridge Technology, The Federal Highway Administration, Washington, D.C.*

80. Goldsworthy, W.B. and Heil, C. (1998). 'Composite Structures are a Snap, in *2nd International Conference on Composites in Infrastructure, Vol. 2 Tucson, AZ.* 1998, pp. 382-396.

81. Fanning, P. and Hollaway, L C. (1993). 'A Case Study in the Design and Analysis for a 5.0 m Deployable Composite Antenna'. *Composites Engineering,* Vol. 3 No. 11, pp. 1007-1023.

Chapter 6

Enhancing the Performance of Masonry Structures

Richard E. Klingner

Ferguson Structural Engineering Laboratory
The University of Texas at Austin

6.1 Introduction

A reader might well ask, "What is the role of masonry, probably humankind's oldest building material, in a text addressing high-performance materials?" The intent of this chapter is first to answer that question directly, and then to provide additional information on the past, present and future of high-performance masonry.

The adjective "high-performance" began to appear almost 20 years ago, in the context of conventional construction, to refer to concrete with specified compressive strengths exceeding 8,000 psi (56 MPa). It rapidly evolved, as users came to understand that high strength was only one of many attributes that might make a concrete suitable for a particular application. Today, "high-performance" refers broadly to a material or construction system, one or more of whose characteristics are intentionally enhanced to make it more suitable for a particular application.

To explore "high-performance" masonry further, we must first ask ourselves, "What are the different ways in which masonry is expected to perform?" Masonry is expected to perform as building envelope; as architecture; and as structure. It must carry out each of these functions in a cost-competitive way, in comparison with other systems and materials. In the remainder of this chapter, how masonry carries out each function is described in more detail; its performance with respect to each function is noted; ways in which that performance has been enhanced are noted;

and suggestions are given for focused research to improve performance further in some areas.

6.2 Performance of Masonry as Building Envelope

Masonry, whether regarded as structural or not, must usually function as part of a building's exterior, or envelope. In that function, it might be required to perform in the following ways:

- resist liquid water;
- control water vapor;
- control the environment inside the envelope (temperature, humidity and noise);
- control damage from fire;
- control damage from hail and wind-borne debris;
- resist or transfer externally applied loads; and
- resist or accommodate differential movement.

In the remainder of this section, masonry's performance in each of those functions is further described.

6.2.1 *Resist liquid water*

When significant quantities of liquid water pass through a building envelope, the envelope has obviously failed, and can also experience hidden problems from loss of thermal insulating effectiveness, formation of mildew or rot, and corrosion. Liquid water does not normally pass through mortar, nor through clay masonry units, nor through the grouted cells of concrete masonry units. It can, however, pass through voids in mortar joints, through cracks between units and mortar, and through gaps between the masonry and other parts of the building envelope. Preventing or sealing these gaps requires:

- proper design of the masonry itself (for example, by making sills from one-piece stone or precast concrete instead of masonry units laid in soldier or rowlock orientation, and by sloping the sills);
- proper design, detailing and installation of the masonry envelope's ventilation and drainage details (weep-holes, flashing and drips);
- proper design, detailing and sealing of gaps in the masonry envelope (movement joints);
- proper design, detailing, and sealing of gaps between masonry and the rest of the envelope;

– proper masonry construction;
– proper maintenance of sealant over the life of the envelope; and
– proper cleaning to avoid the development of openings at interfaces.

Challenges to improving the performance of masonry with respect to water penetration include the following:

– effectively communicate proper masonry design, detailing and construction practices to those responsible for carrying them out and to those responsible for inspecting them;
– effectively communicate proper installation practices for ventilation and drainage details to those responsible for carrying them out and to those responsible for inspecting them;
– eliminate or reduce the need for conventional cleaning of new masonry using innovative coatings or construction practices; and
– increase the effectiveness and design life of sealants, or reduce the envelope's dependence on their performance.

6.2.2 *Control water vapor*

When significant quantities of water vapor are trapped within the thickness of a building envelope, the envelope can have problems with wetting of finishes, loss of thermal insulating effectiveness, formation of mildew or rot, and corrosion. Movement of moisture in vapor form into the envelope should be reduced using appropriate air and vapor barriers, vapor retarders, or both. Moisture in the envelope should be able to escape without damaging the envelope. With the exception of glazed units, most masonry materials themselves permit the passage of water vapor. This passage can be blocked, however, if the surfaces of the masonry assembly are sealed, or if accessory materials are improperly specified or improperly installed.

Prevention of excess moisture in the vapor through the envelope requires:

– minimization of the transmission of water vapor via air leakage; and
– minimization of the formation of condensed water vapor via vapor transmission.

Passage of water vapor through the envelope requires:

– proper placement of vapor barriers, if used; and
– specification of exterior coatings, if used, that are vapor-permeable.

Challenges to improving the performance of masonry with respect to vapor control include the effective communication of proper design, specification and installation practices for vapor barriers to those responsible for carrying them out.

6.2.3 *Control of the environment inside the envelope (temperature, humidity and noise)*

To maintain a productive living and work environment inside a building envelope, temperature, humidity, and noise must be controlled.

6.2.3.1 Control of temperature

Control of temperature requires that the temperature be maintained within a comfortable range, and that the temperature be reasonably uniform inside the building envelope, without hot spots, cold spots, or drafts. This in turn requires that the building envelope be sealed against the passage of air drafts; that it have a sufficiently high thermal insulating value; and that it resist rapid temperature changes. The first objective is probably governed by the above requirements for resistance to the passage of liquid water. The others, however, are not.

The thermal insulating value (R-value) of a building envelope depends on the materials of which it is made, on the thickness of those materials, and on the way in which those materials are arranged. Masonry's inherent R-values are moderate, and the R-value of a masonry assembly can be increased significantly by appropriate placement of insulation.

First, the net R-value of a building envelope is significantly reduced by thermal bridges – places where thermal energy can be transmitted directly through the envelope. For example, if a single-glazed window is placed in an insulated masonry wall, the net R-value of the wall can be reduced from about 8 to about 5. For another example, when the thermal bridging effects of the studs are considered, the net R-value of a wooden stud wall with fiberglass insulation is easily reduced from about 11 to about 8. Reductions due to thermal bridging by steel studs are even greater.

Second, the effective R-value of a building envelope is increased by its thermal mass -- the amount of thermal energy required to change the temperature of the envelope itself. Because of their density, masonry materials generally have high thermal mass, and masonry's effective

R-values are comparable to those found in assemblages made from lighter materials and high-efficiency insulation.

Finally, the thermal performance of a building envelope depends on the extent to which it reduces the building's heat gain or loss. Heat gain depends primarily on the transmission of solar radiation through the envelope. Because thermal transmission is much greater for buildings with transparent envelopes than with opaque ones (even dark, opaque ones), masonry transmits less thermal energy than glass curtain walls. Heat loss depends, among other things, on the building envelope's thermal radiation. Because absorption of thermal radiation is greater for darker materials, and less for lighter ones, masonry's performance with respect to thermal radiation is balanced.

In summary, masonry can be effective in controlling temperature. Challenges in enhancing its effectiveness in that regard include how to increase the thermal insulation value of masonry units.

6.2.3.2 Control of humidity

Humidity is controlled within the building envelope by a suitable HVAC system, and by controlling the ingress of liquid water and the transmission of water vapor through the envelope. Humidity is controlled through the thickness of the envelope (for example, within the cavity of a drainage wall) by control of air and vapor movement through the envelope. Both of these are discussed above.

6.2.3.3 Control of noise

Noise is controlled within the building envelope by controlling sound transmission through the envelope, and absorption within it.

Sound transmission through an envelope depends on the envelope's mass, and on the elimination of openings that might transmit sound directly through the envelope. Because of its mass, masonry's performance with respect to sound transmission is excellent.

Sound absorption depends on the surface porosity and surface texture of the envelope material. With the exception of masonry constructed using special units, masonry's performance with respect to sound absorption ranges from average to poor.

In summary, masonry can be effective in controlling noise. Challenges in enhancing its effectiveness in this regard include finding ways of increasing the surface porosity of masonry, while retaining its durability.

6.2.4   Control damage from fire

Design of buildings for fire in the US follows these basic steps:

(1)   classify the building according to occupancy and use;
(2)   classify the building according to type of construction;
(3)   establish the building's allowable height and area, based on the use and occupancy classification and the type of construction;
(4)   establish each building component's required fire resistance in hours, based on occupancy and material classification;
(5)   establish fire-rating requirements for exterior walls based on fire separation distance;
(6)   establish maximum area of exterior wall openings; and
(7)   verify that each building component's fire resistance in hours equals or exceeds the required fire resistance in hours.

When carrying out the design of exterior or interior walls for fire, the required fire rating is usually ranges from 2 to 4 hours, irrespective of most of the building characteristics addressed in the above steps. A 2-hour rating requires about 4 inches of masonry; a 4-hour rating, about 6 in. Required thickness depends on the type of masonry, and can also be reduced if the building is provided with an automatic sprinkler system. These available resistances, based directly or indirectly on ASTM E 119 fire tests, recognize the good performance of masonry in controlling the spread of fire.

Masonry has other good fire-resistance characteristics, however, that are not directly recognized by ASTM E119 fire tests. Because it is noncombustible and has high thermal mass, masonry offers excellent protection for building elements that are more vulnerable to fire. Finally, it does not emit toxic fumes or contribute to flame spread.

Challenges in enhancing masonry's effectiveness in resisting fire include the following:

−   find ways to increase the fire-resistance rating of a given thickness of masonry;
−   find ways to justify and encourage the use of non-combustible materials such as masonry in compartmentalization of buildings, with or without automatic sprinkler systems; and

- evaluate the fire resistance of masonry in combination with other materials.

6.2.5 *Control damage from hail and wind-borne debris*

Resistance of the building envelope to damage from hail and wind-borne debris depends primarily on the impact resistance of the masonry assembly. Challenges in this area include the need to continue the development and dissemination of effective material specifications for impact resistance of masonry.

6.2.6 *Resist or transfer externally applied loads*

A subsequent section of this chapter addresses the structural functions of masonry. In that section, structural design of masonry is discussed in detail. The structural functions of masonry are also mentioned here, however, because if the masonry does not function as structure, it will probably not function as building envelope, either.

Many masonry components that function as parts of a building envelope are not themselves intended to resist externally applied loads -- they merely transfer those loads to a backup system. That transfer is accomplished by connectors. For the masonry to function properly as building envelope, the connectors themselves must continue to function, and the out-of-plane stiffness of the connectors and the backup system must be compatible with the out-of-plane stiffness of the masonry envelope.

If the connectors or backup system are too flexible compared to the masonry itself, cracks in the masonry may be sufficiently wide to permit the passage of an objectionable amount of water through the building envelope. If too much water passes through the building envelope, serious corrosion problems can occur in the connectors, backup system, or both. Such corrosion problems can eventually jeopardize the structural integrity of the building envelope.

Challenges in enhancing the effectiveness of masonry building envelopes in resisting applied loads are addressed in the section of this chapter dealing with masonry as structure. Challenges in enhancing the effectiveness of masonry building envelopes in transferring applied loads include the following:

- develop user-friendly stiffness requirements for backup systems and connectors so that cracks in the masonry veneer will not permit

an objectionable amount of water to pass through the building envelope; and
- enhance the corrosion resistance of connectors and backup systems to prolong the effective life of masonry veneers.

6.2.7 *Resist or accommodate differential movement*

An ideal building envelope would have no gaps that could permit unintentional air movement or the passage of liquid water. Unfortunately, masonry building envelopes must often have gaps to accommodate differential movement between the envelope and the backup system, and between adjacent elements of the envelope itself. For example, clay masonry veneer (which expands over time) on a concrete frame (which shrinks over time) must be provided with expansion joints to accommodate the differential movement between the veneer and the frame. Masonry envelopes must also be provided with expansion joints (for clay masonry) or control joints (for concrete masonry) to accommodate movement within the envelope itself.

Except for small elements or monolithic construction, it is generally not practical to design masonry envelopes to resist differential movement. Rather, that movement must be accommodated.

Challenges in enhancing the effectiveness of masonry building envelopes in accommodating differential movement include:

- develop standard, user-friendly guidelines for locating and installing movement joints; and
- develop standard, user-friendly details for movement joints.

6.2.8 *Concluding remarks on masonry as building envelope*

Based on forensic experience, most problems with masonry buildings involve the performance of masonry as building envelope, rather than the performance of masonry as structure. This suggests that if masonry is to be added incrementally to university engineering and architecture curricula, then the proper specification and detailing of masonry should be a higher priority than its structural design.

Our challenges in this regard are as follows:

- find effective ways to introduce masonry specification and detailing into undergraduate engineering and architecture curricula;

- encourage more university professors to become involved in the ASTM process, where they can learn about specification and detailing issues in addition to structural design; and
- encourage the continuing education about masonry, of practicing architects, engineers, contractors and building officials.

6.3 Performance of Masonry as Architecture

For more than 10,000 years, masonry has been used in a wide variety of architectural forms, including domes, pyramids, arches, walls, facades, and shells. It is architecturally flexible because it is composed of relatively small, hand-placed units. It is architecturally appealing because of its range of colors, textures, patterns and shapes. This inherent appeal is enhanced by increases in its architectural flexibility and visual attractiveness.

For purposes of this chapter, the term "masonry as architecture" refers to all appearance aspects of masonry – its global appearance in a building, and its local appearance as a composite of units, mortar, grout and accessory materials.

6.3.1 *Global appearance of masonry*

The global appearance of masonry depends primarily on the variety of architectural forms in which it can be used, and also on the variety of architectural details that can be incorporated into those forms.

6.3.1.1 Architectural forms

Masonry is traditionally used in walls. Because it is much stronger in compression than in tension, it is also naturally suited for compressive structural forms such as arches and domes.

Principal challenges in enhancing the global appearance of masonry with respect to architectural forms include the following:

- find ways to increase the variety of architectural forms that can be laid using a relatively small number of different unit sizes, and with little or no cutting of units; and
- develop formwork and scaffolding that will make it easier to construct masonry arches and domes.

6.3.1.2 Architectural details

Masonry details include corbels (masonry that is stepped away from the building with increasing elevation), racks (masonry that is stepped toward the building with increasing elevation); quoins (rectangular protrusions at masonry corners; and bond patterns (orientations of units).

Challenges in enhancing the global appearance of masonry with respect to architectural details include the following:

– find ways to increase the variety of architectural details (such as corbels, racks, quoins, and different bond patterns) that can be laid without cutting units or unduly increasing cost; and
– develop construction techniques or tools that will make it easier to construct masonry architectural details more quickly and reliably.

6.3.2 *Local appearance of masonry*

The local appearance of masonry depends primarily on the color, shape and surface texture of the units and mortar comprising it. Color and shape of units are controlled by manufacturing tolerances. The surface texture of units is controlled by manufacturing processes, and the surface texture of a masonry surface is also controlled by whether the units are laid true to that surface. Finally, local appearance depends on the visual pattern of units and mortar.

Challenges in enhancing the local appearance of masonry include the following:

– find ways to improve the consistency of color of units and mortar;
– find ways to decrease cracking and chipping of masonry units;
– find better ways to control the alignment of units and mortar joints, the variation in thickness of mortar joints, and the variation of masonry walls from level and plumb;
– develop industry-wide standards of acceptance for the installation of masonry (tolerances, joint widths);
– find better ways to control staining, and improve cleaning techniques; and
– find better ways to decrease or eliminate efflorescence.

6.4 Performance of Masonry as Structure

6.4.1 *Introductory remarks to masonry as structure*

Masonry as structure must typically resist loads from gravity and wind, and must occasionally resist extreme loadings such as earthquake. When we teach masonry design as part of engineering curricula, we tend to emphasize structural analysis and calculation, to the detriment of specification and detailing. The ultimate goal of masonry design is to design, not to analyze and calculate. Most masonry can be designed with little or no structural calculation, based on previous experience with similar elements.

In this general context, our challenges in improving the performance of masonry as structure include the following:

– develop simplified design provisions, consistent with the more complex ones, for the design of structural elements that we use often;
– develop user-friendly design aids to take the drudgery out of complex calculations; and
– develop "deemed-to-comply" designs for simple masonry structures.

In specific areas of structural design, some issues should be examined in more detail:

– strength versus allowable-stress versus empirical design;
– behavior of specific masonry elements – gaps in current codes; and
– structural behavior of masonry under extreme loads.

Those issues are addressed in the following subsections.

6.4.2 *Strength design versus allowable-stress design versus empirical design*

Masonry design provisions are intended to produce designs with acceptable probabilities of failure. In general, different design approaches – strength, allowable-stress, and empirical – should produce designs with similar probabilities of failure. To compute a probability of failure requires knowledge of the statistical variability of the load. To compare probabilities of failure is simpler, however, because it requires only a comparison of the strengths of the final designs. In general, elements designed by different approaches should have similar strengths, while allowing for the possibility of some differences. A simpler design method, for example, might be required to produce more conservative results, while a more rigorous and hopefully more accurate design method might be permitted to produce less conservative results.

The strengths of elements designed by different approaches of the 2006 *International Building Code* [1] have been compared in limited contexts within the technical committee structure of the Masonry Standards Joint Committee [2-3]. The following tendencies have been noted:

– In most cases, unreinforced masonry panel walls designed by the strength approach are identical to those designed by the allowable-stress approach. The combined effects of load and resistance factors in strength design cancel, giving a required strength that is identical to that of allowable-stress design.
– Unreinforced masonry bearing walls designed by the strength approach are in general slightly stronger than those designed by the allowable-stress approach, because the critical loading combination for strength design involves *0.9D* rather than *1.0D*, and capacity is governed by flexural tension.
– Reinforced masonry elements designed by the strength approach are in general quite similar to those designed by the allowable-stress approach.

There is need for systematic confirmation of these tendencies.

When strength design of masonry was being developed, one argument sometimes raised in its favor was that masonry elements designed by strength procedures could use less material than corresponding elements designed by working-stress procedures, and hence could be more cost-effective. This issue is complex. It depends on the nature of the element itself, on the accuracy of the strength equation versus the allowable-stress equation for the design of that element, and on non-technical issues as well.

– Some masonry elements are statically determinate. Their capacity is reached as soon as the moment at any point exceeds the corresponding flexural capacity. Unreinforced masonry panel walls, or simply supported masonry beams, are examples of such elements. While strength design offers the possibility of designing for ductile flexural behavior in the case of reinforced masonry, such ductility does not increase the strength of statically determinate elements. By providing more warning of impending collapse, strength design may increase the inherent safety of reinforced masonry elements. It does not, however, permit their size or steel area to be decreased.

– Other masonry elements are statically indeterminate. After such reinforced masonry elements yield in flexure at one section, their capacity can continue to increase if additional flexural capacity exists at other sections. Because elastic moments are usually used for design, however, this reserve capacity is not recognized in design. It unquestionably can decrease the probability of failure of the structure containing those elements, provided that the elements are linked together by reinforcement. While the decreased probability of failure might in theory justify a higher capacity-reduction factor (and hence smaller elements) for strength design than for allowable-stress design, the difficulty of distinguishing between statically determinate and statically indeterminate elements probably means that in practice, capacity-reduction factors cannot be increased, and no savings can be realized in member size or steel area.

– If the formulas used for predicting capacity for strength design are more accurate than the corresponding formulas for allowable-stress design, then their increased reliability can be reflected in higher capacity-reduction factors. In many cases, however, both sets of equations show considerable scatter, particularly for shear capacities, and significant increases in capacity-reduction factors are difficult to justify.

– Strength design equations are usually more obviously related to the actual strength of structural elements, than are their allowable-stress counterparts. Because of this, they are often easier to learn and use correctly by designers who have prior experience with strength design of reinforced concrete. This is an advantage.

– Strength design generally gives more consistent probabilities of failure (collapse) than allowable-stress design. Because allowable-stress design is less consistent, it may sometimes require less material than strength design. Attempts to adjust strength design so that it is always more economical than allowable-stress design should be approached with caution.

Comparisons between analytical design and empirical design are not as clear. Empirical design is generally based on maximum permissible *h/t* ratios, maximum permissible plan aspect ratios of floor diaphragms, and maximum permissible stresses on the gross areas of walls. In the presence of axial load, empirically designed elements may be comparable with their analytically designed counterparts. Empirically designed elements with little or no axial load, however, may be significantly less strong than their analytically designed counterparts. Examples of such elements are parapets and non-bearing walls spanning horizontally between pilasters.

The MSJC has not unanimously accepted the principle that empirically designed elements can be evaluated by analytically computing their strength and comparing that strength with the strength of their analytically designed counterparts. In the opinion of the author, this principle should be accepted, and the adequacy of our empirical provisions should be evaluated using it, particularly for elements with little or no axial load. At the same time, indications of satisfactory performance by masonry elements not meeting analytical requirements, should stimulate code developers to explore possible explanations for such satisfactory behavior, and, if justified, to incorporate those explanations into analytical design provisions.

6.4.3 *Inconsistencies in current design provisions*

Some research needs are related to gaps in our current design provisions. For example, the 2005 *MSJC Code* [2] has inconsistencies and gaps in the following areas:

- minimum and maximum flexural reinforcement;
- moment magnifiers for masonry beam-column;
- the 1/3 stress increase;
- effective shear area for non-rectangular sections;
- effective width in compression around grouted reinforcement;
- effective distribution of bearing stresses under bond beams;
- effective width associated with a prestressing tendon; and
- mixed-approach designs.

In the remainder of this section, each is addressed.

6.4.3.1 Minimum and maximum flexural reinforcement

The allowable-stress provisions of the 2005 *MSJC Code* have no requirements for minimum or maximum flexural reinforcement. Work is currently underway to correct this deficiency, and that work should continue to fruition.

The strength design provisions of the 2005 *MSJC Code* have maximum reinforcement requirements intended to ensure flexural ductility consistent with that assumed in design. The objective of the requirements is to prevent crushing of the compression toe of a flexural element under the design axial load and a critical strain gradient that depends on the seismic force modification coefficient used in design.

The 2005 *MSJC Code* offers the designer the alternative of confined boundary elements. Specific requirements for those boundary elements do not yet exist, however, and must be determined by test. Boundary-element requirements could be developed by focused research.

6.4.3.2 Moment magnifiers for masonry beam-columns

The 2005 *MSJC Code* does not have consistent provisions for addressing the effects of slenderness. In its strength-design provisions, compressive capacities are reduced by a slenderness-related factor. This approach, while not illogical, is not the best way of handling the problem. In theory, moments should be increased by a moment magnifier.

In the allowable-stress provisions, the compressive capacity of unreinforced masonry is decreased by a slenderness-related factor, and is also checked against elastic buckling using a severe penalty factor that rapidly decreases the allowable capacity as the effective eccentricity increases. This approach, while also not illogical, is also not the best way of handling the problem. It was originally proposed as a conservative substitute for a moment magnifier.

Now that moment magnifiers have been commonly accepted in reinforced-concrete design provisions, and the computational aids available to the designer are much more powerful, there seems to be no good reason to avoid the use of moment magnifiers in masonry design provisions. Work should continue to introduce consistent moment magnifiers into masonry design provisions. If necessary, these magnifiers should be checked using focused research.

6.4.3.3 The 1/3 stress increase

ASCE 7-02 [4], cited by the strength design provisions of the 2005 *IMSJC Code*, prescribes reduced load factors for load combinations involving wind or earthquake in addition to live load. The basis for this is probabilistic. It is improbable that the design live load (an upper fractile of the expected statistical distribution of live loads over time) will occur simultaneously with the design wind or earthquake load (upper fractiles of their expected statistical distributions over time).

The allowable-stress provisions of the 2005 *MSJC Code* include a provision permitting that allowable stresses be increased by one-third for loading combinations including wind or earthquake. Unlike the combinations considered above, this increase applies even to combinations of dead load plus wind or earthquake. There is no probabilistic justification for this, because dead load acts all the time. Nor is there any evidence to support the possible contention that the strengths of masonry and reinforcement are increased by one-third under the strain rates commonly associated with wind load. The one-third increase should be viewed as simply an increase in allowable stresses, and a corresponding increase in the probability of failure.

The basic allowable-stress loading combinations of the 2006 *IBC* specifically prohibit the use of the one-third increase. This prohibition has been regarded as unwarranted by some segments of the masonry technical community. The alternative allowable-stress loading combinations of the 2006 *IBC*, in contrast, permit the use of the one-third increase. This inconsistency, and the underlying rationale for the one-third stress increase for masonry could be investigated by focused research.

6.4.3.4 Effective shear area for non-rectangular cross-sections

Shear can in effect control the design of some reinforced masonry elements, by setting the minimum effective depth that is required to avoid the use of shear reinforcement. Masonry beams usually have rectangular cross-sections, for which the computation of shear capacity is reasonably well understood.

Shear can also control the design of some unreinforced masonry elements, such as ungrouted basement walls. For such walls, computation of the shear capacity is not at all well understood. Should the effective shear area include only the face shells, or also some or all of the mortared webs? Determination of this effective shear area can govern the required thickness of the wall. Effective shear areas for non-rectangular cross-sections could be determined by focused research.

6.4.3.5 Effective distribution of bearing stresses under bond beams

The *2005 MSJC Code* offers no guidance for the effective distribution of bearing stresses under bond beams. Appropriate provisions should be developed.

6.4.3.6 Mixed-approach design

The strength and the allowable-stress provisions of the 2005 *MSJC Code* are unclear on whether it is permitted to design a masonry element, such as a wall, as reinforced masonry in one direction and unreinforced masonry in the perpendicular direction. One viewpoint is that because the design of reinforced masonry assumes that the masonry is cracked, and the orientation of that crack is not known, then if the masonry is designed as reinforced in one direction, it should be designed as reinforced in the perpendicular direction as well. An opposing viewpoint is that even though the masonry may be cracked, it can continue to resist out-of-plane loads through mechanisms such as arching action, and that masonry can be designed as reinforced in one direction, and as unreinforced in the perpendicular direction. Focused research is needed to help determine which viewpoint is more correct, and to propose appropriate design provisions.

6.4.4 *Masonry as structure for extreme loadings*

The performance of masonry structures under extreme loading is important for at least two reasons: first, we wish to control losses during extreme loadings; and second, we interpret behavior under extreme loadings as indicative of the reliability of masonry under everyday loadings.

Because extreme loadings are more newsworthy than everyday ones, and because structural testing under conditions representative of extreme loadings is technically more interesting than structural testing under conditions representative of everyday loading, masonry researchers have tended to pay particular attention to the structural behavior of masonry under extreme loadings. While this is essential, it should not distract us from the objective of ensuring reliable, predictable behavior under everyday loads as well. While a structure's resistance to collapse under extreme loadings is important, so is its resistance to damage under overloads. Investigations of the performance of masonry by the TMS

"Investigating Disasters" program and others has shown that masonry performs well in this respect.

6.4.4.1 Earthquake resistance of new masonry

When masonry structural systems are tested under loading conditions representative of strong earthquake, the results are commonly expressed in terms of hysteretic curves showing the variation of base shear as a function of reversed cyclic story drifts. Structural reliability during inelastic response is sometimes expressed in terms of maximum available displacement ductility ratios. Such information, while useful, does not indicate the reliability of a structural system unless it is compared with the probable required story drift.

Masonry structural systems can be designed to respond inelastically by using low quantities of distributed flexural reinforcement, and by carrying out capacity design for shear. These design constraints favor a structural system composed of cantilever walls, lightly coupled by horizontal diaphragms. Such systems typically have a limiting story drift of about 0.8%, governed by gradual deterioration of the compressive stress blocks of the shear walls under repeated, reversed cycles. For most such systems, this is equivalent to a displacement ductility ratio of about 3.

Other structural systems, such as specially detailed reinforced concrete shear walls or masonry wall frames, may have higher available displacement ductility ratios. Because story drifts at yield are kinematically required to be about the same for all reinforced concrete or masonry systems, regardless of element dimensions, structural systems with higher available story drifts generally have greater inelastic deformations, greater damage, higher detailing costs, and possibly unfavorable benefit-cost ratios compared to structural systems with less severe detailing requirements.

Using masonry structural systems with less severe detailing requirements for inelastic response, and lower values of the seismic force-reduction factor, R, can lead to reliable buildings with sufficient stiffness and strength to avoid significant damage under moderate earthquakes. Given this available design solution, it is probably not cost-effective to focus research efforts on incremental enhancements of the seismic performance of masonry under increasingly severe loading histories. Instead, we should accept our current general solution as satisfactory, and investigate the following variations on it for new construction:

- develop a designer-friendly alternative to the bewildering array of classifications for lateral force-resisting systems ("ordinary," "intermediate," "special");
- develop reliable displacement-based design approaches for masonry structural systems without severe detailing requirements for inelastic response;
- develop reliable, user-friendly tools for predicting the inelastic response of masonry structural systems without severe detailing requirements; and
- examine current tools for the performance-based seismic design of reinforced concrete structural systems, and modify them appropriately for masonry structural systems.

6.4.4.2 Earthquake resistance of existing masonry

The 1994 Northridge Earthquake clearly showed the cost-effectiveness of rudimentary seismic rehabilitation techniques for old, unreinforced masonry:

- bracing or removing parapets;
- using through mechanical connectors between walls and horizontal diaphragms;
- bracing walls out-of-plane; and
- verifying the basic integrity of masonry by "shove tests" or other means.

These techniques should be applied throughout the country.

Of equal interest, however, is the earthquake resistance of existing masonry that was designed and constructed in compliance with the criteria of the 1950's, and is now under scrutiny because of the increasingly severe seismic provisions of modern structural design codes. We need reliable tools for estimating the seismic resistance of such structures, and reliable seismic retrofitting techniques for them.

6.4.4.3 Wind resistance of masonry

Examination of the wind resistance of masonry after tornadoes and hurricanes shows that masonry does not appear to have significant problems with wind resistance. Because the actual capacity of masonry veneer is not known, however, its factor of safety against design wind loads is not clear. This capacity depends on the capacity of connectors, and work is needed to verify the strength of existing tie systems and possibly to develop new systems. Work is also needed on the behavior of connections between horizontal diaphragms and the masonry.

6.4.4.4 Concluding remarks on structural performance of masonry under extreme loads

In discussions within the MSJC leading to the development of the current strength design provisions, a repeated topic was how to characterize the structural capacity of masonry – what mean values, and how much scatter, would be appropriate? Apparently valid test data would indicate lower characteristic values of capacity at or near our current allowable stresses. Other test data, apparently of equal validity, would indicate higher characteristic values. A frequent debating point was, "If masonry is as weak as these data suggest, why haven't we seen more failures?"

Technical validity can reside on both sides of this issue. Those questioning the status quo are not necessarily trying to put masonry out of business – they are merely trying to increase our understanding of its behavior. Similarly, those defending the status quo are not necessarily acting irresponsibly – they are merely asking that the historical evidence be considered.

Rational discussion of this issue requires knowledge in the following areas. All are suitable subjects for focused research:

– We need a rational primer on engineering probabilities for non-mathematicians. Developers of design provisions need to know the significance of a combination of statistically distributed design loads and statistically distributed resistances. For a particular design load (and associated scatter), and a particular mean resistance (and associated scatter), what is the probability of failure in a given year? In 10 years? What is the probability of failure of at least some masonry buildings in a city in 10 years? Without this type of understanding, a rational answer to the above question is unlikely. This topic is applicable to all construction materials, not just masonry, and it should be addressed cooperatively by the entire design community.

- We need a clear understanding of all available resistance mechanisms in masonry structural systems. For example, we commonly design unreinforced bearing walls to span vertically. We know that they actually can span in two directions, and that their resistance can exceed that assumed in design. We need to quantify these overstrengths, decide whether they are significant, and if so, adjust our design provisions accordingly.
- We need specific data on the response of masonry structures to extreme loads. For example, large commercial chains, such as Wal-Mart or Target, often use standard designs across the entire country. We could develop standard instrumentation packages for such buildings, install them on buildings in high wind areas, and monitor building performance. Data so obtained could help us better calibrate our design provisions.

6.5   Cost-Competitiveness of Masonry

Regardless of its performance as building envelope, as architecture, and as structure, the viability of masonry as a construction system depends on its cost-competitiveness in the construction marketplace. The cost of masonry can be described in terms of its design cost; its manufacturing cost; its cost at the job-site; its cost in the structure; its initial cost to the owner; its life-cycle cost; and its cost to the environment. In the remainder of this section, each cost is examined in more detail.

6.5.1   *Design and specification cost of masonry*

At first glance, it might seem inappropriate to relate cost-competitiveness to design cost. Closer examination, however, reveals otherwise. In the traditional process of design and construction, design fees are a percentage of the structural cost of the building. The percentage of jobs built according to that process continues to decrease, and to be supplanted by a system in which a designer and the specifier negotiate fees for services. From the owner's viewpoint, a building is an investment tool, and other things being equal, the cost of design and specification should be about the same regardless of the material from which the building is made.

A designer's viewpoint, then, must consider design cost. Construction materials that require relatively low design effort per square foot of usable space are more attractive. Masonry must compete, by this criterion, with other construction materials, including structural steel (in which final design of connections is usually carried out by the steel fabricator); reinforced concrete (in which details of reinforcement are usually prepared by the reinforcement detailer); tilt-up (in which design of reinforcement can be based on standard details disseminated by the tilt-up industry); glass curtain walls; metal panels; exterior insulation and finish systems (EIFS); and metal, vinyl or wood siding. By this criterion, masonry is not as attractive. Relatively few design aids are available for it; its design is viewed as complex, old-fashioned, or difficult in other ways; and it is viewed by some designers as difficult to interface with other materials.

A specifier's viewpoint must also consider cost. Construction materials that require relatively low specification effort per square foot of usable space are more attractive. By this criterion, attractive construction materials include structural steel; reinforced concrete and tilt-up (specification is simple, and standard reference specifications are available), glass curtain walls; metal panels; exterior insulation and finish systems (EIFS); and metal, vinyl or wood siding. By this criterion, masonry is not as attractive. Its specification requires the separate and coordinated specification of units, mortar, grout and accessory materials. Three excellent guide specifications (MasterSpec, SpecText, and the *TMS Annotated Guide*) are available, and the *MSJC Specification* [3] is also excellent. The problem is that many specifiers do not know masonry well enough to use these specification tools effectively.

There is a tremendous need for focused research to develop and promulgate standard specifications for typical masonry construction (veneer on houses, veneer on frame buildings, or low-rise commercial construction).

6.5.2 *Manufacturing cost of masonry units*

Masonry units themselves are not expensive. Gray block (8 x 8 x 16 in.) cost about $1.00. Modular clay units cost about $0.25. Raw materials for masonry units are found within a reasonable distance of most major metropolitan areas in the US.

Challenges to increasing masonry production lie in the time and investment needed to bring new plants on line, and in the increasing complexity of environmental regulations governing production. For example, a modern concrete masonry plant, capable of producing 16 million $8 \times 8 \times 16$ equivalents yearly, requires an investment of about 10 million dollars and a lead time of about 3 years. A modern clay masonry plant, capable of producing 60 million modular equivalents yearly, requires an investment of about 25 million dollars and a lead time of about 5 years. There is a need for focused research to decrease the capital investment and time required to bring new concrete and clay masonry plants on line. Modern environmental regulations governing control of emissions and dust are an expensive reality.

Once operational, modern masonry plants can be very efficient. Concrete masonry plants commonly cure units less than 24 hours. Modern clay masonry plants can use computer-controlled firing cycles lasting 12 hours rather the 30 hours that used to be typical. Increasingly, plants use robotic handling equipment to increase productivity. There is a need for focused research to decrease the production cost of masonry units even more.

6.5.3 Cost of masonry at the job site

Given the cost of masonry units as produced, the cost of masonry at the job site is controlled by the cost of transportation, which depends in turn on the weight of units required for each square foot of wall area. Under current economic conditions, it is ordinarily cost-competitive to ship concrete masonry units within a 150-mile radius of their point of production. This radius increases for concrete and clay units with distinctive appearance or characteristics. Clearly, this radius can be increased by units with decreased density.

There is a need for focused research to decrease the weight of masonry units, or to decrease the thickness of masonry units, so that a wall of the same surface area can weigh less. In addition to decreasing transportation costs, this also offers the potential benefit of reduced seismic forces.

6.5.4 Cost of masonry in the structure

Given the cost of masonry units at the job site, the cost of installed masonry is controlled by the costs of masonry mortar, grout and accessory materials, and by the costs of installing the units plus those other materials to form the masonry assemblage.

The cost of installed units increases whenever units must be cut, or whenever units are broken or chipped in transportation or handling. The cost of installed masonry mortar and grout depends somewhat on the cost of the mortar and grout themselves, but more on the costs of preparing and placing them. It increases whenever materials must be handled in bag form, whenever batching is incorrect, whenever mortar is difficult to use, or whenever mortar is mixed but not used. The cost of installed accessory materials depends to a limited extent on the cost of the materials themselves, but primarily on their ease of installation.

There is a need for focused research in the development of:

– ways to encourage specifiers to use modular design;
– cost-effective ways to decrease breakage and chippage of masonry units;
– masonry mortars with improved performance;
– ways to encourage the use of compatible combinations of mortar and units;
– better techniques for hot- and cold-weather construction;
– more cost-effective ways (such as silo systems) to batch, mix and deliver mortar;
– more cost-effective scaffolding systems;
– cost-effective flashing, insulation and vapor barriers;
– cost-effective ways of protecting masonry during construction;
– cost-effective ways of keeping masonry clean during construction; and
– cost-effective ways of cleaning masonry after construction.

One factor often identified as important in the installation cost of masonry is the cost of masons themselves. The scarcity of masons, widely discussed at the beginning of the 90's, is still with us in many areas. Efforts must continue on recruitment and training of masons. There is a need for focused research on better ways to accomplish this.

6.5.5 *Initial cost of masonry to the owner*

The initial cost of masonry to the owner depends on the cost of in-place masonry, on its cost to specify and design, and also on the cost of financing construction. Many specific needs for focused research are noted in the preceding sections.

One overall need, not included in the above, is for the development of standard wall types with uniform specifications and construction details. For example, we could have a standard residential veneer wall; a rain-screen wall; a standard drainage wall with CMU backup; a standard drainage wall with steel stud backup; a standard fully grouted barrier wall; and a standard partially grouted barrier wall. These 6 basic wall types would represent practically all modern masonry construction. Other types could easily be added.

Development of standard wall types would involve the re-packaging of existing knowledge, rather than the development of new knowledge. Design provisions, specifications, and masonry industry technical notes should be synthesized to give masonry users simple recipes for how to use masonry correctly. A designer should be able to go to a web site and download design procedures, examples, complete specifications, and drawings for each wall type. Finally, constructors should be able to download step-by-step instructions, in words and pictures, and in different languages, showing the proper assembly of each wall type.

6.5.6 *Life-cycle cost of a masonry building*

The life-cycle cost of a building is the present value of its initial cost, plus the present value of the costs incurred over its lifetime, minus the present value of its sale price at the end of its lifetime. Costs incurred over the lifetime of a building include utilities, maintenance, and repairs. If masonry is properly designed, specified and constructed, its maintenance cost is very low compared to that of other envelope materials. For example, it does not need painting.

Focused research is needed to identify ways of documenting and reducing maintenance and repair costs of masonry buildings. For example, focused research on ways to reduce efflorescence would enhance the appeal of masonry, and also reduce the probability of damage due to improper cleaning.

Insurance cost is a significant part of the life-cycle cost of masonry buildings. Insurance costs (for example, for fire) are driven in part by the actuarial risk of loss in masonry buildings, and in part by the by the potential costs of changing current methods for classifying buildings. For example, fire-insurance premiums are the same, or almost the same, for masonry bearing-wall houses as for wood-frame houses, even though masonry is far less flammable. Reasons for this are that much of the loss in residential fires is to the contents of the house rather than to the house itself, and that to charge lower premiums for masonry than for wood-frame houses could reduce the revenue of insurance companies, and would require them to distinguish between types of construction on insurance applications.

Focused research is needed to identify strategies for reducing the cost of insurance premiums for masonry buildings, and to implement those strategies.

Focused research is also needed to update criteria for compiling life-cycle costs for buildings of different materials, and to update the corresponding values for masonry buildings. Updates should reflect historical trends in energy and construction costs. Updates should be used to project the life-cycle cost-competitiveness of masonry buildings, to identify areas in which life-cycle costs could be improved, and to develop research focused on those areas.

6.5.7 *Cost of masonry to the environment*

Another approach to life-cycle cost is the cost of a masonry building to the environment. This is the sum of the energy required to manufacture the raw materials from which the building is made; to construct the building; to operate the building; and to dispose of the building when its useful life is over. This type of life-cycle cost, however, may also include the day-to-day details of the cost of operating the building. For example, buildings with high thermal mass may have lower peak power demands, and therefore lower energy costs.

Focused research is needed to update criteria for compiling life-cycle environmental costs for buildings of different materials, and to update the corresponding values for masonry buildings. Updates should reflect historical trends in energy costs and metering policies. Updates should be used to project the environmental friendliness of masonry buildings, to identify areas in which environmental friendliness could be improved, and to develop research focused on those areas.

6.5.8 *Cost-effectiveness of masonry in niche markets*

Niche markets for masonry include fireplaces, pavers, segmental retaining walls, and landscaping applications. Focused research should continue on potential obstacles to those markets (such as the poor seismic performance of unreinforced masonry chimneys), and on ways of overcoming those obstacles.

6.5.9 *Cost-effectiveness of masonry in new markets*

The masonry technical community has been accused of opposing innovation. In my opinion, this accusation is simplistic. The masonry technical community is a complex array of producers, users and general interest groups. Each of these benefits to some degree from the status quo. While the masonry technical community might benefit collectively from changes to that status quo, it is not accustomed to working together to identify potentially useful changes, to explore their possible repercussions, and to work together to bring them about.

There is a need for the masonry industry to regularly examine the potential of new markets, and to prepare to be competitive in those markets.

– One example of this is post-tensioned masonry. This innovation has only recently arrived in the US, and is growing slowly without mechanisms such as demonstration projects to validate its use.
– Another example is segmental retaining walls, which have increased in popularity with little promotional effort by the masonry industry. The strength of these walls derives primarily from the soil reinforcement, and not from the masonry, which is only a skin. Research by the masonry industry should address ways to optimize the attachment of the face units, to assess their performance under extreme loadings, and to develop new units for this application.

6.6 Summary and Recommendations

6.6.1 *Summary*

This original version of this chapter was prepared as a resource document for the PERM Workshop on Masonry Research, sponsored by the Council for Masonry Research, and held on January 5-6, 2001 in Tempe, Arizona. The objective of the Workshop was to identify research that would enhance the performance of masonry. In this chapter, the ways in which masonry is expected to perform are identified: Masonry performs as building envelope, as architecture, and as structure. It must carry out each of these functions in a cost-competitive way, in comparison with other systems and materials. In this chapter, how masonry carries out each function is described in more detail; its performance with respect to each function is noted; ways in which that performance has been enhanced are noted; and suggestions are given for focused research to improve performance further in some areas.

6.6.2 *Recommendations for focused masonry research*

6.6.2.1 Masonry research to improve the performance of masonry as building envelope

(1) Effectively communicate proper masonry design, detailing and construction practices to those responsible for establishing them and carrying them out.

(2) Effectively communicate proper installation practices for ventilation and drainage details to those responsible for establishing them and carrying them out.

(3) Increase the effectiveness and life of sealants, or reduce the envelope's dependence on them for performance.

(4) Effectively communicate proper design, specification and installation practices for air and vapor barriers to those responsible for establishing them and carrying them out.

(5) Increase the thermal insulation value of masonry units and assemblies.

(6) Find ways of increasing the sound absorption (surface porosity) of masonry, while retaining durability.

(7) Find ways to increase the fire-resistance rating of a given thickness of masonry.

(8) Find ways to justify and encourage the use of non-combustible materials such as masonry in compartmentalization of buildings.

(9) Develop and disseminate effective material specifications for impact resistance of masonry.

(10) Develop user-friendly stiffness requirements for backup systems and connectors so that cracks in the masonry veneer will not permit an objectionable amount of water to pass through the building envelope.

(11) Enhance the corrosion resistance of connectors and backup systems to prolong the effective life of masonry veneers.

(12) Develop standard, user-friendly guidelines for locating and constructing movement joints.

(13) Find effective ways to introduce masonry specification and detailing into undergraduate engineering and architecture curricula.

(14) Find ways to encourage more university professors to become involved in the ASTM process, where they can learn about specification issues.

(15) Encourage the continuing education about masonry, of practicing architects, engineers, contractors, building officials and inspectors.

6.6.2.2 Masonry research to improve the performance of masonry as architecture

(1) Find ways to increase the variety of architectural forms that can be laid using a relatively small number of different unit sizes, and with little or no cutting of units.

(2) Develop formwork and scaffolding that will make it easier to construct masonry arches and domes.

(3) Find ways to increase the variety of architectural details (such as corbels, racks, quoins, and different bond patterns) that can be laid without cutting units or unduly increasing cost.

(4) Develop construction techniques or tools that will make it easier to construct masonry architectural details more quickly and reliably.

(5) Find ways to improve the consistency of color of units and mortar.

(6) Find ways to decrease cracking and chipping of masonry units.

(7) Find better ways to control the alignment of units and mortar joints, the variation in thickness of mortar joints, and the variation of masonry walls from level and plumb.

(8) Find better ways to control staining, and improve cleaning techniques.

(9) Find better ways to decrease or eliminate efflorescence.

6.6.2.3 Masonry research to improve the performance of masonry as structure

(1) Develop simplified design provisions, consistent with the more complex ones, for the design of structural elements that we use often.

(2) Develop user-friendly design aids to take the drudgery out of complex calculations.

(3) Develop "deemed-to-comply" designs for simple masonry structures.

(4)   Confirm the relationships between the strengths of elements designed by the allowable-stress and the strength approach of the 2005 *MSJC Code.*

(5)   Accept the principle that empirically designed masonry elements can be evaluated by analytically computing their strength and comparing that strength with the strength of their analytically designed counterparts. Use that principle to evaluate the adequacy of our empirical provisions, particularly for elements with little or no axial load.

(6)   Use the performance of all masonry, including empirically designed masonry, to identify areas where analytical design provisions may not recognize significant resistance mechanisms, and incorporate those resistance mechanisms into analytical provisions.

(7)   Evaluate the adequacy of our existing analytical design provisions with respect to:

    (a)   minimum and maximum flexural reinforcement;
    (b)   moment magnifiers for masonry beam-columns;
    (c)   the 1/3 stress increase;
    (d)   effective shear area for non-rectangular sections;
    (e)   effective distribution of bearing stresses under bond beams;
    (f)   effective shear area for non-rectangular sections; and
    (g)   mixed-approach designs.

(8)   Develop a designer-friendly alternative to the bewildering array of classifications for lateral force-resisting systems ("ordinary," "intermediate," "special").

(9)   Develop reliable displacement-based design approaches for masonry structural systems without severe detailing requirements for inelastic response.

(10)   Develop reliable, user-friendly tools for predicting the inelastic response of masonry structural systems without severe detailing requirements.

(11)   Examine current tools for the performance-based seismic design of reinforced concrete structural systems, and modify them appropriately for masonry structural systems.

(12)   Apply rudimentary seismic rehabilitation techniques for old, unreinforced masonry throughout the country:

    (a)   brace or remove parapets;

(b) install through mechanical connectors between walls and horizontal diaphragms;

(c) brace walls out-of-plane; and

(d) verify the basic integrity of masonry by "shove tests" or other means.

(13) Develop reliable tools for estimating the earthquake resistance of existing masonry buildings that were designed and constructed in compliance with the criteria of the 1950's, and develop reliable seismic retrofitting techniques for them.

(14) Verify the strength of existing veneer tie systems, and develop new systems if necessary.

(15) Investigate the behavior of connections between floor diaphragms and masonry.

(16) Identify or develop a rational primer on engineering probabilities for non-mathematicians, and use it to estimate probabilities of failure under design loads during different recurrence intervals.

(17) By instrumenting standard buildings around the country, obtain specific data on the response of masonry structures to extreme loads.

6.6.2.4 Masonry research to improve the cost-competitiveness of masonry

(1) Develop and disseminate standard specifications for typical masonry construction (veneer on houses, veneer on frame buildings, or low-rise commercial construction).

(2) Find ways to decrease the capital investment and time required to bring new concrete and clay masonry plants on line.

(3) Find more economical ways to comply with environmental restrictions on emissions and dust.

(4) Find ways to decrease the production cost of masonry units even more.

(5) Find ways to decrease the weight or thickness of masonry units, so that a wall of the same surface area can weigh less.

(6) Develop:

(a) ways to encourage specifiers to use modular design;

(b) cost-effective ways to decrease breakage and chippage of masonry units;

(c)   masonry mortars with improved performance;

(d)   better materials and techniques for grouting;

(e)   better techniques for hot- and cold-weather construction;

(f)   more cost-effective ways (such as silo systems) to batch, mix and deliver mortar;

(g)   more cost-effective scaffolding systems;

(h)   more cost-effective flashing, insulation, and air and vapor barriers;

(i)   more cost-effective ways of protecting masonry during construction;

(j)   more cost-effective ways of keeping masonry clean during construction; and

(k)   more cost-effective ways of cleaning masonry after construction.

(7)   Continue efforts to recruit and train masons, and conduct focused research on better ways to accomplish this.

(8)   Develop standard wall types with uniform specifications and construction details (for example, a standard residential veneer wall; a rain-screen wall; a standard drainage wall with CMU backup; a standard drainage wall with steel stud backup; a standard fully grouted barrier wall; and a standard partially grouted barrier wall).

   (a)   prepare design procedures, examples, specifications and drawings for each wall type.

   (b)   prepare step-by-step instructions, in words and pictures, and in different languages, showing the proper assembly of each wall type.

(9)   Identify ways of reducing maintenance and repair costs of masonry buildings (for example, reducing efflorescence).

(10)   Identify strategies for reducing the cost of insurance premiums for masonry buildings, and implement those strategies.

(11)   Update criteria for compiling life-cycle costs for buildings of different materials, and update the corresponding values for masonry buildings.

(12)   Update criteria for compiling life-cycle environmental costs for buildings of different materials, and update the corresponding values for masonry buildings.

(13) Continue research on potential obstacles to the use of masonry in niche markets (such as chimneys), and on ways of overcoming those obstacles.

(14) Continue to regularly examine the potential of new markets (such as prestressed masonry and segmental masonry retaining walls), and prepare to be competitive in those markets.

6.7 Acknowledgments

The original version of this chapter was prepared by the author under the auspices of The Masonry Society's technical committee on Research. Its first draft was presented and discussed at the PERM Workshop on Masonry Research, held in Tempe, Arizona on January 5-6, 2001, under the sponsorship of the Council for Masonry Research (whose organizational members are the Brick Industry Association, the Expanded Shale, Clay and Slate Institute, the Mason Contractors Association of America, the National Concrete Masonry Association, the National Lime Association, and the Portland Cement Association) and the International Masonry Institute. The project was administered and managed by The Masonry Society. The contents of this chapter reflect the views of the author, who is responsible for the facts and accuracy of the information presented. The contents do not necessarily reflect the views of the sponsoring organizations or The Masonry Society. The author acknowledges the helpfulness of comments made by Workshop participants and others, particularly David Biggs, Gregg Borchelt, Clayford T. Grimm, Lynn Lauersdorf and Patrick Rand.

References

1. *International Building Code*, 2006 Edition, International Code Council, Falls Church, Virginia, 2006.
2. *Building Code Requirements for Masonry Structures (ACI 530-05 / ASCE 5-05 / TMS 402-05)*, American Concrete Institute, American Society of Civil Engineers, and The Masonry Society, 2005.
3. *Specifications for Masonry Structures (ACI 530.1-05 / ASCE 6-05 / TMS 602-05)*, American Concrete Institute, American Society of Civil Engineers, and The Masonry Society, 2005.
4. *Minimum Design Loads for Buildings and Other Structures (ASCE 7-05, Supplement)*, American Society of Civil Engineers, Reston, Virginia, 2005.

Chapter 7

Geosynthetics — Characteristics and Applications

P. E. Stevenson
International Geosynthetic Society, Easley, SC

7.1 Introduction

Geosynthetics refers to a set of materials used to improve the performance of soils or foundations in geotechnical engineering. The International Geosynthetics Society cites its mandate as geotextiles, geomembranes, associated products and related technologies thereby encompassing both synthetic and natural materials.

Geosynthetics of different sorts have been used for thousands of years. They were used in roadway construction in the days of the pharaohs to stabilize roadways and their edges. These early geotextiles were made of natural fibres, fabrics or vegetation mixed with soil to improve road quality, particularly when roads were built on unstable soil. Although modern geosynthetics bear little resemblance to those used in Mesopotamia, the general principles remain the same.

With the recent development of polymers, the development of Geosynthetics has moved along swiftly. Their relatively recent development and acceptance by industry is highlighted by the fact that the International Geosynthetics Society (IGS) was founded in Paris 1983. The worldwide dollar value was very small, not reaching $10,000,000 until 1980 or thereabouts. In 2005 the worldwide value exceeds the billion dollar annual volume mark. The use of geosynthetics in the USA began a rapid expansion when the Federal Highway Administration commissioned the preparation of a manual and a course on the use of geotextiles in transportation applications. This manual and course remain in use and are currently undergoing a fourth revision and update.

7.2 Geotextile Polymers

The primary polymers used in geosynthetics are the olefins, polypropylene, polyethylene and polyester. Almost all geotextiles available in the United States are manufactured from either polyester or polypropylene. Both polymers exhibit suitable physical characteristics and are cost effective in terms of properties, especially tensile, delivered per pound of fiber.

High tenacity polyester fibers and yarns are used in the manufacturing of geotextiles. Polyester has excellent strength and creep properties, and is compatible with most common soil environments.

Polypropylene has low specific gravity, high bulk and is durable. Polypropylene yarns and staple fibers are used in manufacturing woven and nonwoven geotextiles, respectively.

7.3 Geotextile Structures

There are two principal geotextile types, or structures: wovens and nonwovens (Fig. 7.1). Other manufacturing techniques, for example, knitting and stitch bonding, are occasionally used in the manufacture of specialty products.

(a) Nonwoven geotextile   (b) Woven geotextile

Fig. 7.1 Nonwovens and woven geotextile.

7.3.1 *Nonwovens*

Nonwoven geotextiles are manufactured from either staple fibers (staple fibers are short, usually 2.5 to 10 cm in length) or continuous filaments randomly distributed in layers onto a moving belt to form a felt-like "web". The web then passes through a needle loom and/or other bonding machine interlocking the fibers/filaments. Nonwoven geotextiles are highly desirable for subsurface drainage and erosion control applications as well as for road stabilization over wet moisture sensitive soils.

7.3.2 *Wovens*

Weaving is a process of interlacing yarns to make a fabric. Woven geotextiles are made from weaving monofilament, multifilament, or slit film yarns. Slit film yarns can be further subdivided into flat tapes and fibrillated (or spider web-like) yarns. There are two steps in this process of making a woven geotextile: first, manufacture of the filaments and yarns; and second, weaving the yarns to form the geotextile.

Polypropylene slit film fabrics are commonly used for sediment control, i.e. silt fence, and road stabilization applications but are poor choices for subsurface drainage and erosion control applications. Though the flat tape slit film yarns are quite strong, they form a fabric that has relatively poor permeability. Alternatively, fabrics made with fibrillated tape yarns have better permeability and more uniform openings than flat tape products.

Monofilament wovens have better permeability, making them suitable for certain drainage and erosion control applications. High strength multifilament wovens are primarily used in reinforcement applications.

7.4 Geosynthetics Classification

7.4.1 *Classification*

Geosynthetics can be broadly classified into categories based on method of manufacture. The following texts are excerpted from a series of information bulletins prepared by members of the education committee of the International Geosynthetics Society. Figure 7.2 presents the classifications graphically.

Geotextiles (Fig. 7.2(a)) are continuous sheets of woven, knitted or stitch-bonded fibers or yarns. The sheets are flexible and permeable and generally have the appearance of a fabric. Geotextiles are used for separation, filtration, drainage, reinforcement and erosion control applications.

Geogrids (Fig. 7.2(b)) are geosynthetic materials that have an open grid-like appearance. The principal application for geogrids is the reinforcement of soil.

Geonets (Fig. 7.2(c)) are open grid-like materials formed by two sets of coarse, parallel, extruded polymeric strands intersecting at a constant acute angle. The network forms a sheet with in-plane porosity that is used to carry relatively large fluid or gas flows.

Geomembranes (Fig. 7.2(d)) are continuous flexible sheets manufactured from one or more synthetic materials. They are relatively impermeable and are used as liners for fluid or gas containment and as vapour barriers.

Geocomposites (Fig. 7.2(e)) are geosynthetics made from a combination of two or more geosynthetic types. Examples include: geotextile-geonet; geotextile-geogrid; geonet-geomembrane; or a geosynthetic clay liner (GCL). Prefabricated geocomposite drains or prefabricated vertical drains (PVDs) are formed by a plastic drainage core surrounded by a geotextile filter.

Geosynthetic clay liners (GCLs) (Fig. 7.2(f)) are geocomposites that are prefabricated with a bentonite clay layer typically incorporated between a top and bottom geotextile layer or bonded to a geomembrane or single layer of geotextile. Geotextile-encased GCLs are often stitched or needle-punched through the bentonite core to increase internal shear resistance. When hydrated they are effective as a barrier for liquid or gas and are commonly used in landfill liner applications often in conjunction with a geomembranes.

Geopipes (Fig. 7.2(g)) are perforated or solid-wall polymeric pipes used for drainage of liquids or gas (including leachate or gas collection in landfill applications). In some cases the perforated pipe is wrapped with a geotextile filter.

Geofoam blocks (Fig. 7.2(h)) or slabs are created by expansion of polystyrene foam to form a low-density network of closed, gas-filled cells. Geofoam is used for thermal insulation, as a lightweight fill or as a compressible vertical layer to reduce earth pressures against rigid walls.

Geocells (Fig. 7.2(i)) are relatively thick, three-dimensional networks constructed from strips of polymeric sheet. The strips are joined together to form interconnected cells that are infilled with soil and sometimes concrete. In some cases 0.5 m- to 1 m-wide strips of polyolefin geogrids have been linked together with vertical polymeric rods used to form deep geocell layers called geomattresses.

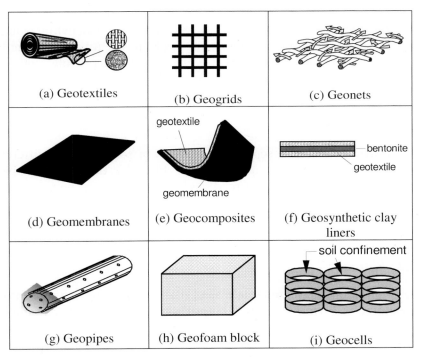

(a) Geotextiles	(b) Geogrids	(c) Geonets
(d) Geomembranes	(e) Geocomposites	(f) Geosynthetic clay liners
(g) Geopipes	(h) Geofoam block	(i) Geocells

Fig. 7.2 Classification of geosynthetics.

7.4.2 *Products*

The range of geosynthetic products and the list of producers that fits into these classifications are very large, continuously changing and there is a climate of continuous development. Several trade associations and professional societies represent the industry worldwide and can provide specific and current product information through their membership or through product directories published by their affiliates.

These associations include the International Geosynthetics Society (IGS), the Geosynthetics Manufacturers Association (GSA) affiliated with Industrial Fabrics Association International (IFAI), European Association of Geotextiles Manufacturers (EAGM), and the European GeoMembrane Association (EGMA).

7.5 Geosynthetics Functions

Geosynthetics include a variety of synthetic polymer materials that are specially fabricated to be used in geotechnical, geoenvironmental, hydraulic and transportation engineering applications. It is convenient to identify the primary function of a geosynthetic as being one of: separation, filtration, drainage, reinforcement, fluid/gas containment, or erosion control. In some cases the geosynthetic may serve dual functions.

7.5.1 *Separation*

The geosynthetic acts to separate two layers of soil that have different particle size distributions as shown in Fig. 7.3 (a). For example, geotextiles are used to prevent road base materials from penetrating into soft underlying subgrade soils, thus maintaining design thickness and roadway integrity. Separators also help to prevent fine-grained subgrade soils from being pumped into permeable granular road bases.

7.5.2 *Filtration*

The geosynthetic acts similar to a sand filter by allowing water to move through the soil while retaining all upstream soil particles. For example, geotextiles are used to prevent soils from migrating into drainage aggregate or pipes while maintaining flow through the system. Geotextile filters are also used below rip rap and other armour materials in coastal and river bank protection systems to prevent soil erosion, as shown in Fig. 7.3(b).

7.5.3 *Drainage*

The geosynthetic acts as a drain to carry fluid flows through less permeable soils. For example, geotextiles are used to dissipate pore water pressures at the base of roadway embankments, as shown in Fig. 7.3(c). or higher flows, geocomposite drains have been developed. These

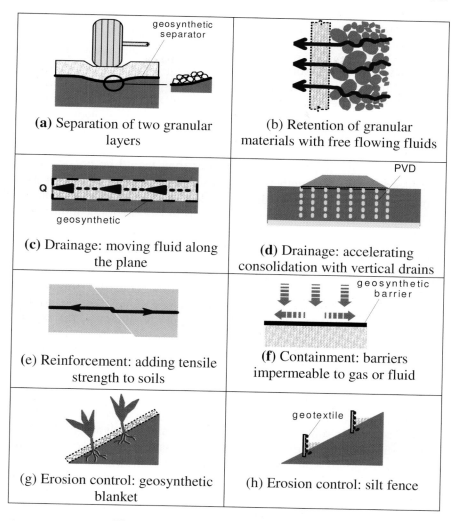

Fig. 7.3 Different functions of geosynthetics.

materials have been used as pavement edge drains, slope interceptor drains, and abutment and retaining wall drains. Prefabricated vertical drains (PVDs) have been used to accelerate consolidation of soft cohesive foundation soils below embankments and preload fills, as shown in Fig. 7.3(d).

7.5.4 *Reinforcement*

The geosynthetic acts as a reinforcement element within a soil mass or in combination with the soil to produce a composite that has improved strength and deformation properties over the unreinforced soil. For example, geotextiles and geogrids are used to add tensile strength to a soil mass in order to create vertical or near-vertical changes in grade (reinforced soil walls), as shown in Fig. 7.3(e). Reinforcement enables embankments to be constructed over very soft foundations and to build embankment side slopes at steeper angles than would be possible with unreinforced soil. Geosynthetics (usually geogrids) have also been used to bridge over voids that may develop below load bearing granular layers (roads and railways) or below cover systems in landfill applications.

7.5.5 *Fluid/Gas (barrier) containment*

The geosynthetic acts as a relatively impermeable barrier to fluids or gases. For example, geomembranes, thin film geotextile composites, geosynthetic clay liners (GCLs) and field-coated geotextiles are used as fluid barriers to impede flow of liquid or gas, as shown in Fig. 7.3(f). This function is also used in asphalt pavement overlays, encapsulation of swelling soils and waste containment.

7.5.6 *Erosion control*

The geosynthetic acts to reduce soil erosion caused by rainfall impact and surface water runoff. For example, temporary geosynthetic blankets and permanent lightweight geosynthetic mats are placed over the otherwise exposed soil surface on slopes, as shown in Fig. 7.3(g). Geotextile silt fences are used to remove suspended particles from sediment-laden runoff water, as shown in Fig. 7.3(h). Some erosion control mats are manufactured using biodegradable wood fibers.

7.5.7 *Other functions*

Geotextiles are also used in other applications. For example, they are used for asphalt pavement reinforcement and as cushion layers to prevent puncture of geomembranes (by reducing point contact stresses) from stones in the adjacent soil, waste or drainage aggregate during installation and while in service. Geotextiles have been used as daily covers to prevent dispersal of loose waste by wind or birds at the working surface of municipal solid waste landfills. Geotextiles have also been used for flexible concrete formworks and for sandbags. Cylindrical geotubes are manufactured from double layers of geotextiles that are filled with hydraulic fill to create shoreline embankments or to dewater sludge.

7.6 Guidelines and Specifications

There are many agencies and organizations world wide involved in the development of guidelines and specifications. In US, ASTM, IGS and GSI (Geosynthetic Research Institute at Drexel University) are the leading organizations. ASTM committee D35 on Geotextiles and Related Products was established in 1984 and has developed many standards during the past 20 years.

GSI was formed in 1986 and constitutes a diverse group of agencies, corporations, companies, partnerships and individuals who actively engage in the design and use of geosynthetics as engineering materials. The mission is to develop and transfer knowledge, assess and critique geosynthetics, and provide service to the member organizations. During the past 20 years, GSI has also developed numerous guidelines and specifications on the testing and application of geosynthetics. Detailed information can be found at www.geosynthetic-institute.org.

IGS was found in 1983 and is a non-profit organization dedicated to the scientific and engineering development of geotextiles, geomembranes, related products and associated technologies. The IGS promotes the dissemination of technical information on geosynthetics through a newsletter (IGS News) and though it's two official journals (Geosynthetics International – www.geosynthetics-international.com and Geotextiles and Geomembranes – www.elsevier.com/locate/geotexmem). Additional information on the IGS and its activities can be obtained at www.geosyntheticssociety.org.

The Federal Highway Administration (FHWA) has championed geosynthetics in the United States since the late 1970's. FHWA commissioned the development of a course and manual on geosynthetics first published in 1981, which remains in use and is under going its fourth revision.

American Association of State Highway and Transportation Officials (AASHTO) develops and publishes standards for construction materials including geosynthetics. It has made two important contributions to the industry. The first is the geosynthetic specification AASHTO M288-

Geotextile Specification for Highway Applications. The second is the quality assurance program known as National Transportation Product Evaluation Program (NTPEP).

In Europe, two parallel organizations the International Organization for Standardization, www.ISO.org (ISO) and European Committee for Standardization, www.cenorm.be (CEN) serve a function parallel to ASTM.

7.7 Applications of Geosynthetics

Geosynthetics has been widely used for a variety of civil constructions. The following is a sampling of the typical applications for geosynthetics and is based on the documents prepared by members of the IGS education committee.

7.7.1 *Geosynthetics in drainage and filtration*

Geosynthetics can be effectively used as drains and filters in civil and environmental works in addition to or in substitution to traditional granular materials. Geosynthetics are easier to install in the field and often cost-effective in situations were granular materials available do not meet design specifications, are scarce or have their use restricted by environmental legislations. Figure 7.4 presents the installation of a geotextile filter in a drainage ditch. As drains, a geosynthetic can be

Fig. 7.4 Installation of geotextile filters.

specified to attend hydraulic characteristics that allow free flow of liquids or gases throughout or across its plane as shown in Fig. 7.3(c) and Fig. 7.3(d).

Geotextile filters have to attend criteria that assure that the base soil will be retained with unimpeded water flow. Available retention criteria establishes that FOS $\leq$ n D_s, where FOS is the geotextile filtration size, which is associated to pore and constriction sizes in the geotextile, n is a number which depends on the criteria used and D_s is a representative dimension of the base soil grains (usually D_{85}, which is the diameter for which 85% in weight of the soil particles are smaller than that diameter).

The filter has also to be considerably more permeable than the base soil throughout the project life time. Therefore, the permeability criterion for geotextiles establishes that $k_G \geq N \, k_s$ where k_G is the geotextile coefficient of permeability, N is a number that depends on the project characteristics (typically varying between 10 and 100) and k_s is the permeability coefficient of the base soil. Clogging criteria require that the geotextile will not clog and that may be based on relations between geotextile filtration opening size and soil particle diameters that should be allowed to pipe through the geotextile or on specific performance tests. For the latter, filtration tests can be carried out in the laboratory to evaluate the compatibility between the soil and a candidate geotextile filter. If properly specified and installed, geosynthetics can provide cost-effective solutions for drainage and filtration in civil and environmental engineering works. Additional information on the use of geosynthetics in such applications and in other fields of geotechnical and geoenviromental engineering can be found at www.geosyntheticssociety.org.

7.7.2   *Geosynthetics in road engineering*

Roads and highway are of extreme importance to the development of any country. Due to systematic traffic of heavy vehicles, climate conditions and mechanical properties of the materials used in their constructions, highway pavements may last considerably less than expected. Figure 7.5(a) shows the damages in a conventional pavement. Figure 7.5(b) shows the use of a geosynthetic in pavement construction. In this sense, geosynthetics can be effectively used to reduce or avoid reflective cracking as shown in Fig. 7.6(a); work as a barrier to avoid

(a) Damages in a conventional   (b) Geosynthetic in pavement
pavement   construction

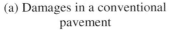

Fig. 7.5 Use of geosynthetic in pavement construction.

pumping of soil fines as shown in Fig. 7.6(b), reduce asphalt cap
thickness as shown in Fig. 7.6(c), and reduce pavement thickness as
shown in Fig. 7.6(d).

The efficiency of the geosynthetics as reinforcement in a pavement
can be estimated by the Efficiency Factor (E):

$$E = \frac{N_r}{N_u}$$

where N_r = number of load repetitions up to failure for the reinforced
pavement. N_u = number of load repetitions up to failure for the
unreinforced pavement. Available data in the literature present values of
E as high as 16, which shows that considerable increases on the
pavement lifetime can be achieved with the use of geosynthetic as
reinforcement or separation. Field observations and research results
confirm the improvements of pavement performance due to geosynthetic
utilization. Figure 7.7 is an illustration of surface rut depth of pavement
with and without geosynthetics. Increase of pavement service life due to
the use of geosynthetic reinforcement If properly specified and installed,
geosynthetics can be cost-effective and improve the performance and
durability of pavements.

Geosynthetics can be effectively used to reinforce unpaved roads
and working platforms on soft soils. If well specified, a geosynthetic can
have one or more of the following functions: separation, reinforcement
and drainage. Geotextiles and geogrids are the most commonly used

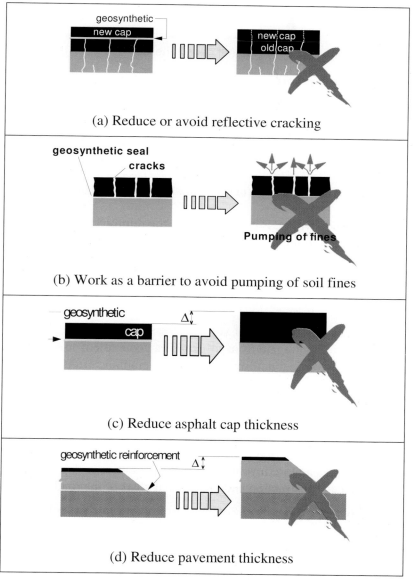

Fig. 7.6 Benefits for the use of geosynthetics in pavement construction.

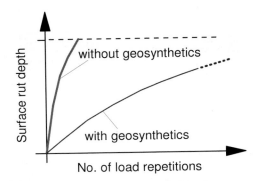

Fig. 7.7 Surface rut depth of pavement with and without geosynthetics.

materials in such works. Figure 7.8 shows the reinforcement mechanisms of geosynthetics in unpaved road. Figure 7.9 show the construction and driving of vehicle on soft clay. Compared to the unreinforced unpaved road, the presence of geosynthetic reinforcement can provide the following benefits:

- Reduces fill thickness
- Widens the spreading of vertical stress increments
- Separates aggregate from soft soil if a geotextile is used
- Reduces vertical deformation due to membrane effect
- Increases soft soil bearing capacity
- Increases the lifetime of the road
- Reduces fill lateral deformation
- Requires less periodic maintenance
- Requires less periodic maintenance
- Generates a more favourable stress distribution
- Reduces construction and operational costs of the road

Influence of geosynthetic reinforcement on unpaved road behaviour. As the depth of the ruts increase the deformed shape of the geosynthetic provides further reinforcement due to the membrane effect. The vertical component of the tensile force in the reinforcement minimizes further vertical deformation of the fill. Several researches in the literature have shown that in a reinforced road a given rut depth will be reached for a number of load repetitions (traffic intensity) larger than in the unreinforced case. This will yield to a greater service life and

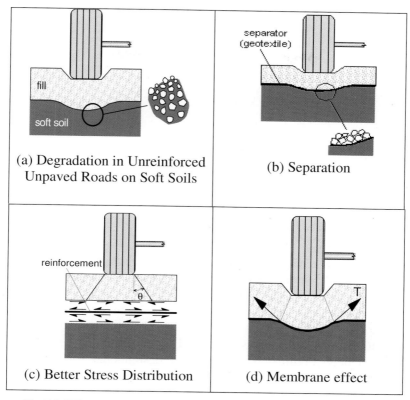

Fig. 7.8 Effect of geosynthetic reinforcement on unpaved road behavior.

(a) Construction of a reinforced unpaved road on soft organic clay

(b) Driving of a Vehicle on a reinforced unpaved road

Fig. 7.9 Use of geosynthetics in unpaved roads.

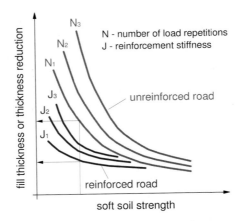

Fig. 7.10 Typical design chart for the geosynthetic reinforcement of unpaved road.

less periodical surface maintenance. Many design methods have been published in the literature, including simple ones involving the use of charts for preliminary analyses. These methods require conventional soil and reinforcement parameters for design purposes under routine conditions. Design charts have also been developed by some geosynthetics manufacturers specifically for preliminary design using their products. Figure 7.10 is a typical design chat for geosynthetics reinforcement of unpaved road. A draining reinforcement material will also accelerate soft soil consolidation, increasing its strength. Drainage of the soft soil can be achieved by using a geotextile, a geogrid and a geotextile or a geocomposite as reinforcement. The stabilisation of the top region of the soft foundation will be beneficial if the road is to be paved in the future reducing construction costs and minimising pavement deformations.

7.7.3 *Geosynthetics in slopes over stable foundations*

Layers of geosynthetic reinforcement are used to stabilize slopes against potential deep-seated failure using horizontal layers of primary reinforcement. The reinforced slope may be part of slope reinstatement and (or) to strengthen the sides of earth fill embankments as shown in Fig. 7.11.

The reinforcement layers allow slope faces to be constructed at steeper angles than the unreinforced slope. It may be necessary to stabilize the face of the slope (particularly during fill placement and compaction) by using relatively short and more tightly spaced secondary reinforcement and (or) by wrapping the reinforcement layers at the face. In most cases the face of the slope must be protected against erosion. This may require geosynthetic materials including thin soil-infilled geocell materials or relatively lightweight geomeshes that are often used to temporarily anchor vegetation.

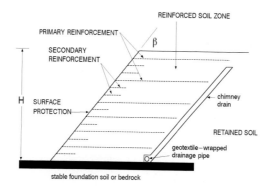

Fig. 7.11 Geosynthetic reinforced soil slope over stable foundation.

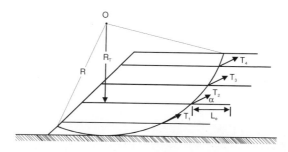

Fig. 7.12 Circular slip analysis of reinforced soil slope over stable foundation.

The location, number, length and strength of the primary reinforcement required to provide an adequate factor-of-safety against slope failure is determined using conventional limit-equilibrium methods of analysis modified to include the stabilizing forces available from the reinforcement. The designer may use a "method of slices" approach together with the assumption of a circular failure surface, composite failure surface, two-part wedge or a multiple wedge failure mechanism. Figure 7.12 shows circular slip analysis of reinforced soil slope over stable foundation. The reinforcement layers are assumed to provide a restraining force at the point of intersection of each layer with the potential failure surface being analysed. A solution for the factor-of-safety using the conventional Bishop's Method of analysis can be carried out using the following equation:

$$FS = \left(\frac{M_R}{M_D} \right)_{unreinforced} + \frac{\sum T_{allow} \times R_T \cos \alpha}{M_D}$$

where M_R and M_D are the resisting and driving moments for the unreinforced slope, respectively, α is the angle of tensile force in the reinforcement with respect to the horizontal, and T_{allow} is the reinforcement maximum allowable tensile strength. Since geosynthetic reinforcement is extensible the designer can assume that the reinforcement force acts tangent to the failure surface in which case $R_T \cos \alpha = R$. The potential failure surfaces must also include those passing partially through the reinforced soil mass and into the soil beyond the reinforced zone as well as those completely contained by the reinforced soil zone. Figure 7.13 shows the use of geosynthetics as a primary reinforcement of a slope and the completed reinforced embankment.

The construction of embankments on soft soils can be a challenging task. The use of geosynthetics to improve embankment stability is one of the most effectives and well-tried forms of the soil reinforcement technique. Figure 7.14 shows four examples for the use of geosynthetics for the improvement of embankment stability over soft soils. In case of

| (a) Primary reinforcement | (b) Completed reinforced embankment |

Fig. 7.13 Use of geosynthetics as a primary reinforcement of a slope and the completed reinforced embankment.

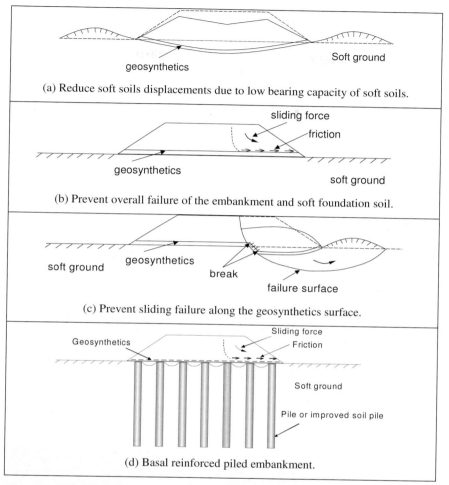

(a) Reduce soft soils displacements due to low bearing capacity of soft soils.

(b) Prevent overall failure of the embankment and soft foundation soil.

(c) Prevent sliding failure along the geosynthetics surface.

(d) Basal reinforced piled embankment.

Fig. 7.14 Use of geosynthetics for the improvement of embankment stability over soft soils.

limited reinforcement effect, the so-called "basal reinforced piled embankment" can be used. Prefabricated piles or improved soil piles can be employed, as shown in Fig. 7.14(d).

The stability level of a reinforced embankment on soft soil can be evaluated by the definition of safety factors (F_s), as shown in Fig. 7.15:

- For overall stability

$$F_s = \frac{M_R + \Delta M_R}{M_D} \geq typically\, 1.2 \sim 1.3$$

where M_D – soil driving moment
M_R – soil resisting moment
ΔM_R – geosynthetic contribution against failure

- For stability against sliding failure

$$F_s = \frac{P_R}{P_A} \geq typically\, 1.5$$

where P_A – active thrust from the fill (from active earth pressures)
P_R: friction force along the fill-reinforcement interface

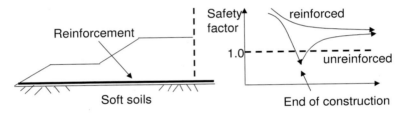

Fig. 7.15 Illustration of safety factor.

Fig. 7.16 Uses of geosynthetics as reinforcement for the improvement of
embankment stability.

The efficiency of geosynthetics as reinforcements of embankments on soft soils can be visualized by the following figures. Figure 7.16 shows applications of geosynthetics for the improvement of embankment stability over soft soils.

7.7.6 *Geosynthetics in walls*

Horizontal layers of geosynthetic reinforcement can be included with retaining wall backfills to provide a reinforced soil mass that acts as a gravity structure to resist the earth forces developed behind the reinforced zone, as shown in Fig. 7.17. Reinforcement types are geogrid, woven geotextile and polyester strap. The local stability of the backfill at the front of the wall is assured by attaching the reinforcement to facing units constructed with polymeric, timber, concrete or metallic wire basket materials comprised of a variety of shapes Fig. 7.18. In North America it has been shown that reinforced soil walls can be constructed for up to 50% of the cost of conventional gravity wall structures.

Analysis and design calculations for reinforced soil walls are related to external, internal, facing and global mechanisms as presented in Fig. 7.19. Global modes refer to instability mechanisms that pass beyond the composite reinforced soil structure. These analyses are routinely handled using conventional slope stability methods of analysis.

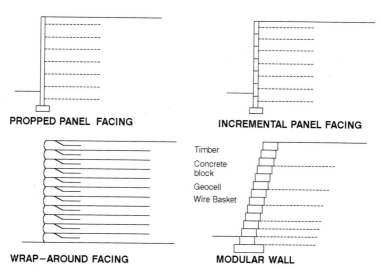

Fig. 7.17 Illustration of different types of reinforced soil wall.

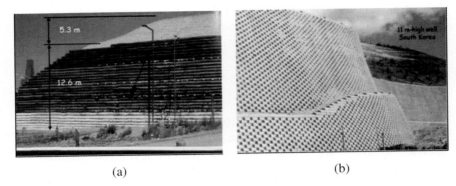

(a)   (b)

Fig. 7.18 Components of modular masonry concrete   (segmental wall) and modular
masonry concrete wall.

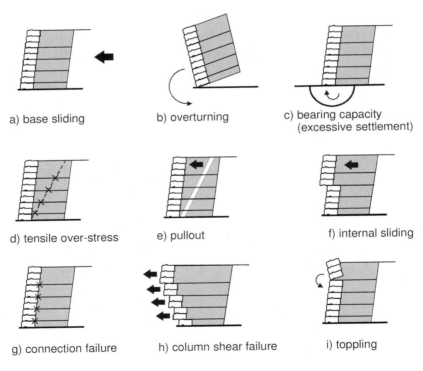

a) base sliding   b) overturning   c) bearing capacity
                                                         (excessive settlement)

d) tensile over-stress   e) pullout   f) internal sliding

g) connection failure   h) column shear failure   i) toppling

a), b), c) external; d), e), f) internal; g), h), i) facing

Fig. 7.19 Design modes for reinforced soil walls.

7.7.7 *Geosynthetics in hydraulic projects*

Hydraulic structures comprise the geosynthetics market segment with arguably the largest growth opportunities. The term "hydraulic structures" includes dams and canals. Hydraulic structures interact with water, which can be one of the more destructive forces in the environment. Geosynthetics are often used to limit the interaction between the structure and water. Geosynthetics can increase the stability of the hydraulic structures. For hydraulic structures, geosynthetics can be used to:

- Reduce or prevent water infiltration through the use of geomembranes
- Reduce or prevent bank erosion of canals through the use of geomembranes liners
- Provide drainage and/or filtration through the use of geotextiles and geonets
- Provide reinforcement for the structure's foundation or the structure itself by using geogrids.

Geomembranes are practically impervious to water infiltration and are commonly used to create a hydraulic barrier on the upstream face of dams. Figure 7.20 shows lining upstream face of a dam and mechanical fastening geomembrane onto a wall. The geomembranes can either be left exposed or covered up using materials such as concrete panels

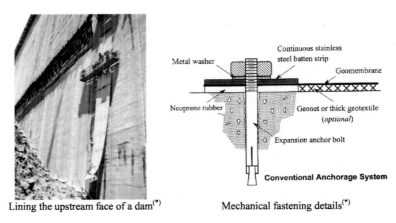

Lining the upstream face of a dam[*]   Mechanical fastening details[*]

Fig. 7.20 Lining upstream face of a dam and mechanical fastening geomembrane onto a wall.

or rip-rap. The use of geomembranes has proven particularly useful in the retrofitting of ageing concrete dams. Exposure can shorten the life-span of the geomembrane due to UV-radiation degradation, but repairs can be made more easily than with covered geomembranes. Covered geomembranes can also be prone to damage, such as puncturing caused by the overlying and/or underlying materials. Geotextiles are often placed underneath, and sometimes over the geomembrane to protect the material from puncturing, serving as cushions to remove stress concentrations.

Leakage through a geomembrane occurs mainly through defects at the seam joints, and puncture holes. Generally, the defects are minimized through CQA/CQC programs at the project site. However, leakage is inevitable especially as the geomembrane begins to age. To protect the structure, geonets or geonet/geotextile geocomposites are typically used as drainage behind the geomembrane. The leak water is collected and deposited downstream through a conduit in the dam or back into the reservoir. The geosynthetic system is affixed to the dam facing by mechanical means, often through the use of anchor bolts and steel batten strips. Gaskets and sealants are used to waterproof the connections and joints. Dams with complicated geometries are more apt to have defects at the seams and joints. The components of the geosynthetic system selected for use with a hydraulic structure are highly project- and site-specific. If properly specified and installed, geosynthetics can be cost-effective and prolong the service life of a hydraulic structure.

7.7.8 Geosynthetics in erosion control

Erosion is a natural process caused by the forces of water and wind. It is influenced by a number of factors, such as soil type, vegetation, and topography and it can be accelerated by various activities that occur on a specific field installation. Uncontrolled erosion processes can cause major damages to existing structures and to the environment. Geosynthetics can be used for erosion control in works such as:

- Slope Protection
- Channels
- Drainage Ditches
- Waterways
- Shoreline Protection
- Reclamation

- Re-vegetation
- Scour Protection
- Rockfall Netting
- Breakwaters
- Weirs
- Embankments

Depending on project and site characteristics, an erosion control work may involve the use of one or more geosynthetic products such as geotextiles, geomats, geonets, geogrids, etc. Figure 7.21 presents the use of a non woven and a geogrid for slope protection and erosion control. Some examples of geosynthetics in erosion control works are presented as follows.

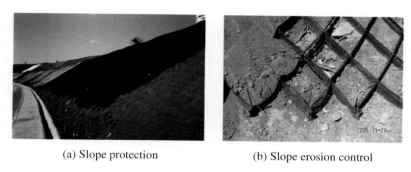

(a) Slope protection   (b) Slope erosion control

Fig. 7.21 Use of geosynthetic for slope protection and erosion control.

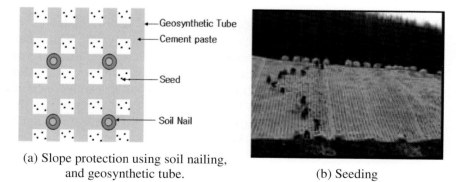

(a) Slope protection using soil nailing, and geosynthetic tube.

(b) Seeding

Fig. 7.22 Slope protection using geosynthetics, soil nailing, rock bolt and seeding.

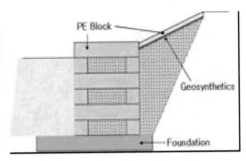

| (a) Protection of river banks using PE block. | (b) Protection of shore lines using PE blocks and panels. |

Fig. 7.23 Protection of river banks and shore lines using PE blocks and panels.

Slope protection is achieving stability using geosynthetics, soil nailing, rock bolt and anchors. The slope is covered with a geosynthetic tube. The geotextile tube is filled with cement paste. Soil Nailing is conducted on slope surface. Vegetation on the geosynthetic protects the cut slope from surface loss due to water or wind. Figure 7.22 shows slope protection using geosynthetics, soil nailing, rock bolt and seeding.

As shown in Fig. 7.23, channel protection was installed using stacked PE Panels. Slope is covered with geosynthetic and PE Block. Vegetation on the surface protects slope from surface loss due to scouring.(*) First three photographs courtesy of Prof. Ennio M. Palmeira.

7.7.9 Geosynthetics in landfills

Geosynthetics are extensively used in the design of both base and cover liner systems of landfill facilities. This includes: _geogrids,_ which can be used to reinforce slopes beneath the waste as well as to reinforce cover soils above geomembranes; _geonets,_ which can be used for in-plane drainage; _geomembranes,_ which are relatively impermeable sheets of polymeric formulations that can be used as a barrier to liquids, gases and/or vapors; _geocomposites,_ which consist of two or more geosynthetics, can be used for separation, filtration or drainage; _geosynthetic clay liners_ (GCLs), which are composite materials consisting of bentonite and geosynthetics that can be used as an infiltration/hydraulic barrier; _geopipes,_ which can be used in landfill applications to facilitate collection and rapid drainage of the leachate to a sump and removal system; _geotextiles,_ which can be used for filtration purpose or as cushion to protect the geomembrane from puncture. The figure below illustrates the extensive multiple uses of geosynthetics in both the cover and the base liner systems of a modern landfill facility.

The base liner system illustrated in the figure above is a double composite liner system. It includes a *geomembrane/GCL* composite as the primary liner system and a *geomembrane/compacted clay liner* composite as the secondary liner system. The leak detection system, located between the primary and secondary liners, is a *geotextile/geonet composite.* The leachate collection system overlying the primary liner on the bottom of the liner system consists of gravel with a network of perforated *geopipes.* A *geotextile* protection layer beneath the gravel provides a cushion to protect the primary *geomembrane* from puncture by stones in the overlying gravel. The leachate collection system overlying the primary liner on the side slopes of the liner system is a *geocomposite* sheet drain *(geotextile/geonet* composite) merging into the gravel on the base. A *geotextile* filter covers the entire footprint of the landfill and prevents clogging of the leachate collection and removal system. The groundwater level may be controlled at the bottom of the landfill by gradient control drains built using *geotextile* filters. Also, the foundation soil below the bottom of the landfill may be stabilized as shown in Fig. 7.24.

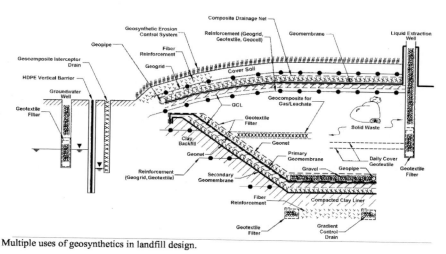

Multiple uses of geosynthetics in landfill design.

Fig. 7.24 Landfill with geosynthetic liners.

Using randomly distributed *fiber reinforcements,* while the steep side soil slopes beneath the liner are reinforced using *geogrids.* The cover system of the landfill illustrated in the figure contains a composite *geomembrane/GCL* barrier layer. The drainage layer overlying the geomembrane is a *geocomposite sheet drain* (composite *geotextile/geonet).* In addition, the soil cover system includes *geogrid, geotextile,* or *geocell* reinforcements below the infiltration barrier system. This layer of reinforcements may be used to minimize the strains that could be induced in the barrier layers by differential settlements of the refuse or by a future vertical expansion of the landfill. In addition, the cover system could include geogrid or *geotextile* reinforcement above the infiltration barrier to provide stability to the vegetative cover soil. *Fiber reinforcement* may also be used for stabilization of the steep portion of the vegetative cover soil. A *geocomposite erosion* control system above the vegetative cover soil is indicated in the figure and provides protection against sheet and gully erosion. The use of *geotextiles* as filters in groundwater and leachate extraction wells is also illustrated in the figure. Finally, the figure shows the use of an *HDPE vertical barrier* system and a geocomposite interceptor drain along the perimeter of the landfill facility. Although not all of the components shown in the figure would be necessarily needed at any one landfill facility, the figure illustrates the many geosynthetic applications that can be considered in landfill design.

7.7.10   *Geosynthetics in wastewater treatment*

Geosynthetics are used in various applications waste water facilities. The most common use is in lagoons operating with anaerobic and aerobic lagoon processes.   Other applications include enhanced evaporation of wastewater and sludge dewatering by permeable geotextile geotubes.

7.7.10.1 Anaerobic lagoons with covers

When wastewater with a reasonably high organic load is kept in a lagoon for several days an active anaerobic sludge accumulates at the bottom of the lagoon. In an uncovered lagoon the anaerobic digestion activity takes place at the base of the lagoon and the activity near the surface tends to be more aerobic.

We can cover these lagoons with a geomembrane floating cover to enhance the anaerobic digestion activity by the exclusion of air (oxygen) to: (a) enable the harvesting of gas (especially methane) which can be used as a fuel, and (b) reduce the effect of odor from the anaerobic activity. Figures 7.25 and 7.26 shows a lagoon covered with a geomembrane.

Fig. 7.25 A lagoon covered with a geomembrane.

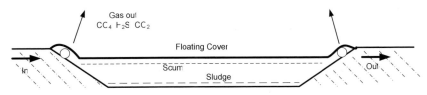

Fig. 7.26 Anaerobic covered lagoon.

Generally these lagoons will take wastewater with BOD of 400 to 5000 kg/cum and the output effluent will have the BOD reduced by 90 to 95%. Detention time is normally 4 –7 days. The anaerobic process is largely self-propelled and the only mechanical input is that required to feed wastewater to the lagoon and force its exit at an overflow outlet. There may be a need for systems to deal with excessive accumulations of sludge (base) and scum (surface under cover) but this will depend on the nature of the wastewater and the dynamics of the system.

7.7.10.2 Aerobic (aerated) lagoons

Aerated systems use either surface aerators or diffuser systems to introduce air into the wastewater and this results in consumption of the organic content of the wastewater, which is mostly released as carbon dioxide. These aerobic systems require considerable mechanical input to operate the aeration system and further work may be need to remove excess sludge from the base from time to time. Typically these systems take wastewater with BOD in the order of 500 to 1500 kg/cum and the output effluent will have the BOD reduced by around 90%. Detention time is normally 4–7 days. Figure 7.27 is an aerated covered lagoon.

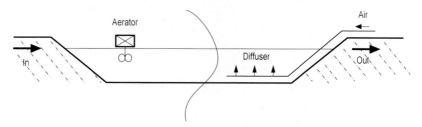

Fig. 7.27 Aerated covered lagoon.

7.7.10.3 Geosynthetics for combined anaerobic and aerobic lagoons

Many wastewater plants make use of anaerobic and aerobic systems as a combined or two-part process. This can be readily achieved in one lagoon using a specially designed geomembrane floating cover. Figure 7.28 illustrates a combined system geomembrane floating cover.

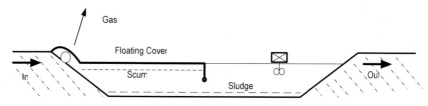

Fig. 7.28 Combined system geomembrane floating cover.

These combined systems have a capacity to take wastewater with BOD of 5000 kg/cum and to achieve an output effluent less than 100 kg/cum.

Total detention times would be in the order of 10 days although some systems use final 'polishing' lagoons or grass filtration/irrigation. These combined systems have the capability for the gas to be used on-site to provide power, which can be used for the aeration energy input.

The applications for geosynthetics in these lagoon systems are essentially associated with the liner system and with the floating cover system but there are many variations that may be chosen according to circumstances. (a) Liner Systems. The liner systems may be a Geosynthetic Clay Liner with soil or concrete cover. However the concrete surface is rather rough and in some cases it will be desirable to use a very smooth HDPE upper liner as this can help to move the sludge to locations from which the excess can be removed. (b) Cover Systems. The more durable and stable cover systems are based on polyethylene materials (PE) but most covers will not allow the use of HDPE for the whole cover because it is too stiff. It is usually necessary to use a hybrid of more flexible PE materials such as mPE-R or LLDPE, which are placed in the flex zones of the cover. Another option is to construct the whole cover in a flexible reinforced geomembrane such as polypropylene or elvalloy. Cover designs may also vary with factors such as the intended operation of the cover with respect to effluent levels, gas collection and associated factors, as well as the construction restrictions, which may limit the cover design options.

7.7.10.4 Enhanced evaporation

A typical black geomembrane with shallow wastewater over it will see the wastewater temperature rise with solar radiation creating an enhanced capacity for evaporation. This is used in wastewater disposal and for salt and mineral extraction processes. A variation of this process can be used in regions with seasonal rainfall and a pronounced dry season. A floating cover over the wastewater will prevent growth of waste volume in the wet season as well as enabling fresh water to be gathered from the cover. In the dry season wastewater can be pumped onto the cover for enhanced evaporation.

7.7.10.5 Sludge dewatering

Geotubes were initially developed as a construction tool enabling the used of dredged sands to build groynes and the like. These filtration properties can also be used to take sludges with high water content and rapidly dry them to a solid state, which allows truck transport without dripping.

7.8 Summary

This chapter mainly describe describes different types of geosynthetics and their applications in civil engineering construction. Almost all geotextiles available in the United States are manufactured from either polyester or polypropylene. There are two principal geotextile types, or structures: wovens and nonwovens. Geosynthetics can be used for one or more purposes of the following functions: separation, filtration, drainage, reinforcement, fluid/gas containment, or erosion control.

References

1. Bathurst, Richard J. "Geosynthetic Classification" International Geosynthetics Society education committee brochure.
2. Bathurst, Richard J. "Geosynthetic Functions" International Geosynthetics Society education committee brochure.
3   Bathurst, Richard J. "Geosynthetics in Walls" International Geosynthetics Society education committee brochure.
4.   Bathurst, Richard J. "Geosynthetics in Slopes over Stable Foundations" International Geosynthetics Society education committee brochure.
5. Gourc, J.P. and Palmeira, E.M "Geosynthetics in Drainage and Filtration" International Geosynthetics Society education committee brochure.
6. Palmeira, E. M. "Geosynthetics in Road Engineering" International Geosynthetics Society education committee brochure.
7. Palmeira, E. M. "Geosynthetics in Unpaved Roads" International Geosynthetics Society education committee brochure.
8. Otani, J and Palmeira, E. M. "Geosynthetics in Embankments on Soft Soils" International Geosynthetics Society education committee brochure.
9. Zornberg, J and Bouazza, M "Geosynthetics in Road Engineering" International Geosynthetics Society education committee brochure.
10. Bouazza, M and Zornberg, J. "Geosynthetics in Landfills" International Geosynthetics Society education committee brochure.
11. Shin, E. C. "Geosynthetics in Erosion Control" International Geosynthetics Society education committee brochure.

12. Sadlier, M. "Geosynthetics in Wastewater Treatment" International Geosynthetics Society education committee brochure.
13. Koerner, R. "Designing with Geosynthetics" Prentiss Hall, 5th Edition
14. ASTM International, www astm.org

Chapter 8

Coatings and Sealers

C. Vipulanandan and M. Isaac
University of Houston, Houston, TX

8.1 Introduction

Reinforced concrete can be degraded by direct chemical attack on concrete or by corrosion of the reinforcements. Coatings and sealers are being increasingly used in improving the performance and increasing the service life of reinforced concrete structures under various environmental conditions. The focus of this chapter is on testing and evaluating the effectiveness of coatings (non-penetrating into concrete) and sealers (penetrating into concrete) in acidic and salty environment in protecting reinforced concrete structural elements. Typically coatings such as epoxies, polyurethanes, acrylics, polymer concrete, latex and cement based composites are being used in protecting concrete against acidic and sulfate attack. Sealers, such as silanes and siloxanes, with and without coatings, are being used to protect the reinforcements in the concrete from salty environments. Since there are number of coatings and sealers in the market place it is essential to evaluate their performance before using them to protect reinforced concrete.

There is more and more interest in protecting new and old reinforced concrete structures exposed to harsh environmental conditions using coatings and sealers. Cement concrete is a highly alkaline material that can easily deteriorate under acidic environments in food factories, chemical plants and wastewater facilities and are usually protected using various types of coating materials. Similarly reinforced concrete structures used in the coastal regions and highway pavements that require seasonal salt application can be protected from chloride attack on steel reinforcements by applying sealers and coatings on the concrete surface.

333

In sewage facilities, sewer pipes lift stations and wastewater treatment plants are corroded due to sulfuric acid produced by sulfuric-acid-producing bacteria (Soebbing, 1996 and Sand, 1994). It has been reported that the concentration of sulfuric acid for the worst case in sewer environments was pH of 0.5 (Islander, 1991). Hence coating system that bonds to concrete and provides protection from microbial sulfuric acid attack would have wide use in the wastewater industry (Redner et al., 1994; Liu and Vipulanandan 1999-2005). Redner et al. (1994) evaluated more than 20 different coating systems in an accelerated test program (in 10% sulfuric acid). Their test results showed that there were no fail-safe coatings. Even coatings that survived the accelerated test had failed in the field (Redner et al., 1994). Liu and Vipulanandan (1999) studied concrete specimens coated with an epoxy coating and immersed in 3% sulfuric acid (pH = 0.45). The results showed that for specimens without holidays in the coating film, the mass gain was only about 1% after three years and the probability of failure increased with the increase in mass for coated concrete in 3% sulfuric acid; hence, predicting the mass change in test specimens of coated concrete was very important in determining the service life of coated concrete and evaluating the effectiveness of coating materials.

Deterioration of reinforced concrete structures is partly caused by the gradual intrusion of chloride ion and moisture into the concrete from deicing salts and in the coastal regions further accelerated by the salt water environment. The corrosion of the reinforcement in bridge elements is due primarily to the presence of chloride ions. As a consequence, a large number of concrete structures, especially bridge elements, are contaminated with chlorides that initiate the corrosion of the reinforcing steel. The resulting presence of chlorides and loss of the alkaline environment causes the embedded steel to lose its surface passivity. Accumulated corrosion products cause cracking of the protective concrete cover. Deterioration caused by the corrosion of reinforcing steel is not limited to bridge decks only. It also affects other bridge members such as abutments, beams, crossbeams, diaphragms, piers, and piles (Pfeifer et al., 1981)). Several methods are used in protecting reinforced concrete surfaces including coating the concrete. Coating materials used in protecting concrete from corrosive salt environments include epoxies, methacrylate, urethane, silicate, siloxane and silanes. The primary function of a surface treatment is to prevent capillary action at the surface, thus preventing the migration of water and chloride ions into the concrete. Surface treatments materials used for protecting reinforced concrete structures include sealers and coatings. The treatments provide either a non-penetrating film (coating) or penetration into concrete pores or pore blockers (sealers) as schematically shown in Fig. 8.1. The general difference between sealers and coatings is that sealers penetrate the concrete surface and block the capillary pores, whereas coatings do not penetrate the concrete surface but form a thick film on the surface.

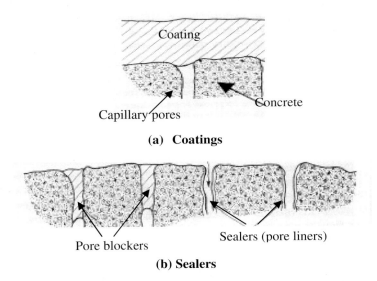

Fig. 8.1 Protecting concrete. (a) Non-penetrating coatings. (b) Penetrating sealers.

Sealers have been used on reinforced concrete surfaces to retard or prevent the migration of chlorides for some time. Applying a sealer to the concrete can be an effective and an inexpensive method of addressing the corrosion problem, thus increasing the service life of a reinforced concrete structures. The practice of sealing has increased appreciably in recent years, and a variety of proprietary sealers are now available. It must be noted that some sealers are not permanent and periodic reapplications may become necessary to maintain the protective properties of the sealers.

Concrete sealers can be divided into two main groups: pore blockers and hydrophobic agents (pore liners). Hydrophobic sealers (such as silanes and siloxanes) penetrate the concrete to some degree and enable the concrete to become water repellant. Pore blocking sealers provide little penetration and form instead a thin film on the concrete's surface. Pore blockers are further distinguished by their ability to partially or fully fill the surface pores, a capability not shared by hydrophobic agents (Wohl et al., 1990).

Concrete sealers work in two ways. First they substantially reduce the absorption of moisture and chloride ions. Second, they allow the progressive internal drying of concrete by reducing the rate of moisture gain from the environment. Hence, the sealer protects the concrete and makes it breathable.

In the 1960's, boiled linseed oil was used for maintenance of existing bridge curbs in Alberta, Canada. Epoxies and other types of sealers were also tried in the 60's. Dehydratine, a black tarry substance, was routinely used on abutments, since the late 1960's (Alberta Transportation Technical Standards, 2001). Coatings such as epoxies and acrylic sealers were a new protective measure against chloride induced corrosion introduced into North America in the mid 70's. The first application in the U.S was on bridges in the Florida Keys in 1982. Penetrating Silane sealers were first used in Alberta on concrete bridge decks in 1986. The first major application in Europe was for the precast concrete segments for the twin bored East Tunnel of the Great Belt Link in Denmark in 1988 (Rostam, 2003).

Highway agencies in the United States and Canada were surveyed about their use of penetrating sealers for protecting cement concrete. One of the first and most complete, independent comparative studies on penetrating sealers for concrete were reported in 1981 (Pfeifer, 1981). In this National Cooperative Highway Research Program (NCHRP) study, total of 21 of sealers of various generic types were subjected to a battery of test procedures developed by the researchers. A study conducted for the Transportation Research Board in 1981 (Pfeifer et al., 1981) researched the protection of concrete bridges against chloride penetration by using various surface treatments (coatings & sealers) representative of all of the chemical types commonly used. Based on the initial screening program, five products with low water absorption, low chloride ion uptake, and good water vapor transmission characteristics, were chosen for further investigation. The five materials chosen were an epoxy, a methyl methacrylate, moisture cured urethane, a silane and polyisobutyl methacrylate. These materials were subjected to further testing to determine the effects of moisture condition of the substrate, coverage rate, and different environmental conditions on the ability to protect against chloride ion intrusion. The five treatments reduced the chloride ion contents by 79 to 97% compared to the uncoated specimens (Mallett, 1994). The study concluded that the epoxy, methyl methacrylate, and the silane were capable of providing added protection to concrete bridge surfaces to reduce intrusion of salt water. This study has become a benchmark for the evaluation of penetrating sealers, at least in a laboratory situation. The procedures used in that report primarily use the ingress of chloride solutions into the concrete (and subsequent chloride ion analyses) as a measure of sealer effectiveness. The NCHRP procedure is similar in principle to AASHTO T259 procedure (AASHTO, 1986) although sample geometry, conditioning, and exposure are different and hence the results from the two methods will be different.

A study conducted by Whiting et al. (1990), surveyed highway agencies in the United States and Canada regarding the use of penetrating sealers. Of the agencies surveyed, 46 U.S. and 9 Canadian agencies have used sealers. The most widely used application of penetrating sealers was on concrete bridge decks. Only about 30% of the respondents were utilizing penetrating sealers in superstructure elements other than decks. The study stated that the respondents noted a variety of problems with the application of penetrating sealers to existing structures. Some of these concerns included the following: drift and evaporation in hot and windy conditions, difficulty in obtaining specified coverage, slippery surfaces, runoff during application, discoloration of concrete, and little or no apparent penetration. The respondents also stated that the performance of the sealers was less than desired. Some indicated that many penetrating sealers were ineffective (or at least not as effective as claimed) in reducing chloride ion infiltration. Other performance problems included: reduction of skid resistance, failure to improve freeze-thaw and scaling resistance, and failure to halt corrosion of reinforcing steel (Whiting et al., 1993).

In contrast to coatings, penetrating sealers allow the concrete to breath since the pores are exposed to the atmosphere. This permits the concrete to dry-out, and with the moisture intake reduction the possibility of corrosion may be lessened. Since most sealers are clear in color, the color of concrete will generally not be affected when applied. In addition, since penetrating sealers are capable of infiltrating well into the surface, they are less affected by the environmental weathering. This can lead to a longer service life when compared to coatings (Kay, 1992).

Sprinkel et al. (1993) studied the performance of sealers as well as overlays and patch materials. The study did not specifically indicate the type of sealer that was evaluated. The investigation concluded that sealers could reduce the infiltration of chloride ions for 5 to 10 years and therefore extended the time until sufficient chloride ions reached the reinforcing steel to initiate corrosion. To ensure adequate skid resistance, sealers should be applied to decks with tined or grooved surfaces. The investigation found that protection provided by sealers varied with tests, exhibiting 0 to 50%, with an average of 32%, reduction in permeability. On the basis of life cycle cost analysis, the most cost effective protection system was determined to be the application of a penetrating sealer (Sprinkel, Weyers, and Sellars, 1991-1993).

A study conducted by the Federal Highway Administration (Whiting et al., 1999) evaluated various corrosion repair techniques and protection systems for prestressed concrete elements. The prestressed concrete specimens were subjected to accelerated corrosive environments to induce corrosion in the steel. Penetrating sealers and coatings were applied to a set of specimens to study their effectiveness. This study concluded that the surface treatments were of limited effectiveness when applied to specimens subjected to active corrosion. In most cases, chlorides continued to penetrate into the concrete, though at a reduced rate. While measurements indicated that corrosion activity was initially reduced after the application of the surface treatment, long-term trends suggest that over time corrosion activity may slowly increase back towards the initial levels. The study also concluded that surface treatments applied to new structures would reduce, but not completely eliminate, the ingress of deleterious substances. If low quality repair materials are used or incorrect construction procedures are employed, corrosion may still occur resulting in the cracking and spalling of the structure. However, in general, application of surface treatments in new construction significantly improves its long-term effectiveness, especially in chloride environments.

Silane sealers penetrate about ½ inch into the concrete and react chemically with concrete to form a layer that resists water and chloride penetration (Bruner, 1992). Siloxane is very similar to silane although not as effective in reducing water and chloride penetration (Bruner, 1992). Silanes and siloxanes are produced from the same basic raw material, silane. Both silanes and siloxanes reduce water and chloride intrusion, but neither change the appearance of concrete. When applied, silane and siloxane repellants penetrate into the substrate and react chemically with the alkalinity therein to form a resin. The resin that forms has no elastomeric properties and is designed to make the capillaries of the substrate smaller than that of a water molecule and hence providing the repellency.

The Alberta Department of Transportation also carried out extensive studies on development of laboratory test procedures for the evaluation of penetrating sealers (Carter et al., 1986; Kottke, 1987; Carter, 1989). While concrete mixtures used for evaluation were similar to those in NCHRP 244, curing was different, and specimens were conditioned prior to testing so as to obtain a pre-selected rate of water absorption for the unsealed control specimens. Tap water was used in the Alberta method, as opposed to a sodium chloride solution, which was used in the NCHRP 244 procedures (Pfeifer, 1981). Additionally, the Alberta procedure allowed for sandblasting the surfaces after applying the sealer, in an attempt to simulate in-service surface abrasion.

Petty (1986) evaluated more than 20 products using both continuous immersion and cyclic soaking in 15-percent sodium chloride solutions. Less than 50 percent of the products tested met specification. Fernandez (1988) applied sealers to oven-dried specimens, then performed water absorption tests and measured depth of penetration. The Ontario Ministry of Transportation utilized a variety of test procedures in their evaluations of penetrating sealers, including depth of penetration, water absorption, water-vapor transmission, chloride absorption, salt scaling, and rapid chloride permeability (Rouings, 1988). Rutkowski (1988) evaluated absorption, resistance to chloride ion penetration, vapor permeability, and impressed current as test procedures for evaluating effectiveness of sealers. A number of field studies on the effectiveness of penetrating sealers have also been reported (Rutkowski, 1988).

Some of the earliest applications were carried out by the Oklahoma Department of Transportation. Smith (1986) reported that by the end of 1985, 245 bridge decks had received a silane treatment. The bridges ranged in age from newly constructed to 15 years when sealers were first applied. Nine bridges were selected for periodic chloride sampling. At the time the report was issued, the data were insufficient to allow conclusions to be drawn as to the effectiveness of the sealers in reducing the rate of chloride accumulation in these decks (Smith, 1986).

Rasoulian et al. (1986) periodically obtained cores from five structures treated with silanes in 1981 and allowed to weather in northern and marine environments for 4 years. Results indicated that the absorption of cores gradually increased with time and was not much less than that of control (unsealed) samples in some cases. Depth of penetration of silanes was found to be 0.1 in. (2 mm) at most (Rasoulian, 1986). Studies in Vermont were similarly disappointing, with silane sealers failing to perform much better than conventional tar-emulsion or linseed oil tested products, though chloride penetration was reduced as compared with untreated concretes in wing walls and median barriers (Fraseoia, 1986; Winters, 1987 a and b).

In a somewhat more extensive study, the Minnesota Department of Transportation evaluated nine products on a bridge deck overlaid in 1983 with low-slump dense concrete. After 3 years of sampling, the most effective products included an oligomeric alkoxy-silane and a penetrating epoxy. Other products, including silanes, fluorosilicates, silanoate, and methyl methacrylate, lost effectiveness within the 3-year test period, as measured by the percentage increase in chloride content compared to untreated sections (Fraseoia, 1986). A similar study was performed by the Pennsylvania Department of Transportation, in which seven sealants were applied to the deck, sidewalks, and parapets of a bridge constructed in 1984. Sampling for chloride content spanned a 4-year period. Only one penetrating sealer (a resin in mineral spirit formulation) was found to be as effective as conventional linseed oil treatment in reducing chloride ion penetration when compared to untreated sections (Whiting, 1990).

To determine how long various sealers extend the service life of bridges affected by chloride corrosion, Zemajtis and Weyers (1996) examined generic sealers applied to steel reinforced concrete bridges in the U.S. It was concluded that the service life of a sealer was affected by various factors, including environmental conditions, traffic wear, penetration depth and ultraviolet (sun) light.

Silanes are monomeric silicon (Si) compounds with four substituents, or groups, attached to the silicon atom (($R\text{-}Si\text{-}R'_3$) where R and R' are organic based). These groups can be the same or different and non-reactive or reactive, with the reactivity being with inorganic (Si) or organic groups. Silane sealers will penetrate into the concrete and make the concrete surface hydrophobic (Perkins, 1997; Klahorst et al., 2004). Silane sealers are considered to be the most widely used products for protecting new and inservice bridge decks. They penetrate and chemically attach to the concrete to reduce water and chloride intrusion. Only as the concrete wears away do the sealers lose effectiveness. Silane sealers are available in three forms: organic solvent, water based, and 100-percent reactible versions. Studies using NCHRP 244 and the Alberta tests showed all versions provide the same basic protection.

An overcoat with a latex coating is expected to block the pores in the concrete surface. High-build latex modified cements are rubbery liquids applied in a relatively thick coating. Cementitious latex modified coatings are relatively impervious, have a similar coefficient of thermal expansion to concrete, and are less expensive than other overlays. The latex additive, with silane sealer, is much more effective at reducing water and chloride intrusion than uncoated concrete.

Ibrahim et al. (1999) studied the effectiveness of concrete surface treatments including sealers and coatings. They evaluated several penetrating sealers and a two-component acrylic coating. The coating was found to be the most effective of the materials investigated in minimizing damage due to sulfate attack after 6 months of sulfate exposure. In addition, it was determined to be effective in reducing the ingress of carbon dioxide. The coating also exhibited considerably lower chloride concentration in comparison to untreated specimens (Ibrahim et al., 1999).

Restoring concrete with non-penetrating coatings requires consideration of the concrete surface conditions (strength, pH, and moisture content) and the porosity of the concrete. The minimum recommended surface strength of concrete for using coatings is in the range of 1.4 to 1.75 MPa (200-250 psi) (Soebbing et al., 1996). There are many factors that affect the performance of coatings in salt water environments, such as chemical and mechanical properties of coating materials, bonding between the concrete surface and the coating material, and the applicability of the coating systems old and new concrete surfaces. Vipulanandan et al. (2001-2005) investigated the effect of water at the concrete surface and concluded that it can react with the coating material and affect the setting and the adhesion of the coating systems (Liu and Vipulanandan, 1999-2003). Coatings can debond and blister if the osmotic pressure under coating films exceeds the tensile adhesion of the coating material. Concrete deterioration can range from slight etching or partial loss of surface cement binder to complete loss of cement binder. Complete binder loss yields exposed coarse aggregates and reinforcing steel which will further accelerate corrosion and cracking and spalling of the concrete. For satisfactory performance, the coating needs to be holiday-free (Vipulanandan et al. 2002, 2005). Many early coating installations did not ensure holiday-free coating which resulted in premature failure of the coatings.

Coatings form an impervious film that bridge over pores and provide an external physical barrier, which protects by slowing down the penetration of liquids and gases. They are designed to control water absorption, vapor transmission, and diffusion of aggressive liquids and gases through the concrete surfaces. Coatings are normally applied by brushing, rolling, or spraying the material onto the surface.

Two commonly employed coatings include cementitious and polymer coating systems. Cementitious coatings allow moisture to escape without debonding or blistering. However, since they do not possess elastomeric properties, they cannot bridge active cracks. Polymer systems consist of epoxies, acrylics, or polyurethanes combined with filler, which provides bulk and thickness. Polymer coatings are hard and durable, but are impervious to vapor transmission.

The characteristics and performance of some common coatings can be summarized as following (Tabatabai, Ghorbanpoor, and Turnquist-Nass, 2005):

- Epoxy coatings provide good adhesion to concrete, exhibit minor shrinkage, and are resistant to light chemical attack.
- Polyurethanes will adhere to dry concrete, are almost shrinkage free, are resistant to light chemical attack, but not to highly alkaline conditions.
- Acrylics display good adherence to concrete and good resistance to alkali, oxidation, and weathering.

Coatings can stay in contact with the concrete and protect it from environmental degradation. Durability of a coating material for concrete structures is as important as its ability to perform in intended applications. Durability is concerned with life expectancy or endurance characteristics of the coating material. A durable coating is one which will withstand, to a satisfactory degree, the effect of service conditions to which it will be subjected. At present, there is no one coating material that is completely immune to physical deterioration. There is only limited information in the literature on the performance of coatings in concrete structural elements and the results are not conclusive on the durability of coating materials.

Most concrete structures in service are subjected to aggressive environmental influences which affect durability. Durability in turn affects structure safety, economy, serviceability, and maintenance. Among all chemicals, sulfates and sulfuric acid are the most harmful compounds for concrete since it is directly attacked by these chemicals. Because sodium sulfate (Na_2SO_4) is the most encountered sulfate in the environment, nearly all of the studies were performed by immersing specimens in the sodium sulfate (Na_2SO_4) solutions. The measurements in these studies mainly focused on strength loss and expansion of concrete, cement mortar, or cement paste. Most of the studies were completed within three to twelve months.

Since several factors in the field can affect the performance of sealers and coatings, it is important to identify the important factors through controlled experiments where important variables are studied one at a time. There are testing protocols to evaluate coatings and sealers for various applications but most of the testing methods are focused on evaluating the coated concrete under accelerated testing conditions. Since several factors in the field can affect the performance of coatings and sealers, it is necessary to test the performance of coated concrete with and without holidays in similar chemical environment of application. Also the bonding strength of coatings to dry and wet concrete is an important property to determine ability of the coatings to adhere to the concrete and protect it from deterioration. The bonding strength of coating is also used as a quality control measure in the field.

8. 2 Test Methods

Based on the application and exposure conditions coated reinforced concrete is subjected to various tests. Under acidic and sulfate environments protecting the concrete is critical and under chloride and moist environment preventing these elements from penetrating and attacking the steel reinforcements is critical. Number of tests has been used to evaluate the performance of coated concrete and these test methods are reviewed with test results reported in the literature.

8.2.1 *Sample preparation*

The method used for preparing the coated concrete for testing will significantly influence the performance of coated concrete. Although there are no standard method for preparing the coated concrete specimens, typically concrete specimen are cured for 28 days, cleaned and dried prior to the application of the coatings. Both concrete cylinders and prism specimens can be used for the study. Concrete surfaces are cleaned by water blasting or sand blasting using pressures of the order of 10 MPa (1500 psi). Typical changes in weight observed during the sample preparation, with and without silane treatment and coating are shown in Fig. 8.2. During the water blasting, relatively small amount of cement and other materials are removed from the surface and the water infiltrates/diffuses into the specimen. The specimens are dried at room condition for one day before spraying the silane coating. After applying the silane sealer, the specimens are left to cure for about 5 days before applying the coating. Leaving the silane coated and uncoated specimens at room condition (23°C and 55% humidity) resulted in loss of weight, indicating moisture loss resulting from breathing through the silane sealer. The weight loss in silane coated specimen continued for 21 days. A acrylic coating was applied using a roller to the concrete with and without silane treatment. The specimens are weighed during the coating cycle accurately to 0.01 grams. Addition of the acrylic coating to the specimens changed the weight of the specimen by over 1 %. The coated specimens are cured at room condition for 21 days or as specified by the coating manufacturer before immersing in water and 15 % NaCl solution for marine environments; and in 3 % or 30 % sulfuric acid in acidic environments.

8.2.2 *Laboratory tests*

Laboratory tests are used in evaluating the effectiveness of sealers and coatings on concrete surface. The tests are based on the potential application for the coatings and sealers and can be broadly classified into marine and acidic/sulfate environments. The tests are as follows:

1. Immersion Test and Drying Test (NCHRPR 244) (marine environment)
2. Chemical Resistance Test (CIGMAT CT-1) (for acidic/sulfate environment)
3. Bonding Test (ASTM D 4541, CIGMAT CT-2, CIGMAT CT-3)

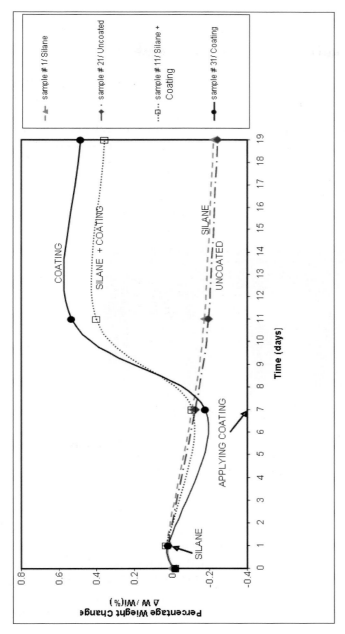

Fig. 8.2 Preparation of coated concrete specimens

8.2.2.1 Immersion test (NCHRP 244)

This test has been recommended for testing the concrete treated with sealers and/or coatings to protect the reinforced concrete from marine environment. As shown in Fig. 8.3, the specimens are immersed in a selected test reagent to the full specimen height in a closed container so that the specimens are exposed totally to the liquid phase. For this test, 76 mm (3-inch) × 152 mm (6-inch) cylindrical cement concrete specimens are used. The 21 days of immersion test is followed by a drying test for another 21 days. Although the original test developed in the early 1980 had no pinholes in the test specimens the recent tests had pinholes (Vipulanandan et al. 2006). In this test, changes in: (1) weight of specimen and (2) appearance of specimen are monitored at regular intervals.

Fig. 8.3 Immersion test with holidays for coated concrete specimens
(modified NCHRP 244).

Water immersion

Typical response of silane treated concrete, coated concrete, silane + coated concrete, and uncoated concrete immersed in water are shown in Figures 8.4 and 8.5. In this case uncoated concrete showed an average increase in weight of 0.61%, 0.72%, and 0.77% in 1, 9 and 21 days of immersion respectively due to water infiltration and diffusion. Silane treated concrete had a 51% less water uptake than the uncoated concrete after 21 days of immersion. Coated concrete had a 9% less water uptake

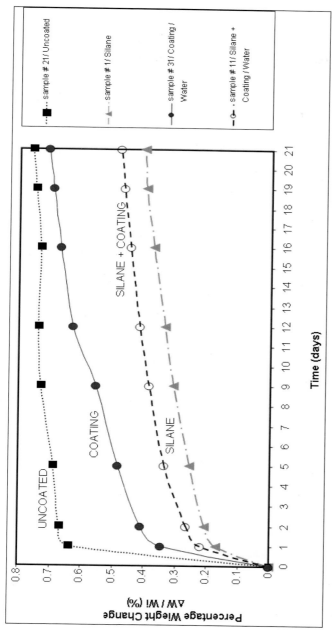

Fig. 8.4 Water uptake by concrete coated with silane, coating, silane + coating combined coating compared to uncoated concrete.

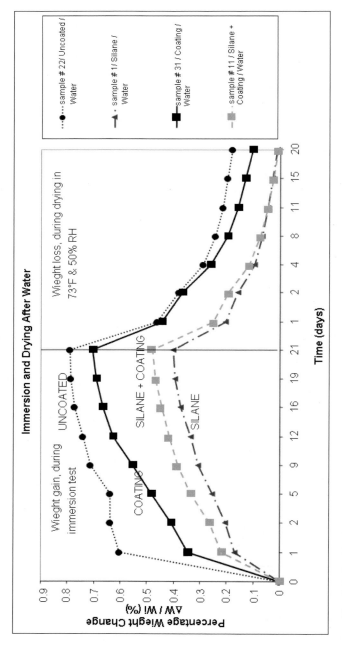

Fig. 8.5   Residual weight change after immersion of specimens (silane, coated concrete, silane + coated concrete) in water for 21 days and drying for 20 days.

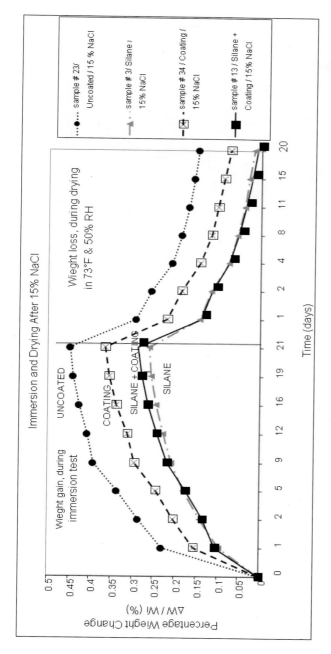

Fig. 8.6 Residual weight changes after immersion of specimens (silane, coated concrete, silane + coated concrete in 15% NaCl solutions.

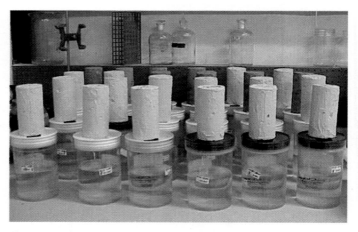

Fig. 8.7   Drying of concrete specimens at the end of the immersion test.

than the uncoated concrete after 21 days of immersion. Silane + coated concrete had a 39% less water uptake than the uncoated concrete after 21 days of immersion. It has been reported that 3 mm pinhole affected the water uptake of the coated concrete but didn't affect the silane treated concrete.

NaCl solution (15%) immersion

Typical results of silane treated concrete, coated concrete, silane + coated concrete, and uncoated concrete immersed in 15% NaCl solution are shown in Fig. 8.6. Uncoated concrete showed an average increase in weight of 0.22%, 0.36% and 0.42% after 1, 9 and 21 days of immersion respectively due to salt infiltration and diffusion. Silane treated concrete had a 38% less sodium chloride uptake than the uncoated concrete after 21 days of immersion and the pinhole did not affect the solution uptake.. Coated concrete had a 9.5%less solution uptake than the uncoated concrete after 21 days of immersion. Silane + coated concrete had a 33.3% less water uptake than the uncoated concrete after 21 days of immersion and the pinhole did affect the solution uptake.

Dry test starts immediately after the immersion test and during this test, specimens are dried at room conditions for 21 days and weight change and visual appearance of the concrete are monitored regularly.

8.2.2.2 Drying test (NCHRP 244)

In order to study the effectiveness of sealers and coatings, drying test at room condition (23°C and relative humidity of 50%) as shown in Fig. 8.7 are performed immediately after the immersion test as outlined in NCHRP 244 Report.

Typical results of silane treated concrete, coated concrete, silane + coated concrete and uncoated concrete during the drying stage after immersion in water and salt solution are shown in Figures 8.5 and 8.6 respectively.

Water

Uncoated concrete showed an average decrease in weight of 0.24%, 0.5%, and 0.58% after 1, 8 and 20 days of drying respectively. Silane coated concrete had a 32.7% less water loss than the uncoated concrete after 20 days of drying. Coated concrete had a 3.3% more water loss than the uncoated concrete after 20 days of drying. Silane + coated concrete had about 19% less water loss than the uncoated concrete after 20 days of drying. Over one immersion and dry cycle the concrete with silane + coating had a weight gain of 0.005%. This performance was even better than the case of silane only (Fig. 8.6). In general, the trends observed can vary based on the silane and coating materials under investigation.

NaCl solution (15%)

Uncoated concrete showed an average decrease in weight of 0.15%, 0.26%, and 0.29% after 1, 8 and 20 days of drying respectively. Silane coated concrete had about 17% less salt loss than the uncoated concrete after 20 days of drying. Coated concrete had about 10 percent more salt loss than the uncoated concrete after 20 days of drying. Silane + coating coated concrete had a 6.8% less salt loss than the uncoated concrete after 20 days of drying. Results of residual weight change after immersion of specimens (silane and coated concrete) in water for 21 days and drying for 20 days are shown in Fig. 8.6. Over one immersion and dry cycle, the coated concrete had a weight gain of 0.01%. This performance was even better than case of silane only (Fig. 8.6) for the case reported. These trends can vary depending on the silane and coating materials.

8.2.2.3   Chemical resistance test (Modified ASTM G 20/CIGMAT CT 1)

In order to study the chemical resistance of coated concrete, ASTM G 20 test was modified. As shown in Fig. 8.8 the specimens are immersed in a selected test reagent to half the specimen height in a closed bottle so that the specimens are exposed to the liquid phase and vapor phase. This method is intended for use as a relatively rapid test to evaluate the acidic/sufate resistance of coated specimens under anticipated service conditions. In this test, 72 mm (3-inch) × 152 mm (6-inch) cylindrical cement concrete specimens are used and coated as mentioned in Section 8.4.1. Dry and wet specimens are coated on all sides and coated concrete is tested. For the test, two radial holes are drilled into the specimen approximately 13 mm (0.5-in.) deep (Fig. 8.8). In this test the changes in (1) diameter/dimension at the holiday level (2) weight of specimen (3) appearance of specimen and (4) pulse velocity (ASTM C 597) of the specimen will be measured at regular intervals.

Typically at least three test reagents are selected for this study and they are (1) deionized (DI) water (pH = 5 to 6) (2) 3% sulfuric acid solution (pH = 0.45; representing the worst reported condition in the wastewater system) and (3) 30% sulfuric acid solution (pH = -0.8; representing accelerated testing conditions). Control tests are performed with no pinholes (holidays).

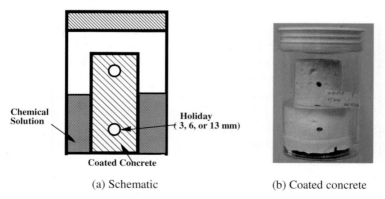

(a) Schematic   (b) Coated concrete

Fig. 8.8 Chemical resistance test on coated concrete specimen.

This test serves as a guide to investigators wishing to compare the relative merits of concrete/brick-coating materials in specific environments. The choice of reagents, concentrations, duration of immersion, temperature of test, and properties to be reported are necessarily arbitrary and should be chosen to reflect conditions known to exist along the concrete/brick structure right-of-way. This test method consists of partially immersion-type test in a closed container where coated pipe specimens are in long-term contact with both the liquid and vapor phase of the test reagent. Specimens exposed in this manner are inspected for visible signs of chemical attack. Other recommended measurements are pulse velocity (ASTM C 597) and calcium leaching from the coated specimens (CIGMAT CH 1-99).

Typically pinholes (holidays) are made on the surface of the coated concrete specimen at points quarter height from the end of the specimens. It shall be made by drilling a radial hole through the coating so that the angular cone point of the drill will fully enter the concrete where the cylindrical portion of the drill meets the concrete surface. Both pinholes (holidays) should lie in the same cylindrical axis. Specimens without intentional pinholes shall also be prepared for testing.

Place the coated concrete specimen vertically in test container and fill the container with the selected reagent so that the liquid level covers one half of the coated concrete/brick specimen up to a point midway between the two intentional holidays. Take the specimen out of the bottle after 30, 180 and 360 days of immersion for inspection and testing. Tests can be continued for longer period of time depending on the information needed to investigate the performance of coated concrete. Observe and report the appearance of specimen after exposure to reagent on the basis of visual examination for evidence of loss of gloss, developed texture, decomposition, discoloration, softening, swelling, injury, bubbling, blistering and cracking as defined in Terminology ASTM D 883. In addition weight, diameter, pulse velocity, calcium, and pH of the solution will be measured. Weight will be measured to 0.01 grams; diameter measued accurate to 0.5 mm (0.02-in); pulse velocity measured to 100 m/s; calcium measured to 5 ppm.

Visual inspection

The coated surfaces are visually inspected regularly and information on blistering, spalling, discoloring and cracking are noted and photographed to understand the failure mechanisms (Vipulanandan et al., 1999-2005).

Failure criteria

In order to evaluate the performance of coatings, the failure criteria for coated concrete are defined as follows:

Criteria 1:   when the diameter of the blister on the coating becomes larger than 25 mm (1 inch);

Criteria 2:   when the length of the crack on the coating is longer than 25 mm (1 inch);

Criteria 3:   multiple blistering, multiple cracking on the coating surface which is not included in Criteria 1 and Criteria 2;

Criteria 4:   weight gain is more than 2% of the initial weight.

Observed coated concrete failure types

The coating failure types observed during the testing of 25 types of coatings included cracking across the pinhole (mainly along the length of the cylindrical specimens), spalling, and blistering around the pinhole. The configurations of the failure types are shown in Fig. 8.9.

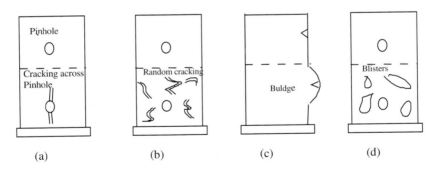

Fig. 8.9 Failure types observed on coated concrete specimens.

Type C1 Failure. Major cracking across the pinhole (Fig. 8.9 (a)). For coated specimens with pinholes, the concrete is directly in contact with sulfuric acid/sulfate through the pinholes. Calcium hydroxide and other calcium complexes react with acid and forms gypsum and ettrigite at different pH levels. Expansion can be observed in the region subjecting the coating to hoop tensile stresses and strains. Initially, small blisters are formed around the pinholes. The blisters continue to grow with the immersion time. Hence, the coating tends to confine the expansion of concrete resulting in tensile stresses and strains in the coating. Pinholes further magnify the stresses in the coating. When tensile stress caused by concrete expansion exceeds the tensile strength of the coating, cracks occur in the direction of the length of the cylindrical specimen. Generally, coated concrete specimens that have larger pinholes cracked first.

Type C2 Failure. Random cracking of coating (Fig. 8.9 (b)). When sulfuric acid/sulfate penetrates through the coating film and reacts with the concrete on the interface of the coating and the concrete substrate or reacts with the primer of the coating, random cracking may occur on the coating film. If the tensile strength of the coating is low or the coating film is very thin (less than 1.5 mm), the coating film will crack at weak points due to concrete expansion. Cracks further develop and cause the coating film to peel off of the concrete substrate. In such cases, coatings fail faster than other failure types.

Type C3 Failure. Blistering around the pinholes (Fig. 8.9 (c)). When concrete expands, coatings deform without cracking, blisters mainly occur around the pinholes. This type of failure was observed in epoxy and polyurethane coated concrete specimens with pinholes in 3% sulfuric acid.

Type C4 Failure. Blistering on the coating surface (Fig. 8.9 (d)). In some cases, blisters are also observed on the coating film. Hard blisters on coating surface are caused by expansion of corroded concrete (or primer) while soft blisters on the coating surface are caused by osmotic pressure.

8.2.2.4 Bonding tests

Both bonding tests, CIGMAT CT-2 test (modified ASTM D 4541) and CIGMAT CT-3 test (modified ASTM C 321) are used to determine the bonding tensile strength of coatings to concrete (Fig. 8.9). Dry concrete specimens are stored in the room environment (temperature 23 ± 2 °C and humidity 50 ± 5 %) and wet concrete specimens were saturated in water for at least 7 days before applying the sealers and/or coatings to prepare specimens for testing.

8.2.2.4.1 CIGMAT CT-2 (modified ASTM D 4541)

The CIGMAT CT-2 (modified ASTM D 4541) test is used to determine the tensile bonding strength of coatings to concrete with/without sealers. Prism concrete specimens are coated in the same manner as the specimens for the immersion test. Bonding strength is determined after specified time of immersion in water or solutions specified otherwise.

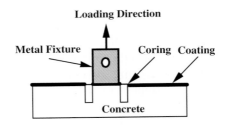

(a)   Test configuration

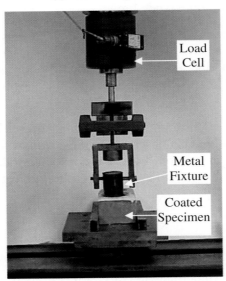

(b)   Test setup

Fig. 8.10 Bonding test configuration CIGMAT CT-2 and set-up
(modified ASTM D 4541).

In this test 51 mm (2-in.) diameter circular area is used for testing (CIGMAT CT-2) (CIGMAT Standards, 2002). Coated concrete blocks, cured in various solutions, are cored using a diamond core drill to predetermined depth to isolate the coating and a metal fixture was glued to the isolated coating section using a rapid setting epoxy. Schematic configuration of the test is shown in Fig. 8.10. After the bonding test, the specimens must be inspected to identify the types of failures as summarized in Table 8.1.

Table 8.1 Types of Failures according to the CIGMAT CT-2 Test.

Failure type	Description	CIGMAT CT 2 test
Type-B1	Substrate failure	
Type-B2	Coating failure	
Type-B3	Bonding failure	
Type-B4	Bonding and substrate failure	
Type-B5	Bonding and coating failure	

8.2.2.4.2 CIGMAT CT-3 (modified ASTM C 321)

The ASTM C 321 test method was modified (CIGMAT CT-3) and used to determine the tensile bonding strength of coatings to concrete with/without sealers. Prism concrete specimens were coated in the same manner as the specimens for the immersion test. Bonding strengths are determined after specified time of curing. In this test, the coating is sandwiched between a pair of rectangular concrete block specimens and then tested for bonding strength (Fig. 8.11). The bonded specimens are cured under water or other solutions up to the time of testing. Compared to CIGMAT CT-2 test (modified ASTM D4541), this is an easier test to perform since no coring or gluing of metal fixture is required.

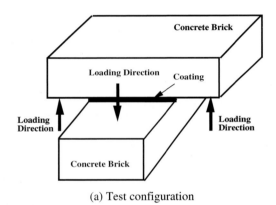

(a) Test configuration

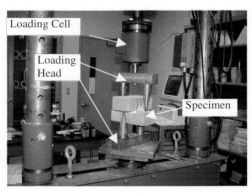

Fig. 8.11 Bonding test setup for CIGMAT CT-3 Test (Modified ASTM C 321 Test).

Over 500 bonding tests have been completed in the CIGMAT laboratory on over 25 coatings with and without sealers. Based on these test results, failures observed in CIGMAT CT-2 and CIGMAT CT-3 tests are summarized in Tables 8.1 and 8.2.

Table 8.2 Types of failures according to the CIGMAT CT-3 test.

Failure type	Description	CIGMAT CT-3 Test
Type B1	Substrate failure	
Type B2	Coating failure	
Type B3	Bonding failure	
Type B4	Bonding and substrate failure	
Type B5	Bonding and coating failure	

Type B1 failure is concrete failure. The failure occurred in concrete due to tension in CIGMAT CT-2 test or bending in CIGMAT CT-3 test. This type of failure is the most desired failure type because the bonding strength between coating and concrete is higher than the tensile strength or flexure strength of concrete.

Type B2 failure is coating failure which is cohesive failure of coating. This type of failure indicates that the tensile strength of the coating is lower than the bonding strength and the tensile strength of concrete.

Type B3 failure is bonding failure where the failure occurs between coating and substrate. This type of failure indicates that the coating has poor bonding strength to concrete substrate. If there is a sealer, then failure can occur between the coating and the concrete surface treated with a sealer.

The three failure types above are the most common observed failure types in the bonding tests. In addition to the three failure types, some other failure types can also occur in the tests.

Type B4 failure is a combined bonding and concrete failure, where the bonding strength of coating to concrete is close to the tensile/flexure strength of concrete.

Type B5 failure is a combined bonding and coating failure, where the bonding strength of coating to concrete is close to the tensile/flexure strength of coating.

8.2.6 *Full-scale test*

8.2.6.1   Hydrostatic pressure test

Applicability of coatings/sealers to concrete surfaces in service must be investigated by stimulating the field conditions. In order to stimulate hydrostatic back-pressure on concrete structures due to the ground water table, it was decided to use concentrically placed concrete pipes to develop the necessary full-scale testing conditions (Vipulanandan et al., 1996). This was achieved by using 900 mm (36 in.) inner pipes and 1600 mm (64 in.) outer pipes with two concrete end plates.

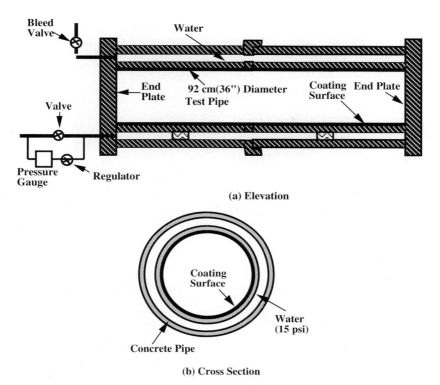

(a) Elevation

(b) Cross Section

Fig. 8.12 Hydrostatic test configuration.

Steel elements were used to support the entire set-up. Inner concrete pipes were representing a concrete surface under a hydrostatic pressure. The total area available for coating was 14 sq. meter (150 sq. ft.). Based on OSHA regulations, a 900 mm (36 in.) diameter pipe is the smallest pipe in which a coating applicator can be allowed to operate within the concrete pipe. Pressure chamber used for the full scale test was designed and built by Hanson (formerly Gifford-Hill & Company), Houston Division, which was representing the American Concrete Pipe Association.

Two different testing conditions were conducted in this test. One was a dry surface condition. Coating was applied to new 900 mm (36 in.) diameter concrete pipe at the Hanson concrete pipe yard in Houston. The coated pipe was then placed in the pressure chamber for hydrostatic pressure testing. This test simulated the new concrete pipes to which

(a)   Outside view of the hydrostatic test

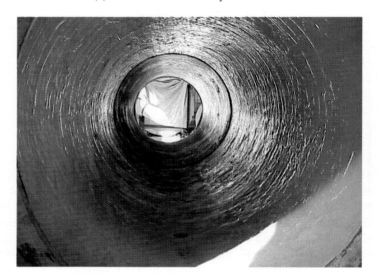

(b)   Coating installed on concrete surface

Fig. 8.13 Hydrostatic test setup and coating installation.

coating was applied on the surface before the pipes were installed. The
other test was a wet surface condition. The 900 mm (36 in.) concrete
pipe was installed in the test chamber and pressurized at 105 kPa
(32 ft of water head) for at least two weeks before applying the coating.

The coating was applied after a water jet blasting of the surface. This test simulated concrete surface under service conditions.

The elevation and cross-section of the hydrostatic test configurations are shown in Fig. 8.12. Figure 8.13 shows one test chamber and a coated surface.

8.2.6.2 In-situ bonding test

In-situ bonding tests on the coating materials can be performed at the end of the hydrostatic test. A 51 mm (2 in.) diameter core drill can be used to core into the concrete surface and isolate the test area, and a metal piece is glued to the coated surface with an epoxy. After two days of curing, the test can be performed using a hydraulic loading system to determine the bonding strength and the type of failure (Fig. 8.14).

8.2.6.3 Water vapor emission test

Water transmission through coating films can be measured in the hydrostatic pressure test chambers. ASTM E 1907 was used in the test. Anhydrous calcium chloride (CaCl2) was used as a desiccant. The desiccant was heated up to 200 °C for 24 hours before use.

Dishes containing $CaCl_2$ (about 2/3 of the depth) were sealed with plastic films. The areas covered by plastic film were 81 in^2 in each test. Two duplicated tests are generally performed under each condition (dry coated and wet coated concrete surfaces). Weight changes are measured after 72 hours. Figure 8.15 shows the test setup for water vapor transmission tests.

8.2.6.4 Measurements

During the hydrostatic test, the coated surfaces are visually inspected regularly and information on blistering, spalling, discoloring, and cracking were noted and photographed.

In the in-situ bonding test, bonding strength from at least two tests can be obtained, and failure types of the bonding tests are recorded. The test results are used to compare with the laboratory test results to determine the effect of the water back pressure on the bonding strength of coatings to concrete substrate.

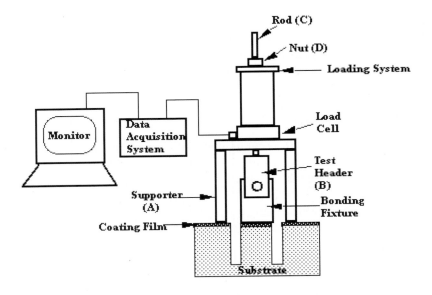

(a) Configuration of in-situ bonding test

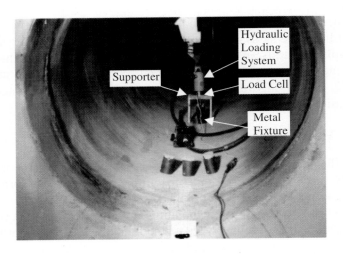

(b) In-situ test setup

Fig. 8.14 Configuration and setup for in-situ bonding test.

Fig. 8.15 Water vapor transmission test setup.

Water vapor transmission tests are performed at the end of the hydrostatic tests. Water vapor transmission rate are calculated according to the weight change of anhydrous calcium chloride after 72 hours. At least two tests are performed on each coating.

8. 3 Modeling Liquid Transport into Coated Concrete

Based on mass transfer concept, models have been developed to better quantify the factors influencing the performance of coated concrete (Liu and Vipulanandan, 2003).

(a) *Physical model*

When coated concrete comes into contact with liquid, liquid penetrated through the coating film into the concrete. Physical models for coated concrete when immersed in liquid can be divided into two categories based on the reactivity of the liquids with the concrete substrate and coatings.

Case 1. *No-Chemical Reaction*

If the liquid did not chemically damage either the coating film or the concrete, such as deionized water (D. I. water) in the present study, the physical model of liquid penetrating into the coated concrete specimens can be represented as shown in Fig. 8.16(a).

In this case, the liquid saturated the contact surface of the coating film. Due to the gradient of the degree of saturation in the coating film, the liquid penetrated through the coating film into the concrete substrate. It was assumed that there was no physical damage to both coating film and concrete.

Case 2. With chemical reaction

If the liquid, such as sulfuric acid solutions, react with the concrete substrate and damages the concrete matrix, the concrete will be attacked layer by layer starting from the interface of the coating film and the concrete substrate. The physical model for this case can be represented as shown in Fig. 8.16(b) where there is a region of reacted concrete and the degree of saturation of the liquid in this concrete substrate is only in the reacted and affected zone as shown in Fig. 8.16(b).

Assumptions in developing the models

To simplify the process of liquid transport in coated concrete, some assumptions are made in developing the models. According to the observed mechanisms, the assumptions for modeling the liquid uptake by coated concrete cylinders are as follows:

(1) degree of saturation of liquid (S) for coating and concrete is defined as $\dfrac{\text{Mass of Liquid Absorbed}}{\text{Solid \quad Volume}}$.

(2) the process can be modeled by second order differential equation (Crank, 1975)

$$\frac{\partial S}{\partial t} = D\left(\frac{\partial^2 S}{\partial x^2} + \frac{\partial^2 S}{\partial y^2} + \frac{\partial^2 S}{\partial z^2}\right); \qquad (8.1)$$

where S is the degree of saturation of the coating or concrete.

(3) the mass transfer coefficient is a constant in the coating film (D_{CT}), un-reacted concrete cylinder (D_{CO}) or reacted concrete region (D_{CO}^r);

(4) coating surface in contact with the liquid is assumed to be saturated;

(5) coating film and concrete surface are in good contact, so there is no accumulation of liquid at the interface;

(6) coating film does not react with liquid;

(b) *Film Model (Model 1)*

The coating thickness on the concrete cylinder was in the range of 1.5 to 2.0 mm for the two epoxy coatings; hence, it can be treated as a thin plane sheet. The degrees of saturation of the coating surface on the liquid side and the interface are S_0^{CT} and S_i^{CT} respectively (Fig. 8.14 (a)).

The mass transport process in one dimension under non-steady state, if the transport parameter is D_{CT}, is given by (Crank, 1975)

$$\frac{\partial S}{\partial t} = D_{CT} \frac{\partial^2 S}{\partial x^2},$$

(8.2)

where

S	=	degree of saturation, g liquid/cm^3 solid;
T	=	time, s;
x	=	dimension in the coating thickness direction, cm;
D_{CT}	=	mass transfer coefficient, cm^2/s.

For a coating film on a porous substrate, the degree of saturation on the interface of the coating film and the substrate vary with time. If a small time interval t to (t + dt) is considered, the process can be assumed to be a pseudo steady state. In this case, Eq. (8.2) in one dimension reduces to

$$\frac{d^2 S}{dx^2} = 0$$

(8.3)

Integrating Eq. (8.3) and applying the conditions at x = 0, S = S_0^{CT} and x = ℓ, S = S_i^{CT}, we have

$$\frac{S - S_0^{CT}}{S_i^{CT} - S_0^{CT}} = \frac{x}{\ell}.$$

(8.4)

The rate of mass transfer F (g liquid/ cm^{2}s) through a plane sheet under steady state is given (Crank, 1975)

$$F = -D_{CT}\left(\frac{dS}{dx}\right). \tag{8.5}$$

Assuming that the degree of saturation on the interface (coating-concrete) varies with time t and can be represented by the following exponential function

$$S_i^{CT} = S_0^{CT}(1 - e^{-\beta^{CT}t}), \tag{8.6}$$

where β^{CT} is a coating material parameter related to the rate of the coating film saturation. The lower the value of β^{CT}, the lower is the rate of saturation in the coating film. In selecting coatings for rehabilitation, low value of β^{CT} will be preferred.

Hence Eq. (8.4) can be written as

$$F_{(t)} = \frac{D_{CT}}{\ell}S_0^{CT}e^{-\beta^{CT}t}, \tag{8.7}$$

The amount of the substance transported through the coating film (in the present case around the cylindrical coated specimen) from time t to (t + dt) is

$$dW_t = 2\pi RhF_{(t)}dt \tag{8.8}$$

The accumulated amount of the substance transported through the coating film from time 0 to t is

$$W_t = 2\pi Rh\int_0^t F_{(t)}dt, \tag{8.9}$$

where   R = radius of concrete specimen, cm;
             h = height of specimen, cm.

(i) That is

$$W_t = 2\pi Rh\int_0^t \frac{D_{CT}}{\ell}S_0^{CT}e^{-\beta^{CT}t}dt, \tag{8.10}$$

Finally, by integrating Eq. (8.10), the following relation is obtained:

$$W_t = \frac{2\pi R h S_0^{CT}}{\beta^{CT}} \frac{D_{CT}}{\ell} \left(1 - e^{-\beta^{CT}t}\right) \tag{8.11}$$

Equation (8.11) (Model 1) is the film model to predict the mass transfer through coating film on cylindrical concrete specimen which results in the measured mass change in the specimens.

The ultimate degree of saturation S_0 was obtained from the experiments on pure coating bulk materials. Relating Eq. (8.11) to the mass change-time relationship of coated concrete, the mass transfer coefficient D_{CT} and parameter β^{CT} for the coatings can be obtained.

(c) Bulk model (Based on concrete)

1. Non-reactive solution (Model 2)

For mass transport in cylindrical specimens, if the degree of saturation is a function of radius and time only, the second order differential equation is:

$$\frac{\partial S}{\partial t} = \frac{1}{r}\frac{\partial}{\partial r}\left(rD_{CO}\frac{\partial S}{\partial r}\right), \tag{8.12}$$

where D_{CO} is the mass transfer coefficient of the substrate.

For liquid transport in a coated concrete cylinder without chemical reaction, the degree of saturation on the concrete surface changes with time. If the degree of saturation at the concrete surface is $\phi(t)$, the solution of the second order differential Eq. (8.12) is given (Crank, 1975) by

$$S = \frac{2D_{CO}}{R}\sum_{n=1}^{\infty}\exp\left(-D_{CO}\alpha_n^2 t\right)\frac{\alpha_n J_0(r\alpha_n)}{J_1(R\alpha_n)}\int_0^t \exp\left(D_{CO}\alpha_n^2 t\right)\phi(t)dt, \tag{8.13}$$

where $J_0(R\alpha_n)$ is the Bessel function of the first kind of order zero, $J_1(R\alpha_n)$ is the Bessel function of the first order and α_n s are the roots of $J_0(R\alpha_n)$.

Assuming $\phi(t) = S_0^{CO}\{1 - \exp(-\bar{\beta}t)\}$, representing the degree of saturation at concrete surface (coating-concrete interface) which approaches a limiting value S_0^{CO}, the sorption-time curve is given by Crank (1975)

$$\frac{W_t^{CO}}{\pi R^2 h S_0^{CO}} = 1 - \frac{2J_1\left\{\left(\bar{\beta}r^2/D_{CO}\right)^{1/2}\right\}\exp\left(-\bar{\beta}t\right)}{\left(\bar{\beta}R^2/D_{CO}\right)^{1/2} J_0\left\{\left(\bar{\beta}R^2/D_{CO}\right)^{1/2}\right\}} \qquad (8.14)$$
$$+ \frac{4}{R^2}\sum_{n=1}^{\infty}\frac{\exp\left(-D_{CO}\alpha_n^2 t\right)}{\alpha_n^2\left\{\alpha_n^2/\left(\bar{\beta}/D_{CO}\right) - 1\right\}}$$

Figure 8.15 shows the liquid uptake curves for different values of parameter $\bar{\beta}R^2/D_{CO}$ (λ) (Crank 1975).

From Fig. 8.15, the solution uptake is determined by the parameter $\bar{\beta}R^2/D_{CO}$ which represents the effects of the coating material properties and the parameter $(D_{CO}t/R^2)^{1/2}$ (x-axis) which represents the time effect.

Let $\lambda_{CO} = \bar{\beta}R^2/D_{CO}$.

Approximating Eq. (8.13) and considering an exponential function of the form

$$\frac{W_t^{CO}}{\pi R^2 h S_0^{CO}} = \left\{1 - \exp\left[-\lambda_{CO}\left(\frac{D_{CO}t}{R^2}\right)^n\right]\right\} \qquad (8.15)$$

where n is a constant.

Equation (8.15) is best fitted to the standard curves in Fig. 8.16 for different λ_{CO} values by using the least-square method. The parameter λ_{CO} and n with coefficient of correlation can be obtained by fitting the curve (Liu and Vipulanandan, 2005). The value of n varied from 0.98 to 1.26 when the value of λ was in the range of 0.01 to 5.

The approximate solution for Eq. (8.17) (below) is in good agreement with the solution suggested by Crank (1975) (Fig. 8.16). The solid curves are the standard curves (Eq. (8.13)) as given by Crank (1975) while the dotted lines are the approximate solution from

Eq. (8.16). The agreement of the fit indicates that the approximate solution can predict the mass increase in the specimens as the exact solution does (Eq. (8.14)) for the assumed degree of saturation ($\phi(t)$) at the concrete interface.

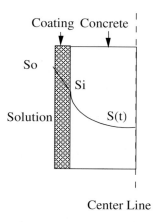

Fig. 8.16 Film model (Model 1) for coated concrete.

2. Reactive Solution (Model 3)

For liquid transport in concrete cylinders with chemical reaction, Eq. (8.14) can be modified as follows (Crank, 1975):

$$\frac{\partial S}{\partial t} = \frac{1}{r}\left\{\frac{\partial}{\partial r}\left(r\frac{D_{CO}}{k+1}\frac{\partial S}{\partial r}\right)\right\}, \qquad (8.16)$$

where k is coefficient of chemical reaction. Eq. (8.16) can be solved analytically as Eq. (8.12). The mass transfer coefficient D_{CO} is replaced by D_{CO}^r where $D_{CO}^r = \dfrac{D_{CO}}{k+1}$. So the solution for liquid transport in coated reactive cylinder substrate is as follows:

$$\frac{W_t^r}{\pi R^2 h S_o^r} = \left\{1 - \exp\left[-\lambda_{CO}^r\left(\frac{D_{CO}^r t}{R^2}\right)^n\right]\right\}, \qquad (8.17)$$

where W_t^r is the mass change, S_O^r is the ultimate degree of saturation in the reacted area, D_{CO}^r is the mass transfer coefficient in reacted area, and λ_{CO}^r is defined as before for the reacted region.

3. Effect of Holiday Sizes

It is difficult to apply coatings on a concrete surface uniformly without any defects. Holidays (Pinholes) and other defects will develop during the coating process (Redner et al. 1994). Under service conditions, the liquid will penetrate through the holidays on the coating surface into the concrete substrate and the interface. The holiday effect on liquid uptake can be clearly seen in Figures 8.17-8.20. Hence, it is important to account for the holiday effect in the modeling.

The effect of holiday sizes on the liquid uptake of the coated substrate can be taken into account by introducing a parameter ξ which is a function of holiday sizes to Eq. (8.11), (8.15) and (8.17). Parameter ξ will be equal to 1 and assumed to reach the limiting value when the

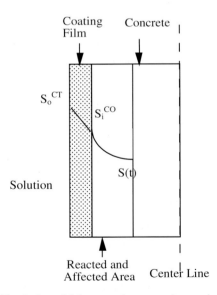

Fig. 8.17 Physical model for coated concrete in reactive solution.

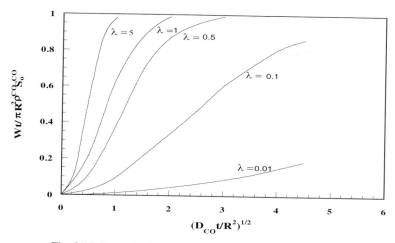

Fig. 8.18 Exact (solid line) solutions for various values of λ.

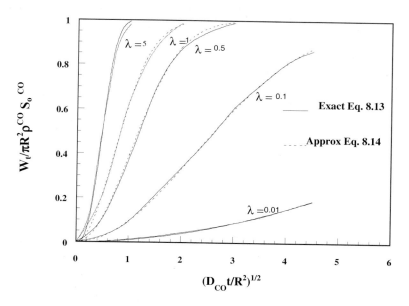

Fig. 8.19 Comparison of exact solution and approximate solution
(Liu and Vipulanandan, 2003)

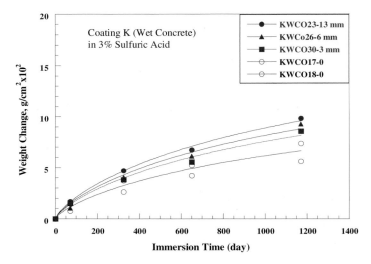

Fig. 8.20 Comparison of film mass transport model (Model 1) prediction to
the experiment data of a coated concrete in 3% sulfuric acid
(symbols are the experiment data, solid lines are model prediction).

holiday size is large; hence parameter ξ is defined as

$$\xi = 1 + \frac{d_h}{k_1 + k_2 d_h} \tag{8.18}$$

where d_h = holiday diameter, cm;

       $k_1,\ k_2$ = parameters related to the coating.

Introducing Eq. (8.18) to Eq. (8.11), (8.15) and (8.17), the equations become

Model 1 $$W_t = \xi \frac{2\pi R h S_0^{CT}}{\beta^{CT}} \frac{D_{CT}}{\ell} \left(1 - e^{-\beta^{CT} t}\right), \tag{8.19}$$

Model 2 $$\frac{W_t^{CO}}{\pi R^2 h S_0^{CO}} = \xi \left\{ 1 - \exp\left[-\lambda_{CO} \left(\frac{D_{CO} t}{R^2} \right)^n \right] \right\}, \tag{8.20}$$

and

Model 3
$$\frac{W_t^r}{\pi R^2 h S_o^r} = \xi \left\{ 1 - \exp\left[-\lambda_{CO}^r \left(\frac{D_{CO}^r t}{R^2} \right)^n \right] \right\}.$$
(8.21)

(d) *Verifications of models*

Film Model (Model 1)

Using Eq. (8.11) the mass change in the test specimens were predicted and compared to the experimental data in Fig. 8.20. The ultimate degree of saturation of the coating (S_0^{CT}) was also determined from pure coating test data . Based on the model predictions, the values of β^{CT} and D_{CT} for the coatings in D. I. water and 3% sulfuric acid can be obtained.

The material parameter (β^{CT}) determines the rate of saturation on the coating-concrete interface and parameters S_0^{CT}, β^{CT} and D_{CT} influence the total uptake of liquid. The material parameters (β^{CT}) for epoxy coatings varied from 0.0020 to 0.0035. The parameter β^{CT} for Epoxy 1 coating was not much affected by the liquid type and the conditions of concrete (dry or wet). This was not the case with Epoxy 2 where β^{CT} changed with liquid type and condition of concrete. The parameter β^{CT} almost tripled from dry concrete in D.I. water to wet condition in acid.

The mass transfer coefficient (D_{CT}) is proportional to the flux of the liquid. For Epoxy 1, the mass transfer coefficients (D_{CT}) are very close in D. I. water and 3% sulfuric acid. The result indicated that solutions did not affect the performance of coating films, but the mass transfer coefficient of the Epoxy 2 coating film had substantially higher value in 3% sulfuric acid than that in D.I. water. Comparing the mass transfer coefficient (D_{CT}) from Model 1 with the mass transfer coefficient (D) from Mebarkia (1995), the results indicate that the film mass transfer coefficients (D_{CT}) (Model 1) were lower than the bulk mass transfer coefficients (D) (Mebarkia 1995) except for Epoxy 2 in 3% sulfuric acid (Vipulanandan et al. 2002).

The comparison of model 1 and experiment data for Epoxy 1 is shown in Fig. 8.20.

The mass transfer coefficients of concrete in D.I. water (DCO) and 3% sulfuric acid (DrCO) was obtained by fitting the cylindrical model (Mebarkia, 1995) to experimental data of uncoated concrete specimens. The values of mass transfer coefficients were 2.55×10^{-10} m^2/s and 3.08×10^{-10} m^2/s in D.I. water and 3% sulfuric acid respectively.

Using Eq. (8.15), Eq. (8.17) and experimental data, the parameters n, λ, and $\overline{\beta}$ can be obtained for different coatings. The values of n, λ, and $\overline{\beta}$ for different coating systems are summarized in several publications (Vipulanandanet al. 2000-2006). The comparisons of model prediction Model 2 (Eq. (8.20) and Model 3 (Eq. (8.21) to experiment data are shown in Figs. 8.21 and 8.22.

The parameter $\overline{\beta}$ represents the saturation rate of the concrete surface. It can reflect the conditions of the interface between the coating and the concrete. Higher $\overline{\beta}$ value means that the coating may not form a good barrier on the concrete surface against solution penetration. Parameter $\overline{\beta}$ varied from 0.86×10^{-8} to 1.46×10^{-8} in D. I. water and from 2.20×10^{-8} to 4.71×10^{-8} in 3% sulfuric acid respectively. The parameter n varied from 0.521 to 0.579 in D. I. water and from 0.493 to 0.761 in 3% sulfuric acid. The values of the parameter n for coated concrete are much less than the value of the parameter n from standard curves. This indicates that the process of liquid penetrating into coated concrete is not a pure diffusion process.

Relating Eq. (8.20) and (8.21) to the experimental data under different holiday sizes, parameters k_1 and k_2 can be obtained. The values of k_1 and k_2 vary with the coating and the type of solutions.

The parameter k_1 indicates the effect of holidays on liquid uptake of coated substrates. The smaller the value of k_1, the more the effect of holidays on the liquid uptake of coated substrates. The value of k_2 indicates the effect of holiday sizes on the liquid uptake of coated substrates. The smaller the value of k_2, the more the effect of holiday sizes on the liquid uptake of coated substrates. The prediction of holiday size effect on mass change of Epoxy 1 coated concrete is shown in Figs. 8.21 and 8.22.

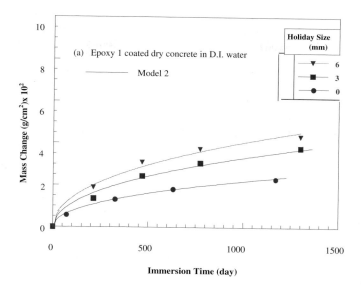

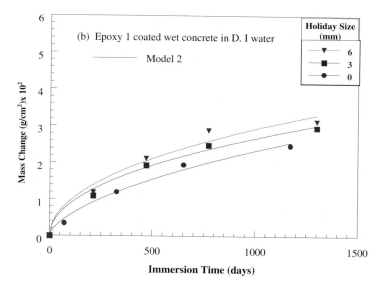

Fig. 8.21 Comparison of the experimental data with Model 2 in D.I. water.

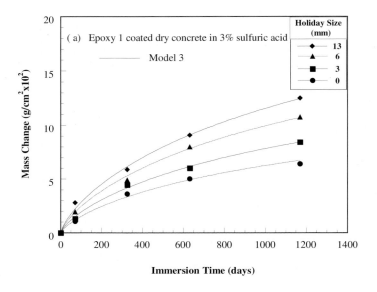

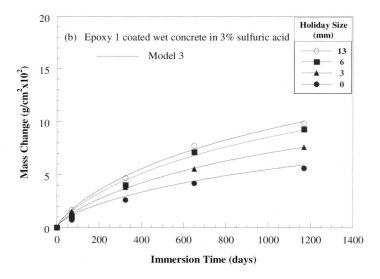

Fig. 8.22 Comparison of the experimental data with Model 3 prediction
in 3% sulfuric acid.

8.4 Conclusion

A combination of full-scale and laboratory tests have been used to evaluate the performance of coatings and sealers for protecting dry and wet reinforced concrete exposed to sulfuric acid and salt environments. In the case of acidic environment commercially available polymer and cement based coatings have been studied for several years. Effectiveness of silanes, with and without coatings, has been studied to determine the chloride penetration into the concrete during immersion and drying. Analytical models were developed to predict weight change and holiday size effect on coated concrete. Based on the literature review, experimental results and analyses, the following observations are advanced:

1) The full-scale hydrostatic test were used to evaluate the applicability of coatings onto concrete under hydrostatic back pressure up to 105 kPa (15 psi) with a moisture emission of $536\ \mu g/(s.m^2)$ $(9.49\ lb/(1000ft^2.24h))$. The test clearly identified the coatings that could be applied to moist concrete structures in service with an external hydrostatic pressure from ground water or storage tanks.

2) Two bonding tests were modified based on the ASTM standards. Based on the test results, five modes of failure have been identified for each type of test. Bonding strength of the coatings varied from 0.25 MPa (36 psi) to 2.50 MPa (360 psi). Both test methods can be used to identify the coating that bond well to dry and wet concrete.

3) Chemical resistance test with total or partial immersion, can effectively evaluate the acid resistance or chloride penetration of the coated concrete. Test results showed that coatings substantially reduce the solution penetration into coated concrete and the holidays affected the performance of coated concrete. This test method can be used to identify the coatings, with or without sealers, to protect the reinforced concrete structures exposed to harsh environmental conditions.

4) The film model and concrete model developed in this study can be used to predict mass increase of coated concrete in reactive and non-reactive solutions. The model parameters were sensitive to the changes in coatings, solutions and holiday sizes. Results showed that the models can better predict the performance of coated substrate in long-term immersion condition.

5) Holidays (pin holes) can substantially increase the mass change in coated concrete. The test results and analyses showed that there is a limiting value for holiday size, beyond which the effect was minimal (depending on coatings, solutions).

Acknowledgement

This project was supported by the Center for Innovative Grouting Materials and Technology (CIGMAT) under grants from the City of Houston, National Science Foundation (CMS-9526094, CMS-9634685), Texas Department of Transportation and various industries.

References

1. AASHTO "Standard Method of Test for Resistance of Concrete to Chloride Ion Penetration." American Association of State Highway and Transportation Officials, Standard Specifications for Transportation Materials and Methods of Sampling and Testing, Part ll-Methods of Sampling and Testing. Washington, D.C, 1986.
2. Alberta Transportation Technical Standards Branch; Section B388: Specification for Concrete Sealers, July, 2001.
3. Bruner, James Jr. "Concrete Surface Treatments – A Selection Guide", Proceedings of the Materials Engineering Conference, Atlanta, GA, August 10-12, 1992. American Society of Civil Engineers, New York, NY, p. 476-482, 1992.
4. Cady, P.D., and Weyers. R.E., Deterioration Rates of Concrete Bridge Decks. Journal of Transportation Engineering, Vol. 110, No. 1, pp. 34-44, 1984.
5. Carter, P.N. "The Use of Penetrating Sealers on Concrete Bridge Decks." In Structural Materials, Proceedings of the Sessions Related to Structural Materials. San Francisco: ASCE Structures Congress, May 1-5, 1989.
6. Carter, P.D. and A.J. Forbes. "Comparative Evaluation of the Waterproofing and Durability of Concrete Sealers." ABTRJRD/ RR-86-09, Final Report 1984-86. Alberta Transportation Research and Development Branch, October 1986.
7. CIGMAT CT-1, CIGMAT Standards, Standard Test Method for Chemical Resistance of Coated or Lined Concrete and Clay Bricks, Center For Innovative Grouting Materials and Technology, University of Houston, Houston, Texas, 2005

8. CIGMAT CT-2, CIGMAT Standards, Standard Test Method for Pull-off Strength of Coatings and Linings, Center For Innovative Grouting Materials and Technology, University of Houston, Houston, Texas, 2005.

9. CIGMAT CT-3, CIGMAT Standards, Standard Test Method for Bonding Strength of Coatings and Linings: Sandwich Method, Center For Innovative Grouting Materials and Technology, University of Houston, Houston, Texas, 2005.

10. CIGMAT CH-1, CIGMAT Standards, Standard Test Method for Determing Calcium using EDTA, Center For Innovative Grouting Materials and Technology, University of Houston, Houston, Texas, 2001.

11. Coti, H., Fernandez-Prini, R. and Gomez, D. "Protective Organic Coatings: Membrane Properties and Performance." Progress in Organic Coatings. Vol. 10, No. 1, 5-33, 1982

12. Crank, J. The mathematics of diffusion, Oxford University Press, 2nd, New York, N.Y. 1975.

13. Ehrich, S., Helard, L., Letourneux, R., Willocq, J. and Bock, E. "Biogenic and Chemical Sulfuric Acid Corrosion of Mortars." Journal of Materials in Civil Engineering, Vol. 11, No. 4, 340-344, 1999.

14. Fattuhi, N. I. and Hughes, B. P. "The Performance of Cement Paste and Concrete Subjected to Sulfuric Acid Attack." Cement and Concrete Research, Vol. 18, No. 4, 545-553, 1988.

15. Fernandez, N. Evaluation of the Performance of Concrete Sealers. Iowa Transportation Department, Materials Section, February 1988.

16. Florida DOT Standard: Specifications for Road and Bridge Construction; Section 413: Sealing Concrete Structure Surfaces.

17. Fraseoia, R.I. Evaluation of Chem-Trete BSM Weather Proofing Solution on 1-89, Fairfax, Vermont. Initial Report 86-5. March 1986.

18. Gospodinove, P. N., Kazandjiev, R. F., Partalin, T. A. and Mironova, M. K. (1999). "Diffusion of Sulfate Ions into Cement Stone Regarding Simulaneous Chemical Reactions and Resulting Effects." Cement and Concrete Research, Vol. 29, No. 10, 1591-1596.

19. Hagen, M, 3. Special Study 367-Extended Evaluation of Selected Experimental Bridge Deck Protective Systems-Concrete Sealers

for Bridge Decks. Final Report. Physical Research Section, Office of Materials, Research, and Standards, Minnesota Department of Transportation, February 1988.

20. Ibrahim, M., Al-Gahtani, A.S., Dakhil, F.H. "Use of Surface Treatment Materials to Improve Concrete Durability", ASCE Journal of Materials in Civil Engineering, Vol. 11, No. 1, pp. 36-40, 1999.

21. Islander, R. L., Devinny, J. S., Mansfeld, F., Postyn, A. and Shih, H. "Microbial Ecology of Crown Corrosion in Sewers." Journal of Environmental Engineering, Vol. 117, No. 6, 751-770, 1991.

22. Kay, T. "Assessment and Renovation of Concrete Structures", Long man Scientific & Technical, Essex, England, 1992.

23. Kim, J. and Vipulanandan, C. "Effect of pH, Sulfate and Sodium on the EDTA titration of Calcium," Cement and Concrete Research, Vol. 33(5), pp. 621-627, 2003.

24. Klahorst, J., Klingner, R. E., Kreger, M. E. and Fowler T.J. Mitigation Techniques for In-Service Structures with premature Concrete Deterioration: Literature Review, Final Report, Research Project 0-4069, TxDOT, 2004.

25. Kottke, E. Evaluation of Sealers for Concrete Bridge Elements. Alberta Transportation and Utilities, Bridge Engineering Branch, August 1987.

26. Liu, J. and Vipulanandan, C. (1999). "Testing Epoxy Coatings for Dry and Wet Concrete Wastewater Facilities." Journal of Protective Coatings and Linings, Vol. 16, No. 12, 26-37, 1999.

27. Liu, J., and Vipulanandan, C. "Tensile Bonding Strength of Epoxy Coatings to Concrete Substrate," Cement and Concrete Research, Vol. 35, pp.1412-1419, 2005.

28. Liu, J., and Vipulanandan, C. "Long-term Performance of Epoxy Coated Clay Bricks in Sulfuric Acid," Journal of Materials in Engineering, ASCE, Vol. 16, No. 4, pp.349-355, 2004.

29. Liu, J. and Vipulanandan, C "Modeling Water and Sulfuric Acid Transport Through Coated Cement Concrete," Journal of Engineering Mechanics, ASCE, Vol. 129(4), pp. 426-437, 2003.

30. Liu, J. and Vipulanandan, C. "Evaluating a Polymer Concrete Coating for Protecting Non-Metallic Underground Facilities from Sulfuric Acid Attack," Journal of Tunnelling and Underground Space Technology, Vol. 16, pp. 311-321, 2001.

31. Mainguy, M. Tognazzi, C., Torrenti, J-M. and Adenot, F. (2000). "Modeling of Leaching in Pure Cement Paste and Mortar." Cement and Concrete Research, Vol. 30, No.1, 83-90, 2000.

32. Martys, N. S. and Ferraris, C. F. (1997). "Capillary Transport in Mortars and Concrete." Cement and Concrete Research, Vol. 27, No. 5, 747-760, 1997.

33. Mallett, G.P. "State of the Art Review: Repair of Concrete Bridges", Thomas Telford, London, 1994

34. Mebarkia, S. and Vipulanandan, C. "Mechanical Properties and Water Diffusion in Polyester Polymer Concrete." Journal of Engineering Mechanics, Vol. 121, No. 12, 1359-1365, 1995.

35. Nguyen, T., Bentz, D. and Byrd, E. "Method for Measuring Water diffusion in A Coating Applied to A Substrate." Journal of Coatings Technology, Vol. 67, No. 844, 37-46, 1995.

36. Perkins, P. H. "Repair, Protection and Waterproofing of Concrete Structures" E & FN SPON, New York, NY, 233 pp., 1997.

37. Petty, D.A. Laboratory Evaluation of Cement Concrete Sealers. Commonwealth of Massachusetts, Department of Public Works, Research and Materials Section, 1986.

38. Pfeifer, D. W. and Scali, M. J. "Concrete sealers for protection of bridge structures." National Cooperative Highway Research Program Report 244 (December 1981).

39. Rasoulian, M., C. Burnett, and R. Desselles. "Evaluation of Experimental Installation of Silane Treatment on Bridges." FHWA/LA-87/207, Final Report. Louisiana Department of Transportation and Development, September 1981-April 1986.

40. Redner, J. A., Hsi, R. P. and Esfandi, E. J. "Evaluating Coatings for Concrete in Wastewater Facilities: An Update." Journal of Protective Coatings and Linings, Vol. 11, No. 12, 50-61, 1994.

41. Rostam, S. "Reinforced Concrete Structures – Shall Concrete Remain the Dominating Means of Corrosion Prevention?," Materials and Corrosion, Vol. 54, pp. 373, 2003.

42. Rouings, R.M. and B. Chojnaeki. A Laboratory Evaluation of Concrete Surface Sealants-Phase H. MI-127. Ontario, Canada: Ministry of Transportation of Ontario, November 1988.

43. Rutkowski, T.S. Evaluation of Penetrating Surface Treatments of Bridge Deck Concretes. Final Report Research File No. 81-5. Research Unit, Materials Section, Division of Highways and

Transportation Facilities, Wisconsin Department of Transportation, November 14, 1988.

44. Sand, W., Dumas, T., Marcdargent, S., Pugliese, A. and Cabiron, J. (1994). "Tests for Biogenic Sulfuric Acid Corrosion in A Simulation Chamber Confirms The On Site Performance of Calcium Aluminate Based Concrete in Sewage Application." ASCE Material Engineering Conference, November 14-16, San Diego, 1994.

45. Soebbing, J. B., Skabo, R. R. and Michel, H. E. (1996). "Rehabilitating Water and Wastewater Treatment Plants." Journal of Protective Coatings and Linings, Vol. 13, No. 5, 54-64, 1996.

46. Smith, M.D. Silane Chemical Protection of Bridge Decks. FHWA/OK 86(4), Final Report. Oklahoma Department of Transportation, December 1986.

47. Sprinkel, M.M., Sellers A.R., Weyers, R.E. "Rapid Concrete Bridge Deck Protection, Repair, and Rehabilitation", SHRP-S-344 Concrete Bridge Protection and Rehabilitation: Chemical and Physical Techniques, National Research Council, Washington, D.C. 1993.

48. Sprinkel, M.M., Weyers, R.E., Sellars, A.R. "Rapid Techniques for the Repair and Protection of Bridge Decks", Transportation Research Record 1304, Highway Maintenance Operations and Research, 1991.

49. Tabatabai, H., Ghorbanpoor A., and Turnquist-Nass A. "Rehabilitation Techniques for Concrete Bridges", Wisconsin DOT Report CEM-050301, pp. 22-30, March, 2005.

50. Tang, L. and Nilsson, L. "A Study of The Quantitative Relationship between Permeability and Pore Size Distribution of Hardened Cement Pastes." Cement and Concrete Research, Vol. 22, No. 4, 541-550, 1992.

51. Texas DOT Material Specifications; Section 5. DMS-8110, Coatings for Concrete, Section 9. DMS-8140, Concrete Surface Treatment (Penetrating), Austin, Texas.

52. Thomas, N. L. "The barrier Properties of Paint Coatings." Progress in Organic Coatings, Vol. 19, No. 2, 101-121, 1991.

53. Vipulanandan, C. and Liu, J. "Coatings for concrete surfaces: Testing and Modeling," Chapter 21, Handbook of Environmental Degradation of Materials, Editor Myer Kutz, William Andrew Publishing, Norwich, NY, 2005.

54. Vipulanandan, C. and Liu, J. "Film Model for Coated Cement Concrete," Cement and Concrete Research, Vol. 32(4), pp. 1931-1936, 2002.

55. Vipulanandan, C. and Liu, J. "Glass-Fiber Mat Reinforced Epoxy Coating for Concrete in Sulfuric Acid Environment," Cement and Concrete Research, Vol. 32(2), pp. 205-210, 2002.

56. Vipulanandan, C. and Liu, J. "Performance of Polyurethane-Coated Concrete in Sewer Environment," Cement and Concrete Research, Vol. 35, pp.1754-1763, 2005.

57. Whiting, D. "Penetrating Sealers for Concrete: Survey of Highway Agencies Transportation Research Record 1284, pp. 79-84, 1990.

58. Whiting, D., Ost, B., Nagi, M., Cady, P. "Methods for Evaluating the Effectiveness of Penetrating Sealers", SHRP-S/FR-92-107 Condition Evaluation of Concrete Bridges Relative to Reinforcement Corrosion Volume 5, National Research Council, Washington D.C. 1993.

59. Whiting, D., Tabatabai, H., Stejskal, B., Nagi, M. Rehabilitation of Prestressed Concrete Bridge Components by Non-Electrical (Conventional) Methods. Federal Highway Administration Publication FHWA-RD-98-189, McLean, VA, February 1999.

60. Winters, P. Product Evaluation P87-31-lnitial Report. Vermont Agency of Transportation, July 15, 1987a.

61. Winters, P. Product Evaluation P87-30-Final Report. Vermont Agency of Transportation, June 30, 1987b.

62. Wohl, R. L., and LaFraugh, R. W. Criteria for the Selection of Penetrating Hydrophobic Sealers Used in the Repair of Concrete Parking Decks. In Building Deck Waterproofing, ASTM STP 1084 (L. E. Gish, ed.), American Society for Testing and Materials, Philadelphia, Pa., pp. 75-82, 1990.

63. Zemajtis, J. and Weyers, R. E. "Concrete Bridge Service Life Extension Using Sealers in Chloride-Laden Environments," Transportation Research Record, Vol. n1561, pp.1-5, 1996.

64. Zemajtis, J., Weyers, R. E., and Sprinkel, M. "Corrosion Protection Service Life of Low-Permeable Concrete with a Corrosion Inhibitor," Transportation Research Record, Vol. n1642, No. 98, pp.51-59, 1998.

Chapter 9

Smart Materials and Structures

D. D. L. Chung
Composite Materials Research Laboratory
University at Buffalo, State University of New York

G. Song, N. Ma and H. Gu
Smart Materials and Structures Laboratory
Department of Mechanical Engineering
University of Houston

9.1 Introduction

Whether a structural material is load bearing or not in a structure, its strength and stiffness are important. Although purely structural applications dominate, combined structural and nonstructural applications are increasingly important as smart structures and electronics become more common. Such combined applications are facilitated by smart materials, which include multifunctional structural materials and non-structural functional materials [1]. The functions include sensing, which is relevant to smart structures, structural vibration control, traffic monitoring and structural health monitoring. They also include damping, thermal insulation, electrical grounding, resistance heating (say, for deicing), controlled electrical conduction, electromagnetic interference shielding, lateral guidance of vehicles, thermoelectricity, piezoresistivity, etc.

In addition to mechanical properties, a structural material may be required to have other properties, such as low density (light weight) for fuel saving in the case of aircraft and automobiles, for high speed in the case of race bicycles, and for handleability in the case of wheelchairs and armor. Another property that is often required is corrosion resistance, which is desirable for the durability of all structures, particularly automobiles and bridges. Yet another property that may be required is the ability to withstand high temperatures and/or thermal cycling as heat may be encountered by the structure during operation, maintenance or repair.

A relatively new trend is for a structural material to be able to serve functions other than the structural function, so that the material becomes multifunctional (akin to killing two or more birds with one stone, thereby saving cost and simplifying design). An example of a non-structural function is the sensing of damage. Such sensing, also called structural health monitoring, is valuable for the prevention of hazards. It is particularly important to aging aircraft and bridges. The sensing function can be attained by embedding sensors (such as optical fibers, the damage or strain of which affects the light throughput) in the structure. However, the embedding usually causes degradation of the mechanical properties and the embedded devices are costly and poor in durability compared to the structural material. Another way to attain the sensing function is to detect the change in property (e.g., the electrical resistivity) of the structural material due to damage. In this way, the structural material serves as its own sensor and is said to be self-sensing. Such multifunctional structural materials are also referred to as intrinsically smart materials. Intrinsic smartness is to be distinguished from extrinsic smartness, which relies on embedded or attached devices rather than the structural materials themselves in order to attain a certain non-structural function.

9.2 Self-Sensing Materials

Smart structures are those that have the ability to sense certain stimuli and are able to respond to them in an appropriate fashion, somewhat like a human being. Smart structures are important because of their relevance to hazard mitigation, structural vibration control, structural health monitoring, transportation engineering and thermal control. Sensing is the most fundamental aspect of a smart structure. A structural composite which is itself a sensor is multifunctional.

This section focuses on structural composites for smart structures. It addresses cement-matrix and polymer-matrix composites. The smart functions addressed include strain sensing (for structural vibration control and traffic monitoring) and damage sensing (both mechanical and thermal damage in relation to structural health monitoring).

9.2.1 *Self-sensing cement-matrix composites*

The electrical resistance of strain-sensing concrete (without embedded or attached sensors) changes reversibly with strain, such that the gage factor (fractional change in resistance per unit strain) is up to 700 under compression or tension. The resistance (DC/AC) increases reversibly upon tension and decreases reversibly upon compression, owing to fiber pull-out upon microcrack opening (<1 μm) and the consequent increase in fiber-matrix contact resistivity [2]. The concrete contains as low as 0.2 vol.% short carbon fibers, which are preferably those that have been surface treated [3]. The fibers do not need to touch one another in the composite. The treatment improves the wettability with water. The presence of a large aggregate or of damage decreases the gage factor, but the strain-sensing ability remains sufficient for practical use. Strain-sensing concrete works even when data acquisition is wireless. The applications include structural vibration control and traffic monitoring.

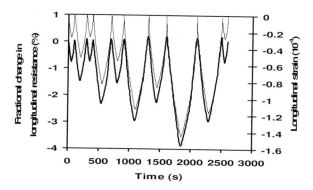

Fig. 9.1 Curves of fractional change in longitudinal resistance (thick curve) versus time and of longitudinal strain (thin curve) versus time during repeated compression at various strain amplitudes for a carbon fiber reinforced cement [4].

Figure 9.1 shows the fractional changes in the longitudinal resistance for carbon fiber (0.5 vol.%) silica-fume cement paste at 28 days of curing during repeated uniaxial compressive loading at a series of various strain amplitudes [4]. The strain essentially returns to zero at the end of each cycle, indicating elastic deformation. The resistance decreases reversibly upon compression. Figure 9.2 [4] shows that the resistivity at the peak strain varies essentially linearly with the peak strain, thereby allowing strain sensing by resistance measurement. Figure 9.3 [4] shows that the gage factor (fractional change in resistance per unit strain) approaches 300, which is high compared to the value of 2 that is typical of metallic strain gages. However, the gage factor decreases with increasing strain amplitude, as shown in Fig. 9.3 [4]. Upon uniaxial tension, the resistivity increases, as shown in Fig. 9.4 [5] and Fig. 9.5 [5] for the longitudinal and transverse directions respectively. In the absence of the fibers, the resistance varies much less upon loading and the effect is much less reversible.

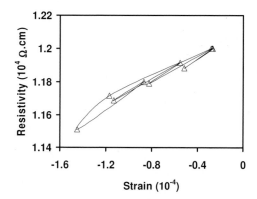

Fig. 9.2 Variation of the resistivity at the peak strain under compression versus the strain amplitude for carbon fiber reinforced cement [4]. The data points are connected with a line drawn to indicate the order in which the various strain amplitudes are imposed. The order and data correspond to those of Fig. 9.1.

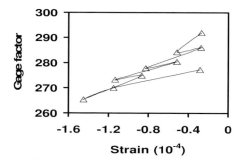

Fig. 9.3 Effect of strain amplitude on the gage factor for carbon fiber reinforced cement [4]. The data points are connected with a line drawn to indicate the order in which the various strain amplitudes are imposed. The order and data correspond to those of Fig. 9.1.

Moisture in the form of free water increases the electrical conductivity and decreases the gage factor of carbon fiber reinforced cement [5]. In addition, it increases the variability of the gage factor with the strain amplitude and with the strain history. Piezoresistivity involves electrical conduction across the interface between the fiber and

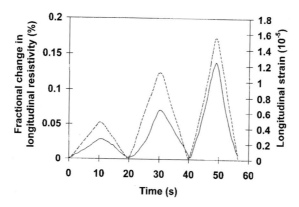

Fig. 9.4 Variation of the fractional change in longitudinal electrical resistivity with time (solid curve) and of the longitudinal strain with time (dashed curve) during dynamic uniaxial tensile loading at increasing stress amplitudes within the elastic regime for carbon fiber reinforced cement [5].

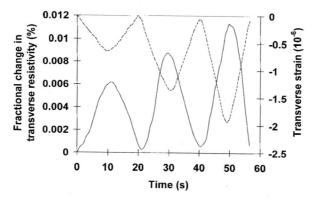

Fig. 9.5 Variation of the fractional change in transverse electrical resistivity with time (solid curve) and of the transverse strain with time (dashed curve) during dynamic uniaxial tensile loading at increasing stress amplitudes within the elastic regime for carbon fiber reinforced cement [5].

the cement matrix [2]. The diminished piezoresistivity due to excessive moisture is because of the water at the fiber-matrix interface interfering the electronic conduction across the interface. In spite of the lower piezoresistive performance, the piezoresistivity remains strong and the relationship between resistivity and strain remains quite linear [6].

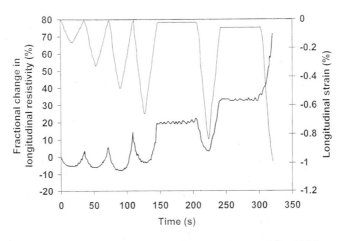

Fig. 9.6 Variation of the fractional change in longitudinal resistivity (thick curve) with time and of the longitudinal strain (thin curve) with time during uniaxial compression at progressively increasing stress amplitudes for carbon fiber reinforced cement [7].

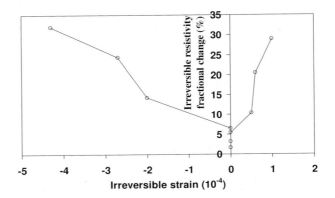

Fig. 9.7 Relationship of the irreversible resistivity fractional change with the irreversible strain for carbon fiber reinforced cement under uniaxial compression [7]. Negative strain is in the longitudinal direction; positive strain is in the transverse direction.

Carbon fiber reinforced cement is also capable of sensing damage, as damage causes the resistance to increase irreversibly, as shown in Fig. 9.6 [7], which is obtained during uniaxial compression at progressively increasing strain amplitudes. The higher is the strain amplitude, the more is the damage, and the more is the irreversible resistance increase at the end of a loading cycle. The irreversible resistance increase correlates with the irreversible strain, as shown in Fig. 9.7 [7].

That both strain and damage can be sensed simultaneously through resistance measurement means that the strain/stress condition (during dynamic loading) under which damage occurs can be obtained, thus facilitating damage origin identification.

9.2.2 *Self-sensing polymer-matrix composites*

Polymer-matrix composites for structural applications typically contain continuous fibers such as carbon, polymer and glass fibers, as continuous fibers tend to be more effective than short fibers as a reinforcement. Polymer-matrix composites with continuous carbon fibers are used for aerospace, automobile and civil structures. (In contrast, continuous fibers are too expensive for reinforcing concrete.) Due to the fact that carbon fibers are electrically conducting, whereas polymer and glass fibers are not, carbon fiber composites are predominant among polymer-matrix composites that are intrinsically smart.

Polymer-matrix composites containing continuous carbon fibers are important structural materials owing to their high tensile strength, high tensile modulus and low density. They are used for lightweight structures such as satellites, aircraft, automobiles, bicycles, ships, submarines, sporting goods, wheelchairs, armor and rotating machinery (such as turbine blades and helicopter rotors).

Carbon fibers are electrically conductive, while the polymer matrix is electrically insulating (except for the uncommon situation in which the polymer is an electrically conducting one). The continuous fibers in a composite laminate are in form of layers called laminae. Each lamina comprises many bundles (called tows) of fibers in a polymer matrix. Each tow consists of thousands of fibers. There may or may not be twist in a tow. Each fiber has a diameter typically ranging from 7 to 12 μm. The tows within a lamina are typically oriented in the same direction, but tows in different laminae may or may not be in the same direction. A laminate with tows in all the laminae oriented in the same direction is said to be unidirectional. A laminate with tows in adjacent laminae oriented at an angle of 90° is said to be crossply. In general, an angle of 45° and other angles may also be involved for the various laminae, as desired for attaining the mechanical properties required for the laminate in various directions in the plane of the laminate.

Within a lamina with tows in the same direction, the electrical conductivity is highest in the fiber direction. In the transverse direction in the plane of the lamina, the conductivity is not zero, even though the polymer matrix is insulating. This is because there are contacts between fibers of adjacent tows. In other words, a fraction of the fibers of one tow touch a fraction of the fibers of an adjacent tow here and there along the length of the fibers. These contacts result from the fact that fibers are not perfectly straight or parallel (even though the lamina is said to be unidirectional), and that the flow of the polymer matrix (or resin) during composite fabrication can cause a fiber to be not completely covered by the polymer or resin (even though, prior to composite fabrication, each fiber may be completely covered by the polymer or resin, as in the case of a prepreg, i.e., a fiber sheet impregnated with the polymer or resin). Fiber waviness is known as marcelling. Thus, the transverse conductivity gives information on the number of fiber-fiber contacts in the plane of the lamina.

For similar reasons, the contacts between fibers of adjacent laminae cause the conductivity in the through-thickness direction (direction perpendicular to the plane of the laminate) to be non-zero. Thus, the through-thickness conductivity gives information on the number of fiber-fiber contacts between adjacent laminae.

Matrix cracking between the tows of a lamina decreases the number of fiber-fiber contacts in the plane of the lamina, thus decreasing the transverse conductivity. Similarly, matrix cracking between adjacent laminae (as in delamination) decreases the number of fiber-fiber contacts between adjacent laminae, thus decreasing the through-thickness conductivity. This means that the transverse and through-thickness conductivities can indicate damage in the form of matrix cracking.

Fiber damage (as distinct from fiber fracture) decreases the conductivity of a fiber, thereby decreasing the longitudinal conductivity (conductivity in the fiber direction). However, owing to the brittleness of carbon fibers, the decrease in conductivity due to fiber damage prior to fiber fracture is rather small.

Fiber fracture causes a much larger decrease in the longitudinal conductivity of a lamina than fiber damage. If there is only one fiber, a broken fiber results in an open circuit, i.e., zero conducitivity. However, a lamina has a large number of fibers and adjacent fibers can make contact here and there. Therefore, the portions of a broken fiber still contribute to the longitudinal conductivity of the lamina. As a result, the decrease in conductivity due to fiber fracture is less than what it would be if a broken fiber did not contribute to the conductivity. Nevertheless, the effect of fiber fracture on the longitudinal conductivity is significant, so that the longitudinal conductivity can indicate damage in the form of fiber fracture.

The through-thickness volume resistance of a laminate is the sum of the volume resistance of each of the laminae in the through-thickness direction and the contact resistance of each of the interfaces between adjacent laminae (i.e., the interlaminar interface). For example, a laminate with eight laminae has eight volume resistances and seven contact resistance, all in the through-thickness direction. Thus, to study the interlaminar interface, it is better to measure the contact resistance between two laminae rather than the through-thickness volume resistance of the entire laminate.

Measurement of the contact resistance between laminae can be made by allowing two laminae (strips) to contact at a junction and using the two ends of each strip for making four electrical contacts. An end of the top strip and an end of the bottom strip serve as contacts for passing current. The other end of the top strip and that of the bottom strip serve as contacts for voltage measurement. The fibers in the two strips can be in the same direction or in different directions. This method is a form of the four-probe method of electrical resistance measurement. The configuration is illustrated in Fig. 9.8 for a crossply laminate. To make sure that the volume resistance within a lamina in the through-thickness direction does not contribute to the measured resistance, the fibers at each end of a lamina strip should be electrically shorted together by using silver paint or other conductive media. The measured resistance is the contact resistance of the junction. This resistance, multiplied by the area of the junction, gives the contact resistivity, which is independent of the area of the junction and just depends on the nature of the interlaminar interface. The unit of contact resistivity is $\Omega \cdot m^2$, whereas that of volume resistivity is $\Omega \cdot m$.

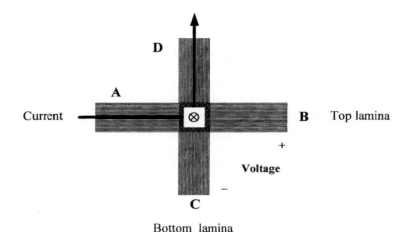

Fig. 9.8 Specimen configuration for measurement of contact electrical resistivity
between laminae.

The structure of the interlaminar interface tends to be more prone to change than the structure within a lamina. For example, damage in the form of delamination is much more common than damage in the form of fiber fracture. Moreover, the structure of the interlaminar interface is affected by the interlaminar stress (whether thermal stress or curing stress), which is particularly significant when the laminae are not unidirectional (as the anisotropy within each lamina enhances the interlaminar stress). The structure of the interlaminar interface also depends on the extent of consolidation of the laminae during composite fabrication. The contact resistance provides a sensitive probe of the structure of the interlaminar interface. The self-sensing of strain (reversible) has been achieved in continuous carbon fiber epoxy-matrix composites without the use of embedded or attached sensors [8,9]. Upon flexure, the tension surface resistance increases reversibly (Fig. 9.9, due to decrease in the penetration of the surface current) while the compression surface resistance decreases reversibly (Fig. 9.10, due to increase in the current penetration) [8]. The effect of flexure on the surface current penetration is consistent with the observation that the through-thickness resistance increases upon uniaxial tension (Fig. 9.11) and decreases upon uniaxial compression (Fig. 9.12) [8].

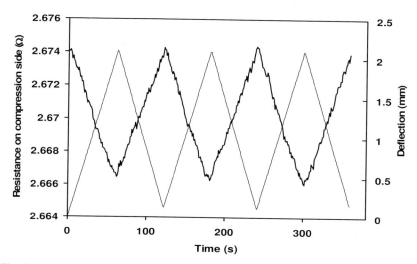

Fig. 9.9 Compression surface resistance (thick curve) during deflection (thin curve) cycling at an maximum deflection of 2.098 mm (stress amplitude of 392.3 MPa) for a 24-lamina quasi-isotropic continuous carbon fiber epoxy-matrix composite [8].

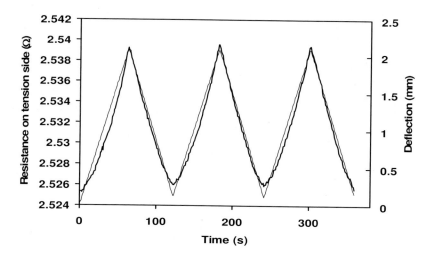

Fig. 9.10 Tension surface resistance (thick curve) during deflection (thin curve) cycling at an maximum deflection of 2.098 mm (stress amplitude of 392.3 MPa) for a 24-lamina quasi-isotropic continuous carbon fiber epoxy-matrix composite [8].

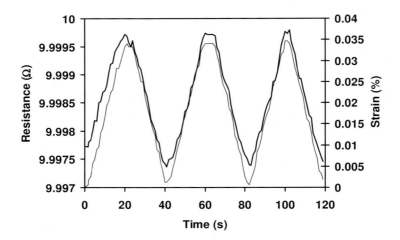

Fig. 9.11 Through-thickness resistance (thick curve) during uniaxial tension (strain shown by the thin curve) at a stress amplitude of +17.5 MPa for a 24-lamina quasi-isotropic continuous carbon fiber epoxy-matrix composite [8].

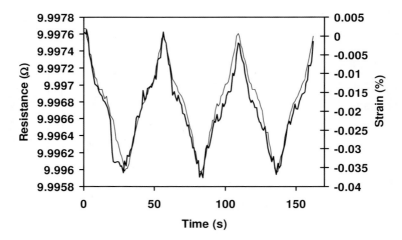

Fig. 9.12 Through-thickness resistance (thick curve) during uniaxial compression (strain shown by the thin curve) at a stress amplitude of -17.4 MPa for a 24-lamina quasi-isotropic continuous carbon fiber epoxy-matrix composite [8].

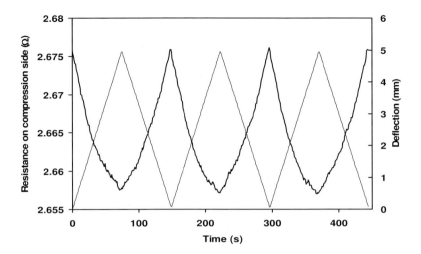

Fig. 9.13 Compressive surface resistance (thick curve) during deflection (thin curve) cycling at an maximum deflection of 4.945 mm (stress amplitude of 996.2 MPa) for a 24-lamina quasi-isotropic continuous carbon fiber epoxy-matrix composite [8].

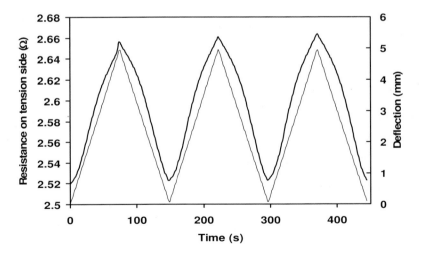

Fig. 9.14 Tensile surface resistance (thick curve) during deflection (thin curve) cycling at an maximum deflection of 4.945 mm (stress amplitude of 996.2 MPa) for a 24-lamina quasi-isotropic continuous carbon fiber epoxy-matrix composite [8].

The self-sensing of damage (whether due to stress or temperature, under static or dynamic conditions) has been achieved in continuous carbon fiber polymer-matrix composites, as the electrical resistance of the composite changes with damage. Minor damage under flexure is indicated by the curve of the resistance of the tension or compression surface vs. deflection becoming nonlinear, as shown in Figs. 9.13 and 9.14 for the compression and tension surface resistances respectively. Upon major damage, all resistances abruptly and irreversibly increase.

Damage in the form of delamination or interlaminar interface degradation is indicated by the through-thickness volume resistance (or more exactly the contact resistivity of the interlaminar interface) increasing due to decrease in the number of contacts between fibers of different laminae. Major damage in the form of fiber breakage is indicated by the longitudinal volume resistance increasing irreversibly.

9.3 Self-Actuating Materials

Self-actuating materials refer to structural materials that provide strain or stress in response to an input such as an electric field or a magnetic field. In the case of an electric field, the phenomenon pertains to either the reverse piezoelectric effect or electrostriction. In the case of a magnetic field, the phenomenon pertains to magnetostriction.

Sensing allows a structure to know its situation, whereas actuation is a way of allowing a structure to respond to what has been sensed. Thus, the presence of both self-sensing and self-actuating abilities enables a structural material to be really smart.

Self-actuating structural materials have not yet been well developed, although there are reports of self-actuation in the form of the reverse piezoelectric effect in carbon fiber (short) cement and in carbon fiber (continuous) polymer-matrix composite.

9.4 Self-Healing Materials

Self-healing refers to the ability of a structural material to heal or repair itself automatically upon the sensing of damage. This ability enhances safety, which is particularly needed for strategic structures. A self-healing cement-based material that involves the embedment of macroscopic tubules of an adhesive in selected locations of the cement-based material has been reported. The tubule fractures upon damage of the structural material, thereby allowing the adhesive to ooze out of the tubule. The adhesive fills the crack in its vicinity, thus causing healing. This technology suffers from the structural degradation due to the embedment of the tubules and the inability of a given tubule location to provide healing after the first time of damage infliction. Once the tubule has broken and the adhesive has solidified, the tubule cannot provide the healing function any more.

The problem with tubules may be circumvented by using a microencapsulated monomer, i.e., microcapsules containing a monomer. Upon breaking of the microcapsule, the monomer oozes out and meets the catalyst that is present in the structural material outside the microcapsule. Reaction between the monomer and the catalyst causes the formation of a polymer, which fills the crack. This method of self-healing has been shown to a limited extent in polymers, but not in cement-based materials, due to the pores in cement-based materials acting as sinks for the polymer. Furthermore, the microcapsule method suffers from the high cost of the catalyst, which needs to be able to promote polymerization at room temperature (without heating). It also suffers from the toxicity of the monomer.

9.5 Extrinsically Smart Materials and Structures

Extrinsically smart materials are non-structural materials that provide certain non-structural functions. Extrinsically smart structures are structures containing embedded or attached devices; the devices rather than the structural materials render the smartness. Due to the large variety of devices, extrinsic smartness is more common than intrinsic smartness. However, the devices are associated with high cost, low durability, poor repairability, limited functional volume and frequently structural performance degradation.

9.5.1 *Sensing using optical fibers*

Optical fibers are well-known for communication use, but they are also used as sensors. An optical fiber is a fiber that is highly transparent to light and that allows the light to travel inside the fiber even when the fiber is bent. For transparency, glass is commonly used. The ability for an optical fiber to guide a light beam is because of total internal reflection, which results from the fact that the fiber has a core and a concentric cladding, such that the refractive index is higher for the core than the cladding.

When an optical fiber is embedded in a structure, the fiber can sense the strain and damage of the structure. This is because the light throughput and other characteristics of the light beam traveling through the fiber are affected by the strain and damage of the fiber. The strain and damage of the structure relate to those of the optical fiber, provided that the fiber is well bonded to the structure. The use of optical fibers requires a light source (a laser or a light-emitting diode) and a light detector.

9.5.2 *Sensing and actuation using piezoelectric devices*

A piezoelectric device is one which converts between mechanical energy and electrical energy. It is a type of transducer. The conversion from mechanical energy to electrical energy (i.e., from strain/stress to electrical current/voltage) is based on the direct piezoelectric effect and allows application in strain/stress sensing and energy harvesting (i.e., converting mechanical vibration energy to electricity). The conversion from electrical energy to mechanical energy is based on the reverse piezoelectric effect and allows application in actuation.

Piezoelectric materials are dielectric materials, including ceramics and polymers. The most common piezoelectric material is lead zirconotitanate (abbreviated PZT, a $PbZrO_3$-$PbTiO_3$ solid solution).

The severity of the piezoelectric effect is described by the piezoelectric coupling coefficient d, which is defined as the electric polarization divided by the applied stress. In general, d is a tensor quantity, because the polarization and stress can be in different directions. A low value of the relative dielectric constant enhances the direct piezoelectric effect, due to the relationship between the polarization and the relative dielectric constant.

9.5.3 *Actuation using electrostrictive and magnetostrictive devices*

Electrostriction refers to the strain resulting from applying an electric field. Magnetostriction refers to the strain resulting from applying a magnetic field. These effects commonly stem from the interaction of the electric/magnetic field with the electrons in the solid and the consequent effect on the shape of the atoms in the solid. They can also be due to change in the orientation of electric/magnetic dipoles in the solid. In addition, electrostriction can be due to bonds between ions changing in length. These strain effects mean that electrostriction and magnetostriction can be used for actuation.

An electrostrictive material is a dielectric material, akin to a piezoelectric material. Compared to the reverse piezoelectric effect, electriction is advantageous in its smaller dependence on the history of electric field application. A magnetostrictive material is a magnetic material, most commonly a ferromagnetic material.

9.5.4 *Actuation using the shape-memory effect*

The shape-memory effect refers to the ability of a material to transform to a phase having a twinned microstructure (the phase called the martensite) that, after a subsequent plastic deformation, can return the material to its initial shape when heated, as illustrated in Fig. 9.15. If the shape-memory alloy (SMA) is constrained from recovering (say, within a composite material), a recovery stress is generated. The recovery stress builds up upon heating and it increases with increasing prior strain of the martensite. The return in shape upon heating can be used for actuation that is activated by heat.

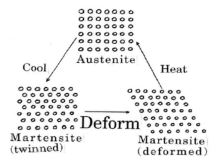

Fig. 9.15 The shape-memory effect illustrated in the crystal lattice level.

Another manifestation of the shape-memory effect involves the use of stress rather cooling to transform the material to martensite. Accompanying this change in phase is elastic deformation that exceeds the elasticity of ordinary alloys by a factor of 10 or more. Upon removal of the stress, the martensite changes back to the original phase and the strain (shape) returns to the value prior to the martensitic transformation [10]. This phenomenon is known as superelasticity or pseudoelasticity, and is illustrated in Fig. 9.16. It can be used for providing a nearly constant stress when strained (typically between 1.5% and 7%). The near constancy of the stress during unloading is exploited in orthodontal braces, where the shape-memory alloy (SMA) is used for the archwire, which applies forces according to the unloading plateau in order to restore the teeth to their proper location. The large hysteresis between loading and unloading in Fig. 9.16 means that a significant part of the strain energy put into the SMA is dissipated as heat. This energy dissipation provides a mechanism for vibration damping.

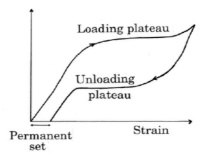

Fig. 9.16 Stress-strain curve of an SMA, showing a small permanent set
after unloading.

9.5.5   *Actuation using magnetorheological fluids*

Magnetorheology refers to the phenomenon in which the rheological behavior changes upon application of a magnetic field. A magnetorheological fluid (abbreviated MR fluid) is a non-colloidal dispersion of fine magnetic (ferromagnetic or ferrimagnetic) particles in a liquid medium. The most common particles are iron, which is ferromagnetic. It exhibits shear thinning (i.e., a decrease in the viscosity upon increase in the shear strain rate). Upon application of a magnetic field, the magnetic dipole moments of different particles align, resulting in columns of particles in the direction of the magnetic field (Fig. 9.17). Thus, the viscosity of the fluid increases with increasing magnetic field at any shear strain rate.

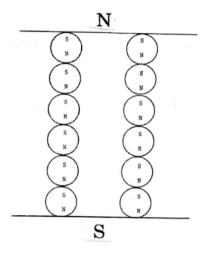

Fig. 9.17 Columns of magnetic particles in a magnetorheological fluid in the presence of a magnetic field in the direction of the columns.

9.6 Applications of Smart Materials and Structures

Applications of smart materials and structures include structural vibration control, traffic monitoring, border monitoring, weighing (including weighing in motion), room occupancy monitoring (for building facility management), building security (intruder detection), structural health monitoring, electric power generation (thermoelectricity and piezoelectricity), actuation (providing stress and strain), modulus control, yield stress control and damping control.

In general, practical implementation requires the use of hardware and software that are associated with the acquisition (wire or wireless), processing and storage of data. In case of electrical measurement, electrical contacts need to be designed in terms of their material, dimensions, shape, spacing, positions and protection. In case of the use of optical fibers, the coupling between the light source and the fiber and between the fiber and the light detector need to be designed to reduce the coupling loss.

The following sections address applications of shape memory alloys (Sec. 9.5.4) for structural vibration control (Sec. 9.7 and 9.8) and of piezoceramics (Sec. 9.5.2) for structural health monitoring (Sec. 9.9). The applications covered are those that are relatively established.

The shape memory alloys are used for both passive and active forms of structural vibration control, as described in Sec. 9.5 and 9.6 respectively.   The passive form refers to energy dissipation without actuation. The active form refers to suppression of the vibration by using actuation that is synchronized with the vibration.

9.7   Application of Passive Shape Memory Alloy (SMA) Devices in Structural Vibration Control

9.7.1 *Overview*

A passive structural control using SMA takes advantage of the SMA's damping property to reduce the vibration response of the structure subjected to severe loading. SMA can be effectively used for this purpose via two mechanisms, namely isolation and energy dissipation [11]. Examples of isolation are shown in Figs. 9.11 and 9.12.

In a ground isolation system, an SMA isolator is installed between a super-structure and the ground to form an uncoupled system.   The isolator filters the seismic motion transferred from the ground to the super-structure, so that the damage of the super-structure is reduced. On the other hand, the martensitic or austenitic SMA dampers greatly absorb the vibration energy by using the energy dissipation mechanism.

Although both isolation and energy dissipation take advantage of the damping capacity of SMAs, they are different in design and function. An SMA isolator is capable of providing variable stiffness to the structure according to the excitation levels, in addition to energy dissipation and restoration of the original position after unloading.

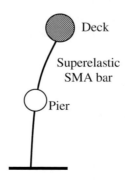

Fig. 9.11. Schematic of the SMA isolation device for highway bridges [12].

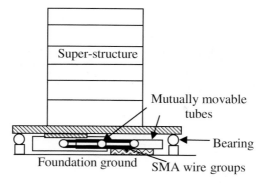

Fig. 9.12 Schematic of the SMA isolation system for buildings [13,14].

Superelastic SMAs are suitable for an isolator design. On the other hand, an SMA energy dissipation system is mainly used to reduce the dynamic response of the structures by direct energy dissipation [15-17]. Both martensitic and superelastic SMAs are suitable for this purpose.

9.7.2 *SMA isolation devices*

Reported SMA isolation systems include SMA bars for highway bridges (Fig. 9.11) [12], SMA wire re-centering devices for buildings (Fig. 9.12) [13,14], an SMA spring isolation system (Fig. 9.13) [18,19] and an SMA tendon isolation system for a shear frame structure with multiple degrees of freedom (Fig. 9.14) [20].

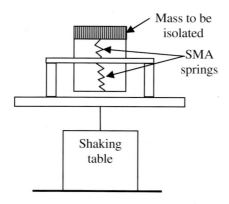

Fig. 9.13 Schematic of the SMA spring isolation device [18].

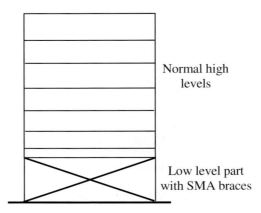

Normal high
levels

Low level part
with SMA braces

Fig. 9.14 Schematic of the SMA tendon isolation system for a MDOF structure [20].

It should be emphasized that only the superelastic SMA has been studied for isolation, due to its zero residual strain upon unloading. To improve the damping capacity of isolation devices, the martensitic SMA elements can be added into the superelastic SMA based isolation system. The re-centering device by Dolce *et al.* [13,14] is a good example of combination the superelastic and martensitic SMAs.

9.7.3 *SMA energy dissipation devices*

SMA energy dissipation devices exist in the form of braces for framed structures [13,14,22-29], dampers for cable-stayed bridges [30] and for simply supported bridges [31,32], anchorages for columns [26,33], beam-column connections [34], and retrofitting devices for historic buildings, as shown in Fig. 9.15 [21,35].

9.7.3.1 SMA braces for frame structures

The SMA wire braces are installed diagonally in frame structures. As the frame structures deform under excitation, the SMA braces dissipate energy through the stress-induced martensitic transformation (for the superelastic SMAs) or the martensitic reorientation (for the martensitic SMAs). Several different scaled-down prototypes of the SMA braces were designed, implemented and tested. The examples include the brace consisting of 210 loops of Nitinol wire installed on a six-story two-bay by two-bay steel frame, as shown in Fig. 9.16 [22], the re-centering

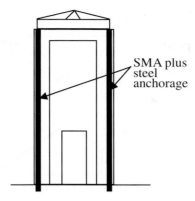

Fig. 9.15 Schematic of retrofit of a bell tower using SMA anchorage [21].

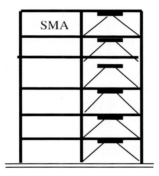

Fig. 9.16 Schematic of the setup of the SMA brace reinforced frame structure [4].

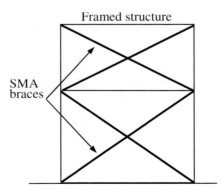

Fig. 9.17 Schematic the SMA braces for a two-story steel frame [22].

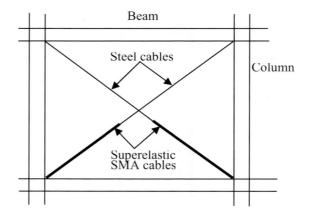

Fig. 9.18 Schematic the SMA braces for a frame structure [26].

braces tested on a real two-story building [13,14], and the hybrid braces made of rigid segments (mostly steel wire) and the Nitinol wires (illustrated in Figs. 9.17 and 9.18) [22,24,26]. The SMA braces have self-centering capability, high stiffness for small displacements and good energy dissipation. The effectiveness of the SMA braces for the vibration suppression is dependent on the pre-strain and the geometry of the SMA wire brace [26,28].

9.7.3.2 SMA dampers for bridges

Both superelastic and martensitic SMAs can be used as damper elements for bridges. Li *et al.* [30] numerically studied an application of a superelastic SMA damper for vibration reduction of a cable-stay bridge. This SMA-cable damper system is illustrated in Fig. 9.19. The numerical simulation shows that the proposed SMA damper can effectively suppress the cable's vibration.

In [31], the testing and simulation analysis results of a full-scale superelastic SMA bar restrainer (Fig. 9.20) used for seismic retrofit of a multi-span simply supported bridge were reported. The results have shown that the SMA restrainer more effectively reduced the relative hinge displacements at the abutment and it provided a large elastic deformation range in comparison with conventional steel restrainer cables. In addition, the SMA restrainer effectively limits the motion of the bridge decks.

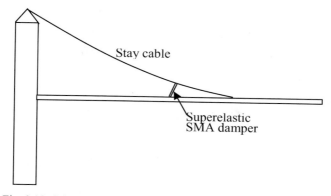

Fig. 9.19 Schematic of the SMA damper for a cable-stay bridge [30]

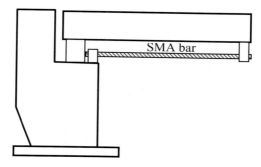

Fig. 9.20 Schematic of the setup of SMA restrainer for
a simple-supported bridge [31].

Casciati *et al.* [1] studies an application of a large martensitic Nitinol bar as a seismic protection device in a bridge. The finite element analysis used shows the applicability of the martensitic Nitinol bar in energy dissipation in relation to both the static and dynamic response to strong earthquakes. The analysis conclusion is in agreement with the experimental results.

9.7.3.3   SMA column anchorage and SMA beam-column connection

Connectors or connections in various structures are prone to damage during an earthquake. SMAs have been used to protect a beam-column connection or a ground-column connection. An SMA-made anchorage

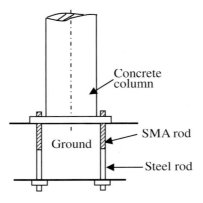

Fig. 9.21 Schematic of SMA bar anchorage for a column [26].

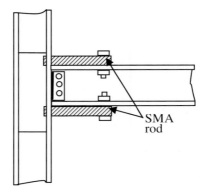

Fig. 9.22 Schematic of SMA connector for steel structures [34].

(Fig. 9.21) or auxiliary (Fig. 9.22) is able to not only reduce the relative motion between the two connected parts by dissipating energy, but also tolerate a relatively large deformation without a connection failure.

Pulsating tension loading tests and numerical simulations were conducted on an exposed-type column base with anchorages (Fig. 9.21), each of which is made of a Nitinol rod and a steel rod [26]. The work shows that this type of anchorage is very effective in dissipating energy and reducing vibration of the column. In addition, it shows that the Nitinol segment of the anchorage can recover its original length after cyclic loadings.

The loading tests of the two full-scale SMA enhanced steel beam-column connections (the shaded parts in Fig. 9.22) demonstrate that the connections exhibit a stable and repeatable hysteresis for rotations up to 4% to absorb the vibration energy [34]. Also it is shown that the SMA enhancement is able to sustain up to 5% strain without permanent damage.

9.8 Application of Active SMA Devices for Structural Vibration Control

9.8.1 *SMAs for active structural frequency tuning*

For a structure vibrating at its resonant frequency, the vibration can be reduced by actively tuning the resonant frequency of the structure. Upon heating, SMA actuators embedded or installed in structures will increase the stiffness of the host structure, so that the natural frequency of the structures can be actively tuned. By active frequency tuning, the vibration control for the structure can be achieved. This is the basic principle of SMAs for semi-active structural vibration control. For example, McGavin and Guerin [36] reported a proof-of-concept experiment in which the frequency of a steel structure is adjusted by using SMA wire actuators in real time. They achieved about 32% change of the natural frequency.

9.8.2 *Shape restoration using superelastic SMAs*

There is a specific type of application of superelastic SMA wires for structural control purpose that is different from the aforementioned examples. This application uses the shape restoration property of superelastic SMA wires. For example, Sakai *et al.* [37] researched self-restoration of a concrete beam using superelastic SMA wires. The experimental results show that the mortar beam with SMA wires recovers almost completely after incurring an extremely large crack.

In recent work [38] at University of Houston, a more efficient way to use superelastic SMA wires in the form of stranded cables to achieve a relatively large restoration force was developed. Shown in Figs. 9.23 and 9.24 is a concrete beam (24 in × 4 in × 6 in) reinforced with fourteen 1/8-inch diameter superelastic stranded cables using the method of post tensioning to achieve a 2% pre-strain. Each cable has seven strands and each strand has seven superelastic wires. Special clamps are used to hold the superelastic strands/cables without slippage. After loading at

11,000 lb, a large crack appeared (Fig. 9.23). Upon subsequent unloading, the crack closed (Fig. 9.24) under the elastic restoration force of the superelastic SMA cables. This research also demonstrates that the stranded cable provides a new and effective way to use SMAs for civil applications.

Fig. 9.23 Large crack in the beam before activation of the SMA.

Fig. 9.24 Crack width reduced in the beam after activation of the SMA.

9.9 Application of piezoelectric material to structural health monitoring

In recent years, piezoelectric materials have been successfully applied to the structural health monitoring of composite structures, metallic structures and concrete structures. Extensive theoretical and experimental research has been conducted. The piezoelectric-based health monitoring approach is a nondestructive evaluation (NDE) method that is suitable for health monitoring of inaccessible in-situ civil structures without using additional expensive and bulky equipment.

According to the principle of the testing method, piezoelectric-based health monitoring can be classified into three major categories: (a) impedance domain-based approach, (b) vibration-characteristic-based approach and (c) Lamb wave-based approach.

9.9.1 *Impedance-based approach*

The impedance-based qualitative health monitoring technique is a real-time structural damage detection method. Due to the electromechanical coupling property of piezoelectric materials, the measured electrical impedance is directly related to the mechanical impedance, and will also be affected by the presence of damage. Sun *et al.* [39] conducted an automated real-time health monitoring system on an assembled truss structure. Chaudhry *et al.* [33] conducted local area health monitoring of aircraft by measuring the impedance of piezoelectric transducers. Tseng *et al.* [41] presented the results of an experimental study for the detection and characterization of damages using PZT transducers on aluminum specimens. The impedance characteristics of the PZT transducer are extracted to detect damage. Besides metallic structures, concrete structures are also suitable for the impedance-based health monitoring method. Soh *et al.* [42] surface-bonded piezoceramic (PZT) patches to carry out health monitoring during the destructive load testing of a prototype reinforced concrete (RC) bridge.

9.9.2 *Vibration-characteristic-based approach*

The vibration-characteristic approach utilizes the piezoelectric actuator to generate certain wave to propagate in the structure and compares the structural vibration-characteristic parameters (modal shape, model frequency, damping, stiffness, etc.) or vibration-characteristic response curves (frequency response, time response, transfer function, etc.) with those of the healthy state to detect damage.

Piezoelectric-based active sensing system is commonly utilized in the health monitoring of civil structures. In the piezoelectric-based active sensing system, one piezoelectric transducer is used as actuator to send excitation waves. Other distributed piezoelectric transducers are used as sensors. The crack or damage inside the structure acts as a stress relief during the wave propagation path. The amplitude of wave and the transmission energy will decrease due to the existence of a crack. The decrease in value of the transmission energy will be correlated with the degree of the damage inside. Shown in Fig. 9.25 is such a system with two embedded piezoelectric transducers. The piezoelectric-based active sensing has the following advantages. 1. This method can be operated at a very low cost; 2. The piezoelectric-based active sensing has high sensitivity in local areas which is good for locating damage; 3 Health monitoring can be conducted on the in-situ concrete structure without using extra bulky equipment. Besides, a piezoelectric actuator can be excited in a broad frequency range which is ideal for health monitoring. The piezoelectric-based active sensing method is an effective and economical approach for health monitoring of large civil concrete structure.

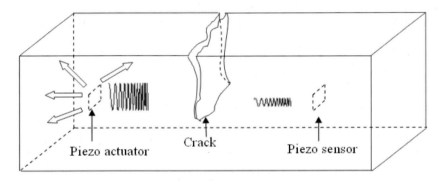

Fig. 9.25 Piezoelectric-based active sensing system.

The piezoelectric-based active sensing method has been successfully applied to the health monitoring of civil structures such as composite structures and concrete structures. The delamination of a composite is a critical issue because it can cause failure of the composite structure. One consequence of delamination in a composite structure is a change in its stiffness. This change in stiffness will degrade the modal frequencies of the composite structure. Okafor [41] conducted theoretical and experimental studies to investigate the effect of delamination on modal frequencies of composite beams with a piezoelectric actuator. Experiments results showed that the third and fourth modal frequencies degrade significantly with increasing delamination size. Saafi *et al.* [44]

proposed the active damage interrogation technique to conduct health monitoring of composite reinforced concrete structures. This system uses an array of piezoelectric transducers (PZT) attached or embedded within the structure for both actuation and sensing. Experimental results showed a distinct difference between the transfer function of the healthy reinforced concrete specimens and that of the damage reinforced concrete specimens. Song et al. [45,46] developed smart aggregates based on piezoelectric materials for health monitoring of two full-size concrete bent-caps (specimens W1 and W2). The experimental setup for the testing of the concrete bent-cap is shown in Fig. 9.26.

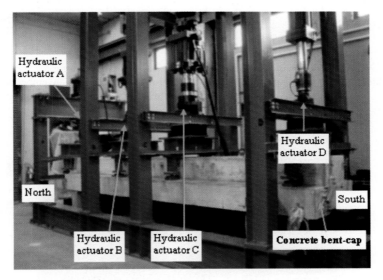

Fig. 9.26 Experimental setup for the testing of the concrete bent-cap.

For the health monitoring of the first concrete bent-cap specimen W1, the smart aggregates are positioned in a planar configuration at one end. For the health monitoring of the second concrete bent-cap specimen W2, the smart aggregates (a total of 10) are placed at the spatial position shown in Fig. 9.27. The smart aggregates were manufactured by embedding a small PZT patch (0.8 cm $\times$ 0.8 cm $\times$ 0.0267 cm) in a small (about 2.5 cm^3) concrete block before curing. The smart aggregates were placed at the desired position in the concrete bent cap before casting, as shown in Fig. 9.28. Both specimens W1 and W2 were subjected to

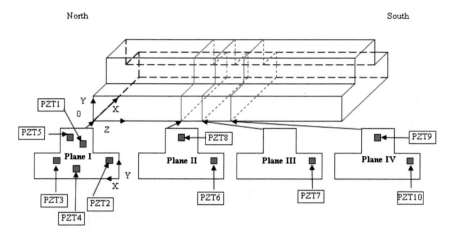

Fig. 9.27 The location of 10 smart aggregates (labeled by PZT1-10) for the structural health monitoring of concrete bent-cap W2. The 10 smart aggregates are embedded in 4 different planes (labeled by Plane I-IV).

Fig. 9.28 The location of smart aggregates before casting.

destructive tests with loadings applied by the four hydraulic actuators. During the experiments, the crack growth on both specimens was monitored by microscopy (MS) and LVDTs with results shown in Figs. 9.29 and 9.30, respectively.

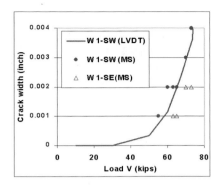

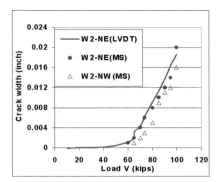

Fig. 9.29 Crack width measured by microscopy (MS) and LVDT for specimen W1. (SW: southwest location; SE: southeast location)

Fig. 9.30 Crack width measured by microscopy (MS) and LVDT for specimen W2. (NE: northeast location; NW: northwest location)

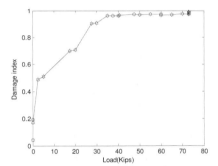

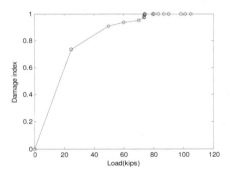

Fig. 9.31 Damage index curve vs. load (specimen W1) of PZT2

Fig. 9.32 Damage index curve vs. load (specimen W2) of PZT2

With wavelet packet analysis, a damage index (DI) was defined to present the damage status of a concrete structure with embedded smart aggregates [34]. When the DI value is zero, there is no damage. When the DI value is one, the structural is completely damaged. For specimens W1 and W2, damage index curves are shown in Figs. 9.31 and 9.32, respectively. From the damage index curve of specimen W1, as shown in Fig. 9.31, the damage index is close to the critical value when the load is around 40 kips (276 MPa), while at the same load value, from the LVDT and MS result of W1, as shown in Fig. 9.29, the crack width has just begun to increase on the surface. From the experimental result of the second specimen W2, the damage index reaches the critical point around 74 kips (510 MPa), as shown in Fig. 9.32, while at the same load value the crack width on the surface has just begun to increase, as shown in Fig. 9.30. This means that severe cracks have happened before the crack width begins to increase on the surface. For both specimens, the critical points in the proposed method are earlier than the critical point deduced from LVDT and microscopy results for both concrete bent-caps. This implies that the PZT sensors are more sensitive than LVDTs or microscopy for health monitoring [47,48].

9.9.3 *Lamb-wave-based approach*

Lamb waves are the most commonly used plate waves for health monitoring of plate-like structures. The propagation of a Lamb wave depends a great deal on the selected frequency and the material thickness. The Lamb-wave-based approach has been successfully applied to the health monitoring of composite structures, metallic structures and steel-reinforced concrete structures.

- *Health monitoring of composite structures.* Delamination is a major concern for in-service composites. Su *et al.* [49] applied four distributed piezoelectric transducers to generate and monitor the ultrasonic Lamb wave with a narrowband frequency in quasi-isotropic carbon fiber/epoxy (CF/EP) laminates. Toyama *et al.* [50] investigate the effects of transverse cracking and delamination on the S_0 mode velocity in carbon fiber reinforced plastic (CFRP) cross-ply laminates. Experimental results showed that both the stiffness and the velocity decreased as the transverse crack density increased. Metallic structures are common in many areas of engineering, particularly in aerospace, ground and sea transportation.
- *Health monitoring of metallic structures.* Due to aging, fatigue and erosion of metals, there can be cracks, holes or other types of damage inside or on the surface of a metallic structure. Staszewski *et al.* [51,52] proposed a laser-based Lamb wave sensing approach for a rectangular metallic plate by using two piezoceramic disks as actuator and sensor. The study showed the potential of the method for simple and rapid detection of the location of damage in a structure. Tua *et al.* [53] proposed an approach to locate and determine the extent of linear cracks in homogeneous aluminum

plates based on the flight time of Lamb wave by using piezoelectric actuators and sensors. Hiber-Huang transform is used to process the sensor signal to determine accurate flight time and also estimate the orientation of the crack. In situ health monitoring is important in the maintenance of metallic structures. Rajic *et al.* [54] proposed an *in situ* health monitoring system based on a piezoceramic wafer element to the detection of specimen fatigue under cyclic loading condition. Lamb waves propagating through the beam test specimen are sensed using small surface-mounted piezoelectric transducers, then analyzed for indications that relate to the onset of fatigue. The study and the experimental results show the great potential for developing similar automated *in situ* structural health monitoring systems for application to operational structures such as aircraft.

- *Health monitoring of concrete structures.* Steel-reinforced concrete (RC) is widely used in civil infrastructure. Steel reinforced concrete (RC) structures usually serve under harsh environments. Wang *et al.* [55] studied the Lamb-wave-based health monitoring of both fiber-reinforced composites and steel-reinforced concrete. The piezoelectric sensor network is installed in selected rebars in areas such as the deck, the columns of bridges, and the footing area of columns for the purpose of health monitoring. Experimental results showed that cracks or debonding damage in RC structures can be detected by the proposed built-in active sensing system.

References

1. Chung DDL. Composite Materials, Springer, 2003.
2. Wen S, Chung DDL. Model of piezoresistivity in carbon fiber cement. Cem Concr Res 2006;36(10):1879-1885.
3. Fu X, Lu W, Chung DDL. Ozone treatment of carbon fiber for reinforcing cement. Carbon 1998;36(9):1337-1345.
4. Wen S, Chung DDL. Piezoresistivity-based strain sensing in carbon fiber reinforced cement. ACI Mater J 2007;104(2):171-179.
5. Wen S, Chung DDL. Uniaxial tension in carbon fiber reinforced cement, sensed by electrical resistivity measurement in

longitudinal and transverse directions, Cem Concr Res 2000; 30(8): 1289-1294.

6. Wen S, Chung DDL. Unpublished result.

7. Wen S, Chung DDL. Electrical-resistance-based damage self-sensing in carbon fiber reinforced cement. Carbon 2007; 45(4): 710-716.

8. Wang S, Chung DDL. Self-sensing of flexural strain and damage in carbon fiber polymer-matrix composite by electrical resistance measurement. Carbon 2006; 44(13): 2739-2751.

9. Wang S, Chung DDL. Negative piezoresistivity in continuous carbon fiber epoxy-matrix composite. J Mater Sci 2007; 42(13): 4987-4995.

10. Duerig, T. W., Melton, K.N., Stockel, D. and Wayman, C.M. Engineering aspects of shape memory alloys. Butterworth Heinemann, London, 1990.

11. Saadat, S., Salichs, J., Noori, M., Hou, Z., Davoodi, H., Bar-on, I., Z., Suzuki, Y. and Masuda, A. "An overview of vibration and seismic application of NiTi shape memory alloy", *Smart Materials and Structures*. 11, 2001, pp. 218-229.

12. Wilde, K., Gardoni, P. and Fujino, Y. "Base isolation system with shape memory alloy device for elevated highway bridges", *Engineering Structures*. 22, 2000, pp. 222-229.

13. Dolce, M., Cardone, D. and Marnetto, R. "Implementation and testing of passive control devices based on shape memory alloys", *Earthquake Engineering and Structural Dynamics*. V. 29, 2000, pp. 945-968.

14. Dolce, M., Cardone, D. and Marnetto, R. "SMA re-centering devices for seismic isolation of civil structures", *Proceedings of SPIE*, V.4330, 2001, pp. 238-249, 2001.

15. Dolce, M. and Cardone, D. "Mechanical behavior of shape memory alloys for seismic application II. Austenite NiTi wires subjected to torsion", *International Journal of Mechanical Sciences*, V. 43, 2001, pp. 2656-2677.

16. Gandhi, F. and Wolons, D. "Characterization of the pseudoelastic damping behavior of shape memory alloy wires using complex modulus", *Smart Materials and structures*. 8, 1999, pp.49-56.

17. Ip, K. H. "Energy dissipation in shape memory alloy wire under cyclic bending", *Smart Materials and Structures*. 9, 2000, pp. 653-659.

18. Khan, M. M. and Lagoudas, D. "Modeling of shape memory alloy pseudoelastic spring elements using Preisach model for passive vibration isolation", *Proceedings of SPIE*. V. 4693, 2002, pp. 336-347.

19. Mayes, J.J., Lagoudas, D. and Henderson, B.K. "An experimental investigation of shape memory alloy pseudoelastic springs for passive vibration isolation", *AIAA Space 2001 Conference and Exposition*, Albuquerque, NM, 2001.

20. Corbi, O. "Shape memory alloys and their application in structural oscillations attenuation". *Simulation Modeling Practice and Theory*, V. 11, 2003, pp. 387-402.

21. Indirli, M., Castellano, M.G., Clemente, P., and Martelli, A. "Demo application of shape memory alloy devices: The rehabilitation of S. Georgio Church Bell Tower", *Proceedings of SPIE*. V. 4330, 2001, pp. 262-272.

22. Clark, P., Aiken, I., Kelly, J., Higashino, M. and Krumme, R. "Experimental and analytical studies of shape memory alloy dampers for structural control", *Proceedings of SPIE*, V. 2445, 1995, pp. 241-251.

23. Han, Y.L., Li, Q.S., Li, A.Q., Leung, A.Y.T. and Lin, P.H. "Structural vibration control by shape memory alloy damper", *Earthquake Engineering and Structural Dynamics*, V. 32, 2003, pp. 483-494.

24. Saadat, S., Noori, M., Davoodi, H., Zhou, Z., Suzuki, Y. and Masuda A. "Using NiTi SMA tendons for vibration control of coastal structures". *Smart Materials and Structures*. V. 10, 2001, pp. 695-704.

25. Davoodi, H., Just, F. A, Saffar, A. and Noori, M. Building system using shape memory alloy members, United States Patent, Patent No.: US 6,170,202, B1, 2001.

26. Tamai, H. and Kitagawa, Y. "Pseudoelastic behavior of shape memory alloy wires and its application to seismic resistance member for building". *Computational Materials Science*. V. 25, 2002, pp. 218-227.

27. Sun, S. and Rajapakse, R. K. N. D. "Dynamic response of a frame with SMA bracing", *Proceedings of SPIE*. V. 5053, 2003, pp. 262-270.

28. Seelecke, S., Heintze, O. and Masuda A. "Simulation of earthquake-induced structural vibration in systems with SMA

damping elements", *Proceedings of SPIE*. V.4697, 2002, pp. 238-245.

29.   Duval, L., Noori, M. N., Hou, Z., Davoodi, H. and Seelecke, S. Random vibration studies of an SDOF system with shape memory restoring force. Physica B, 275: 138-141, 2000.

30.   Li, H., Liu M. and Ou, J. P. "Vibration mitigation of a stay cable with one shape memory alloy damper", *Structural Control and Health Monitoring*, V.11, 2004, pp. 1-36.

31.   DesRoches, R. and Delemont, M. "Seismic retrofit of simply supported bridges using shape memory alloys". *Engineering Structures*, V. 24, 2002, pp. 325-332.

32.   Casciati, F., Faravelli, L. and Petrini L. "Energy dissipation in shape memory alloy devices. Computed-Aided Civil and Infrastructure Engineering", V.13, 1998, pp. 433-442.

33.   Tamai, H., Miura, K., Kitagawa, Y. and Fukuta, T. "Application of SMA rod to exposed-type column base in smart structural system", *Proceedings of SPIE*. V. 5057, 2003, pp.169-177.

34.   Leon, R. T., DesRoches, R., Ocel, J. and Hess, G. "Innovative beam column using shape memory alloys", *Proceeding of SPIE*. V. 4330, 2001, pp. 227-237.

35.   Croci, G., Bonci, A. and Viskovic, A. "The use of shape memory alloys devices in the basilica of St Francis in Assisi", *Proceedings of the Final Workshop of ISTECH Project*, pp. 117-140, Ispra, Italy, 2000.

36.   McGavin, G. and Guerin, G. "Real-time seismic damping and frequency control of steel structures using Nitinol wire", *Proceedings of SPIE*. V. 4696, 2002, pp. 176-184.

37.   Sakai Y., Kitagawa Y., Fukuta T., and Iiba M. "Experimental study on enhancement of self-restoration of concrete beams using SMA wire", *Proceedings of SPIE*. V.5057, 2003, pp. 178-186.

38.   Mo, Y. L., Song, G., and Otero, K. "Development and testing of a proof-of-concept smart concrete structure", *Proceeding of Smart Structures Technologies and Earthquake Engineering (SE04)*, Osaka, Japan, 2004.

39.   Sun F. P., Chaudhry Z, Rogers C A and Majmundar M., "Automated real-time structure health monitoring via signature pattern recognition", Proceedings of the SPIE -Smart Structures and Materials,. V.2443, 1995, pp. 236-247.

40. Chaudhry, Z., Joseph T., Sun F. and Rogers C. A., "Local-area health monitoring of aircraft via piezoelectric actuator/sensor patches", *Proceedings of the SPIE-Smart Structure and Materials Conference*, V. 2443, 1995, pp. 268-276.

41. Tseng K. K-H and Naidu A.S.K. "Non-parametric damage detection and characterization using smart piezoceramic material", *Smart Materials and Structures*, V. 11, No.3, 2002, pp. 317-329.

42. Soh C. H., Tseng K.K., Bhalla S. and Gupta A. "Performance of smart piezoceramic patches in health monitoring of a RC bridge", *Smart Materials and Structures*, V. 9, No. 4, 2000, pp. 533-542.

43. Okafor A.C., Chandrashekhara and Jiang Y.P., "Delamination prediction in composite beams with built-in piezoelectric devices using modal analysis and neural network", *Smart Materials and Structures*, V.5, No. 3, 1996, pp. 338-347.

44. Saafi M. and Sayyah T., "Health monitoring of concrete structures strengthened with advanced composite materials using piezoelectric transducers", *Composites Part B: Engineering*, V.32, No. 4, 2001, pp. 333-342.

45. Song G., Gu H., Mo Y. L., Hsu T., Dhonde H. and Zhu, R.R.H. "Health monitoring of a concrete structure using piezoceramic materials", Proceedings of SPIE – The International Society for Optical Engineering, V. 5765, Smart Structures and Materials , 2005, pp. 108-119.

46. Song G., Mo Y. L., Otero K. and Gu H., "Develop intelligent reinforced concrete structures using shape memory alloys and piezoceramics", the Third International Conference on Earthquake Engineering: New Frontier and Research Transformation, October 18-20, 2004, Nanjing, P.R.China.

47. Song G., Gu H., Mo Y. L., Hsu T., Dhonde H. and Zhu, R.R.H., "Health monitoring of a reinforced concrete bridge bent-cap using piezoceramic materials", Proceedings of the Third European Conference on Structural Control, 3ECSC,12-15 July ,2004, Vienna University of Technology, Vienna, Austria.

48. Song G., Olmi C. and Gu H., "An overheight collision detection and evaluation system for bridge girder using piezoelectric transducer", Second International Conference on Structural Health Monitoring of Intelligent Infrastructure, Nov. 16-18, 2005, Shenzhen, P. R. China.

49. Su Z., Ye L. and Bu X. "A damage identification technique for CF/EP composite laminates using distributed piezoelectric transducers", Composite Structures, V. 57, 2002, pp. 465-471.

50. Toyama N., Noda J. and Okabe T. "Quantitative damage detection in cross-ply laminates using Lamb wave method", *Composites Science and Technology,* V. 63, No.10, 2003, pp. 1473-1479.

51. Staszewski W. J., Lee B.C., Mallet L and Scarpa F. "Structure health monitoring using scanning laser vibrometry: I. Lamb wave sensing", *Smart Materials and Structures*, V. 13, No. 2, 2004, pp. 251-260.

52. Staszewski W. J., Lee B.C., Mallet L and Scarpa F. "Structure health monitoring using scanning laser vibrometry: II. Lamb wave for damage detection", *Smart Materials and Structures* , V. 13, No. 2, 2004, pp. 261-269

53. Tua P.S. Quek S. T. and Wang Q. "Detection of cracks in plates using piezo-actuated lamb waves", *Smart Materials and Structures*, V.13, No. 4, 2004, pp. 643-660

54. Rajic N., Galea S.C. and Chiu W.K. ,"Autonomous detection of crack initiation using surface-mounted piezotransducers", *Smart Materials and Structures* ,V. 11, No. 1, 2002, pp.107-114

55. Wang C. S., Wu F. and Chang F. Structural health monitoring from fiber-reinforced composites to steel-reinforced concrete *Smart Materials and Structures,* V. 10, No. 3 , 2001, pp. 548-552

Index